Fortran 95 Language Guide

Springer
London
Berlin
Heidelberg
New York
Barcelona
Budapest
Hong Kong
Milan
Paris
Santa Clara
Singapore
Tokyo

Wilhelm Gehrke

Fortran 95
Language Guide

 Springer

Wilhelm Gehrke
Regional Computing Centre
University of Hannover
Hannover
Germany

ISBN 3-540-76062-8 Springer-Verlag Berlin Heidelberg New York

British Library Cataloguing in Publication Data
Gehrke, Wilhelm, 1940-
   Fortran 95 language guide
   1.FORTRAN (Computer program language)
   I.Title
   005.1'33
ISBN 3540760628

Library of Congress Cataloging-in-Publication Data
Gehrke, Wilhelm, 1940-
   Fortran 95 language guide / Wilhelm Gehrke
      p.   cm.
   Includes index.
   ISBN 3-540-76062-8 (pbk.)
   1. Fortran 90 (Computer program language)   I. Title.
   QA76.73.F25G433   1996   96-28622
   005.13'3 - - dc20   CIP

Apart from any fair dealing for the purposes of research or private study, or criticism or review, as permitted under the Copyright, Designs and Patents Act 1988, this publication may only be reproduced, stored or transmitted, in any form or by any means, with the prior permission in writing of the publishers, or in the case of reprographic reproduction in accordance with the terms of licences issued by the Copyright Licensing Agency. Enquiries concerning reproduction outside those terms should be sent to the publishers.

© Springer-Verlag London Limited 1996
Printed in Great Britain

The use of registered names, trademarks etc. in this publication does not imply, even in the absence of a specific statement, that such names are exempt from the relevant laws and regulations and therefore free for general use.

The publisher makes no representation, express or implied, with regard to the accuracy of the information contained in this book and cannot accept any legal responsibility or liability for any errors or omissions that may be made.

Typesetting: camera ready by author
Printed and bound at the Athenæum Press Ltd., Gateshead, Tyne and Wear
34/3830-543210 Printed on acid-free paper

PREFACE

Fortran has been and will be the most important programming language for the development of engineering and scientific applications. The outmoded revision FORTRAN 77 of the International Standard was replaced in 1991 by a completely new Fortran language, its name was Fortran 90. This revision is still a powerful tool, in fact it is closer to the state of the art of high level problem oriented programming languages than other famous languages that are used for the same area of application.

The next revision Fortran 95 of the International Standard will be published in the second half of 1996. It is a relatively minor enhancement of Fortran 90 with a small number of new features and mainly devoted to clarifications, corrections, and interpretations of Fortran 90.

The major new language features in Fortran 95 are:
- FORALL statement and construct,
- Pure and elemental user-defined subprograms,
- Implicit initialization of derived type objects, and
- Initial association status for pointers.

The minor new language features in Fortran 95 are:
- Comments in namelist input data,
- Minimal output field width for formatted numeric output,
- Intrinsic SIGN function may distinguish −0 and +0,
- New intrinsic function CPU_TIME returns processor time,
- New intrinsic function NULL returns disconnected pointer,
- References to certain pure functions in specification expressions,
- Nested WHERE constructs and masked ELSEWHERE statements,
- Modifications to intrinsic functions CEILING, FLOOR, MAXLOC, and MINLOC,
- Automatic deallocation of allocatable arrays, and
- Generic identifier in END INTERFACE statement.

Some of these new features have been added to keep Fortran aligned with HPF (High Performance Fortran) which is the Fortran-based language for the data parallel programming model for (massively) parallel computer architectures.

The ISO Fortran committee WG5 is already planning for the next major revision of the Fortran language. The target publication date is 2001. The list of suggested topics in preparation for Fortran 2000 contains "exception handling", "interoperability with C", "allocatable structure components" and "parameterized data types", and "object-oriented programming".

This "Fortran 95 Language Guide" is a comprehensible description of the complete Fortran 95 programming language as it is defined in the final draft [1]. There are only few modifications to the language that could be described in a self-contained section. Most of the new features and the large number of minor modifications and of corrections of defects to Fortran 90 is scattered throughout the language such that it was necessary to prepare this "Fortran 95 Language Guide" as a completely revised edition of the "Fortran 90 Language Guide" [4].

The "Fortran 95 Language Guide" is intended to serve as a language reference manual for programmers, as teaching material for introductory courses in Fortran programming, and as a help for experienced Fortran programmers migrating to Fortran 95. It is not intended to replace any textbook for beginners, any textbook on programming methodology, any annotated standard document, or the standard document itself as a reference for compiler writers. The guide concentrates on the language as a programmers' tool and abstains from personal, historical, and philosophical comments and interpretations.

Layout

The following conventions are used throughout this guide:

Upper-case	ABC	indicate a Fortran keyword which must be written as given.
Lower-case	abc	in a syntax rule indicate a language element which is to be inserted by the programmer.
Lower-case	xyz	in an example indicate a name invented by the programmer.
Special characters	+ *	of the Fortran character set must be written as given.
Square brackets	[ ]	in a syntax rule enclose an optional language element, which may be used or omitted.
Dot sequence	...	in a syntax rule indicate that the preceding optional language element enclosed in brackets may be repeated as necessary.
Braces	{ }	in a syntax rule enclose several language elements; one of them must be selected.
Ampersand	&	at the end of a syntax line indicates continuation.

This is the font for normal text.
This is the font for formal syntax.
```
This is the font for examples.
```
This is the font for definitions.
Terms written such are either emphasized or are defined elsewhere in this guide.

The formal meta language used within this guide supports the precise description of single language features. Note that it is not the formal meta language used in the standard document [1].

Sources

[1]   ISO WG5, *Draft International Stamdard Fortran*, N1176, 1996
[2]   Gehrke, *Fortran 95–Handbuch*, 950 pages, 1996, in preparation
[3]   RRZN, *Fortran 95, Ein Nachschlagewerk*, 375 pages, 1996, in preparation
[4]   Gehrke, *Fortran 90 Language Guide*, Springer, 1995

Technical Terms

The following terms are assumed as known or are defined elsewhere in this guide; in any case, reading the guide will be easier if the reader is aware of them:

block: A sequence of executable statements which is a part of an executable construct.

data object: A variable or a constant.

data entity: A data object, the result of the evaluation of an expression, or the result of a function reference.

definition: A derived type definition defines a derived data type. A subprogram definition defines a user-defined subroutine or function. A variable or a record of an internal file are defined if they have a valid value.

double precision real: The double precision real type is a special case of the real type.

Fortran processor: The computing system consisting of hardware and software by which programs are transformed for use on that system.

parent object: A subobject is a part of a parent object.

presence: An optional dummy argument is present if an actual argument is associated with it which is either a present dummy argument of the caller or which is not a dummy argument of the caller.

reference: A "data object reference" is the appearance of the data object name or data object designator where its value is required during program execution. A "subprogram reference" is the appearance of the subprogram name, of an operator symbol, or of the assignment symbol where the execution of the subprogram is required during program execution. A "module reference" is the appearance of a module name in a USE statement.

variable: A named variable is a scalar or an array object that has an own name. An unnamed variable is an array element (scalar), an array section (array), a structure component (scalar or array), or a character string (scalar).

The standard document [1] uses several concepts and technical terms which are only needed for the precise description of other technical terms. We try to avoid such technical terms within this guide. For example: the definition of the terms "explicit interface" and "implicit interface" in the standard document is fairly poor and hard to read. The **Fortran 95** experts are familiar with these concepts and they don't bother about the details of the definition in the standard document. But my impression is that the casual or even the "normal" **Fortran** programmer will soon forget the precise definitions of such terms. Therefore, we don't use such terms and concepts in this guide and describe the full details where they are needed.

Because **standard-conformance** is an important feature, this term completes our short glossary.

A *program* is standard-conforming if it uses only those forms and relationships described in the standard document and if the program has an interpretation according to the standard.

A *program unit* is standard-conforming if it can be included in a program in a manner that allows the program to conform to the standard.

Acknowledgements

I would like to thank Katrin, Jens and Meike for their support and for their endless patience during the preparation of this book.

Hannover
May 1996
W. G.

CONTENTS

1	SOURCE FORM	1-1
	1.1 Fixed Source Form	1-2
	1.2 Free Source Form	1-4
	1.3 Embedding of Program Lines by INCLUDE	1-6
	1.4 Classification of Fortran Statements	1-7
	1.5 Statement Ordering	1-8
2	**TYPE CONCEPT**	**2-1**
	2.1 Intrinsic Types	2-1
	2.1.1 Integer Type	2-1
	2.1.2 Real Type and Double Precision Real Type	2-2
	2.1.3 Complex Type	2-3
	2.1.4 Logical Type	2-4
	2.1.5 Character Type	2-4
	2.2 Derived Types	2-5
	2.2.1 Derived Type Definition	2-5
	2.2.1.1 Type Component Definition	2-7
	2.2.1.2 Private and Public Derived Type Definitions	2-12
	2.2.2 Structure Objects	2-13
3	**LEXICAL TOKENS**	**3-1**
	3.1 Scoping Units	3-1
	3.2 Keywords	3-1
	3.3 Names	3-2
	3.4 Operators and Assignment Symbol	3-3
	3.5 Statement Labels	3-3
	3.6 Literal Constants	3-3
	3.6.1 Integer Literal Constants	3-4
	3.6.2 Real Literal Constants	3-4
	3.6.3 Double Precision Real Literal Constants	3-5
	3.6.4 Complex Literal Constants	3-6
	3.6.5 Logical Literal Constants	3-6
	3.6.6 Character Literal Constants	3-7
	3.6.7 Binary, Octal, and Hexadecimal Literal Constants	3-8

4 Data Objects   4-1
    4.1   Constants 4-2
    4.2   Variables 4-3
    4.3   Scalars 4-4
         4.3.1   Character Substrings 4-4
    4.4   Arrays 4-5
         4.4.1   Inner Structure of Arrays 4-8
    4.5   Structure Components 4-9
    4.6   Automatic Variables 4-11
    4.7   Association 4-12
         4.7.1   Name Association 4-12
         4.7.2   Pointer Association 4-14
         4.7.3   Storage Association 4-14
    4.8   Definition Status 4-16

5 POINTERS   5-1
    5.1   Pointer Processing 5-2
         5.1.1   Creation of Pointer Targets 5-2
         5.1.2   Association Status 5-3
         5.1.3   Deallocation of Pointer Targets 5-5
         5.1.4   Nullification of Pointer Associations 5-6

6 ARRAY PROCESSING   6-1
    6.1   Array Declaration 6-1
         6.1.1   Explicit-Shape Arrays 6-2
         6.1.2   Assumed-Shape Arrays 6-2
         6.1.3   Assumed-Size Arrays 6-3
    6.2   Reference and Use 6-4
         6.2.1   Whole Arrays 6-4
         6.2.2   Array Elements 6-5
         6.2.3   Array Sections 6-6
                 6.2.3.1   Subscript-Triplet 6-9
                 6.2.3.2   Vector-Subscript 6-11
                 6.2.3.3   Array Sections of Substrings 6-12
    6.3   Memory Management and Dynamic Control 6-13
         6.3.1   Automatic Arrays 6-13
         6.3.2   Allocatable Arrays 6-14
         6.3.3   Array Pointers 6-16
    6.4   Construction of Array Values 6-18
    6.5   Operations on Arrays 6-19
         6.5.1   Array Expressions 6-19
         6.5.2   Array Subprograms 6-21
         6.5.3   Array Assignments 6-21

7	**EXPRESSIONS**		**7-1**
	7.1 Numeric Intrinsic Expressions		7-3
	7.2 Relational Intrinsic Expressions		7-8
		7.2.1 Numeric Relational Intrinsic Expressions	7-8
		7.2.2 Character Relational Intrinsic Expressions	7-9
	7.3 Logical Intrinsic Expressions		7-10
	7.4 Character Intrinsic Expressions		7-12
	7.5 Defined Expressions		7-14
		7.5.1 Defined Operators and Extended Intrinsic Operators . .	7-14
		7.5.1.1 Nonextended Defined Operator	7-17
		7.5.1.2 Extended Defined Operator	7-18
		7.5.1.3 Extended Intrinsic Operator	7-18
	7.6 Common Rules for Expressions		7-19
		7.6.1 Precedence of Operators	7-19
		7.6.2 Interpretation of Expressions	7-20
		7.6.3 Evaluation of Expressions	7-21
	7.7 Special Expressions		7-23
		7.7.1 Constant Expressions	7-24
		7.7.2 Initialization Expressions	7-25
		7.7.3 Specification Expressions	7-26
8	**ASSIGNMENTS**		**8-1**
	8.1 Intrinsic Assignment Statements		8-1
		8.1.1 Numeric Assignment Statement	8-3
		8.1.2 Logical Assignment Statement	8-3
		8.1.3 Character Assignment Statement	8-4
		8.1.4 Assignment Statement for Derived Types	8-5
	8.2 Defined Assignment Statements		8-6
		8.2.1 Nonextended Defined Assignment	8-8
		8.2.2 Extended Defined Assignment	8-9
	8.3 Pointer Assignment Statement		8-10
	8.4 Masked Array Assignments		8-11
		8.4.1 WHERE Statement	8-11
		8.4.2 WHERE Construct	8-12
		8.4.3 Common Rules for Masked Array Assignments	8-15
	8.5 Indexed Assignments		8-17
		8.5.1 FORALL Statement	8-17
		8.5.1.1 Execution of the FORALL Statement	8-17
		8.5.2 FORALL Construct	8-19
		8.5.3 Common Rules for Indexed Assignments	8-21
9	**DECLARATIONS AND SPECIFICATIONS**		**9-1**

9.1	Attributes		9-2
	9.1.1	ALLOCATABLE Attribute	9-3
	9.1.2	DATA Attribute	9-4
	9.1.3	DIMENSION Attribute	9-4
	9.1.4	EXTERNAL Attribute	9-5
	9.1.5	INTENT Attribute	9-5
	9.1.6	INTRINSIC Attribute	9-6
	9.1.7	OPTIONAL Attribute	9-7
	9.1.8	PARAMETER Attribute	9-7
	9.1.9	POINTER Attribute	9-8
	9.1.10	PRIVATE Attribute	9-8
	9.1.11	PUBLIC Attribute	9-9
	9.1.12	SAVE Attribute	9-9
	9.1.13	TARGET Attribute	9-10
9.2	Type Declaration Statements		9-11
	9.2.1	INTEGER Statement	9-14
	9.2.2	REAL Statement	9-14
	9.2.3	DOUBLE PRECISION Statement	9-15
	9.2.4	COMPLEX Statement	9-15
	9.2.5	LOGICAL Statement	9-16
	9.2.6	CHARACTER Statement	9-16
		9.2.6.1 Character Length	9-17
	9.2.7	TYPE Declaration Statement	9-18
9.3	Attribute Specification Statements		9-19
	9.3.1	ALLOCATABLE Statement	9-19
	9.3.2	DATA Statement	9-19
		9.3.2.1 Implied-DO	9-22
	9.3.3	DIMENSION Statement	9-23
	9.3.4	EXTERNAL Statement	9-24
	9.3.5	INTENT Statement	9-24
	9.3.6	INTRINSIC Statement	9-25
	9.3.7	OPTIONAL Statement	9-25
	9.3.8	PARAMETER Statement	9-25
	9.3.9	POINTER Statement	9-26
	9.3.10	PRIVATE Statement	9-27
	9.3.11	PUBLIC Statement	9-27
	9.3.12	SAVE Statement	9-28
	9.3.13	TARGET Statement	9-30
9.4	Additional Specification Statements		9-30
	9.4.1	COMMON Statement	9-31
	9.4.2	EQUIVALENCE Statement	9-36
		9.4.2.1 EQUIVALENCE and COMMON	9-39

	9.4.3 IMPLICIT Statement	9-40
	9.4.4 NAMELIST Statement	9-43

10 EXECUTION CONTROL   10-1

- 10.1 GO TO Statements ... 10-2
  - 10.1.1 Unconditional GO TO Statement ... 10-2
  - 10.1.2 Computed GO TO Statement ... 10-2
- 10.2 IF Statements ... 10-3
  - 10.2.1 Arithmetic IF Statement ... 10-3
  - 10.2.2 Logical IF Statement ... 10-4
- 10.3 IF Construct ... 10-5
  - 10.3.1 Simple IF Constructs ... 10-6
  - 10.3.2 Nested IF Constructs ... 10-9
- 10.4 CASE Construct ... 10-10
  - 10.4.1 Simple CASE Constructs ... 10-13
- 10.5 DO Construct ... 10-14
  - 10.5.1 DO Statement ... 10-15
  - 10.5.2 Do-Termination Statement ... 10-15
  - 10.5.3 Forms of DO Constructs ... 10-16
  - 10.5.4 Execution of a DO Construct ... 10-17
    - 10.5.4.1 Additional Details about Count Loops ... 10-18
    - 10.5.4.2 Additional Details about WHILE Loops ... 10-19
    - 10.5.4.3 Additional Details about Endless Loops ... 10-20
    - 10.5.4.4 CYCLE Statement and EXIT Statement ... 10-20
  - 10.5.5 Nested DO Constructs ... 10-21
- 10.6 Nested Constructs ... 10-22
- 10.7 CONTINUE Statement ... 10-23
- 10.8 STOP Statement ... 10-23
- 10.9 CALL, END, and RETURN Statements ... 10-24

11 INPUT/OUTPUT   11-1

- 11.1 Records ... 11-1
- 11.2 Files ... 11-2
- 11.3 File Attributes of External Files ... 11-2
  - 11.3.1 File Names ... 11-3
  - 11.3.2 Access Methods ... 11-3
    - 11.3.2.1 Sequential Access ... 11-3
    - 11.3.2.2 Direct Access ... 11-3
  - 11.3.3 Form of a File ... 11-5
  - 11.3.4 File Position ... 11-5
- 11.4 Units ... 11-6
- 11.5 Preconnected Units and Predefined Files ... 11-7

11.6 Input/Output Statements 11-8
    11.6.1 Input/Output Specifiers 11-8
        11.6.1.1 UNIT= Specifier 11-9
        11.6.1.2 FMT= Specifier 11-9
        11.6.1.3 NML= Specifier 11-9
        11.6.1.4 REC= Specifier 11-10
        11.6.1.5 ADVANCE= Specifier 11-10
        11.6.1.6 End-of-Record Condition and EOR= Specifier 11-10
        11.6.1.7 IOSTAT= Specifier 11-11
        11.6.1.8 Error Conditions and ERR= Specifier 11-12
        11.6.1.9 End-of-File Condition and END= Specifier . . 11-13
        11.6.1.10 SIZE= Specifier 11-14
    11.6.2 Input/Output Lists 11-14
        11.6.2.1 Implied-DO 11-17
    11.6.3 Data Transfer Statements 11-19
        11.6.3.1 Formatted Input/Output 11-21
        11.6.3.2 Unformatted Input/Output 11-24
        11.6.3.3 List-Directed Input/Output 11-25
        11.6.3.4 Internal Input/Output 11-30
        11.6.3.5 Namelist Input/Output 11-33
        11.6.3.6 Nonadvancing Input/Output 11-38
        11.6.3.7 Printing 11-41
    11.6.4 File Status Statements 11-42
        11.6.4.1 OPEN Statement 11-42
        11.6.4.2 CLOSE Statement 11-46
        11.6.4.3 INQUIRE Statement 11-47
    11.6.5 File Positioning Statements 11-52

12 FORMATS   12-1

12.1 Format Specification 12-1
    12.1.1 Format Specification in FORMAT Statement 12-1
    12.1.2 Character Format Specification 12-2
12.2 Interaction between Input/Output List and Format 12-3
    12.2.1 Repeat Specification, Groups of Edit Descriptors 12-4
    12.2.2 Reversion of Format Control 12-4
12.3 Edit Descriptors 12-5
    12.3.1 A Edit Descriptors 12-8
    12.3.2 B Edit Descriptors 12-9
    12.3.3 BN and BZ Edit Descriptors 12-10
    12.3.4 Character Constant Edit Descriptors 12-11
    12.3.5 Colon Edit Descriptor 12-11
    12.3.6 D Edit Descriptor 12-12

    12.3.7   E Edit Descriptors 12-14
    12.3.8   EN Edit Descriptors 12-14
    12.3.9   ES Edit Descriptors 12-15
    12.3.10 F Edit Descriptor 12-16
    12.3.11 G Edit Descriptors 12-16
    12.3.12 I Edit Descriptors 12-17
    12.3.13 L Edit Descriptor 12-18
    12.3.14 O Edit Descriptors 12-19
    12.3.15 P Edit Descriptor, Scale Factor 12-20
    12.3.16 Sign Control Edit Descriptors 12-22
    12.3.17 Slash Edit Descriptor 12-22
    12.3.18 Tabulator Edit Descriptors 12-23
    12.3.19 X Edit Descriptor 12-24
    12.3.20 Z Edit Descriptors 12-25

13 PROGRAM UNITS AND SUBPROGRAMS   13-1
13.1 Main Program 13-1
13.2 Modules 13-3
    13.2.1   USE Statement 13-4
13.3 Block Data Program Units 13-6
13.4 Subprograms 13-8
    13.4.1   User-Defined Functions (except Statement Functions) . 13-10
             13.4.1.1   Function Definition 13-11
             13.4.1.2   Explicit Invocation, Function Reference 13-16
             13.4.1.3   Operator Functions 13-18
    13.4.2   User-Defined Subroutines 13-18
             13.4.2.1   Subroutine Definition 13-19
             13.4.2.2   Explicit Invocation, CALL Statement 13-21
             13.4.2.3   Assignment Subroutines 13-23
    13.4.3   External Subprograms 13-23
    13.4.4   Internal Subprograms 13-25
    13.4.5   Module Subprograms 13-27
    13.4.6   Dummy Subprograms 13-28
    13.4.7   Statement Functions 13-29
             13.4.7.1   Statement Function Definition 13-29
             13.4.7.2   Invocation of a Statement Function 13-31
    13.4.8   Interface Blocks 13-32
    13.4.9   Overloaded Generic Subprogram Names 13-36
    13.4.10 Additional Entry Points, ENTRY Statement 13-37
    13.4.11 Return from the Invoked Subprogram 13-39
13.5 Internal Program Communication 13-42
    13.5.1   Argument Lists 13-42

		13.5.1.1	Dummy Argument List	13-42
		13.5.1.2	Actual Argument List	13-44
	13.5.2	Argument Association		13-45

              13.5.2.1 Data Objects as Dummy Arguments 13-47
              13.5.2.2 Implicit Association of Two Dummy Arguments 13-48
              13.5.2.3 Length of Character Dummy Arguments . . . 13-49
              13.5.2.4 Scalar Arguments 13-51
              13.5.2.5 Dummy (Argument) Arrays 13-51
              13.5.2.6 Dummy (Argument) Pointers 13-52
              13.5.2.7 Sequence Association 13-52
              13.5.2.8 Assumed-Size Arrays 13-55
              13.5.2.9 Assumed-Shape Arrays 13-55
              13.5.2.10 Restrictions on the Association of Data Entities 13-55
              13.5.2.11 Dummy Subprograms 13-56
              13.5.2.12 Asterisk Dummy Arguments 13-57
      13.5.3 Optional Dummy Arguments 13-57
      13.5.4 Dummy Argument with INTENT Attribute 13-58
      13.5.5 Common Blocks 13-59

14 INTRINSIC SUBPROGRAMS   14-1

14.1 Intrinsic Functions 14-1
      14.1.1 Table of Intrinsic Functions 14-3
14.2 Intrinsic Subroutines 14-9
14.3 Intrinsic Subprogram Reference 14-9
14.4 Intrinsic Subprogram Definitions 14-10
      14.4.1 Descriptions 14-13

A CHARACTER SETS AND COLLATING SEQUENCES   A-1

A.1 Processor-Dependent Character Sets A-1
A.2 ASCII Character Set A-1

B MODELS FOR NUMBERS   B-1

B.1 Models for Integers B-1
B.2 Models for Reals B-1
B.3 Models for Bit Manipulation B-2

C DECREMENTAL LANGUAGE FEATURES   C-1

C.1 Deleted Language Features C-1
C.2 Obsolescent Language Features C-2

D INDEX   D-1

1 SOURCE FORM

A Fortran program is a collection of *program units*. It consists of one *main program* and any number of *modules, external subprograms*, and *block data* program units. Each of these program units is a sequence of program lines consisting of Fortran *statements, INCLUDE lines*, and/or *comments*. The characters in these program lines form *lexical tokens*. The following rules govern the form of the program lines in a program unit.

A **program line** is a sequence of characters. All program lines in a particular program unit must be written either in *fixed source form* or in *free source form*. The character positions of a program line are counted from the left to the right beginning with 1.

With few exceptions, a program line must contain only characters of the Fortran *character set*. The exceptions are *literal constants, character constant edit descriptors*, and *comments*, which may contain certain other characters of the *processor character set*.

The **Fortran character set**, which is embedded in the processor character set, consists of the following characters:

Letters:	A B C D E F G H I J K L M N O P Q R S T U V W X Y Z
Digits:	0 1 2 3 4 5 6 7 8 9

Special characters:

		Blank		:	Colon
=	Equals	+	Plus	−	Minus
*	Asterisk	/	Slash	(	Left Parenthesis
)	Right Parenthesis	,	Comma	.	Point
'	Apostrophe	!	Exclamation Point	"	Quote
%	Percent	&	Ampersand	;	Semicolon
<	Less Than	>	Greater Than	?	Question Mark
$	Currency Symbol	_	Underscore		

If the **Fortran** processor supports lower-case letters in the source program, they are equivalent to corresponding upper-case letters except in literal constants, in character constant edit descriptors, and in certain parameters of the OPEN statement and the INQUIRE statement.

A **statement label** may be used to identify a statement. Other statements in the same program unit or subprogram may refer to such a labeled statement. Therefore, the statement label must be unique within a *scoping unit*; that means, a particular statement label may identify exactly one statement in a scoping unit.

A statement label is a sequence of 1 – 5 digits, one of which must be nonzero. Leading zeros are not significant. Any statement which is not part of another statement may be preceded by a statement label. But only FORMAT statements and *executable statements* may be referred to by their statement labels.

Multiple statements: A program line may contain more than one statement. Two statements may be separated by a ";" character. But a semicolon which appears in a literal constant or in a character constant edit descriptor is not a statement separator. Two or more consecutive semicolons with or without intervening blanks have the same interpretation as a single statement separator. A semicolon appearing as the last nonblank character in a program line or as the last nonblank character preceding a comment is ignored.

A **comment** serves only documentation purposes and has no effect on the interpretation of a program unit. A comment may contain any characters of the processor character set. Lines containing only blanks are comment lines. Comment lines may appear anywhere before the END statement of a program unit.

1.1 Fixed Source Form

Each program line has *exactly* 72 characters if the line contains only characters of type *default character*; otherwise the line length is processor-dependent.

An **initial line** is the first or the only program line of a Fortran statement. Position 1 – 5 must be blank or must contain a statement label. Position 6 must contain a blank or the digit zero. The statement occurs anywhere between position 7 and 72 inclusively. If the statement is longer than 66 characters, it may be continued on a *continuation line*. Note that the END statement of a program unit must not be continued; and a statement whose initial line appears to be a program unit END statement must not be continued.

An initial line may be continued on the next line which is no comment line. The **continuation line** must have in position 6 any character of the Fortran character set except blank or zero. The positions 1 – 5 must be blank. The Fortran processor processes position 7 of a continuation line as if it follows position 72 of the continued line. A continuation line may be continued. A statement must not have more than 19 continuation lines. Note that a ";" character in position 6 indicates a continuation line (and does not separate two statements).

A **statement label** may occur in position 1 – 5 of a program line. Embedded blanks may appear anywhere within a statement label, but these blank(s) are ignored by the Fortran processor. Only the first statement of an initial line may be labeled.

Blanks are normally insignificant. They may be inserted freely for better readability of the source program. Exceptions: blanks in literal constants, in character edit descriptors, and a blank in position 6 of any program line are significant.

1.1 Fixed Source Form

A line containing the character "C" or the character "*" in position 1 is a **comment line**. A comment line may appear anywhere in a program unit, even between an initial line and a continuation line of a statement or between two continuation lines.

An "!" character initiates a comment except when it appears in position 6, in a character literal constant or in a character constant edit descriptor. Such a comment ends in position 72 of the line.

Summary of fixed source form

Position(s)	Character	Interpretation
1	C or * or !	Indicates a comment line
1 – 5, 7 – 72	!	Initiates a comment
1 – 5	Digit(s)	Statement label
6	No 0 and no blank	Indicates a continuation line
7 – 72	Fortran character set	Fortran statement
7 – 72	;	Separates 2 statements
1 – 72	Only blanks	Comment line

```
C0000000011111111112222222222333333333344444444...556666666666777
C2345678901234567890123456789012345678...890123456789012
      P R O G R A M  pascal_triangle            ! initial line
      INTEGER basis (13); CHARACTER fmt *10     ! multipl stmt
      DATA basis /13*1/, null   /23/            ! initial line
                                                ! comment line
      WRITE (*, 1111)                           ! initial line
C     writes the first 3 elements                 comment line
     1  basis(13),                              ! contin. line
     2  basis(12), basis(13)                    ! contin. line
 1111 FORMAT (5X,'Pascal ',                     ! initial line
     1         'triangle',5X,I6/22X,I6,I5)      ! contin. line
      DO20i=12,2,-1                             ! initial line
      DO    j=i,12                              ! initial line
         basis(j) = basis(j) + basis(j+1)       ! initial line
         END DO                                 ! initial line
      null = null - 2; WRITE (fmt, 22) null     ! multipl stmt
 22   FORMAT ('(',I2,'X,13I5)')                 ! initial line
      WRITE (*, fmt) (basis(k), k=i-1,13)       ! initial line
   20 CONTINUE                                  ! initial line
      E N D P R O G R A M  pascal_triangle      ! initial line
```

1.2 Free Source Form

Each program line has *maximal* 132 characters if the line contains only characters of type *default character*; otherwise the line length is processor-dependent. A statement may occur anywhere in the program line.

Character literal constants, character constant edit descriptors, and comments may contain any characters of the processor character set but no control character.

An **initial line** is a program line which is no blank line and has no "!" in its first nonblank character position.

A statement may be **continued** on the next line. The character "&" as the last nonblank character in a line (and not as a part of a comment) indicates that the statement continues with the next statement line. If the first nonblank character of that line is an "&", the statement continues with the next character; otherwise the statement continues with position 1 of that line. A statement must have not more than 39 continuation lines. A program line must not contain an "&" as the only nonblank character or as the only nonblank character followed by an "!".

If a character literal constant or a character constant edit descriptor is continued, an "&" must be the last character of the line (that is, no comment must follow) and the first nonblank character of the next noncomment line must be an "&". If the continuation is in any other context, a comment may follow after the "&".

If a *lexical token*, such as a keyword, name, constant, operator, statement label, "=>", "(/", or "/)", is separated, the first nonblank character of the continuation line must be an "&" followed by the rest of the lexical token.

A **statement label** may occur after a semicolon.

Blanks must not appear within lexical tokens except within character constants. But blanks must be used as separators between keywords, names, constants, and statement labels and subsequent keywords, names, constants, and statement labels. Two or more blanks used as separators have the same interpretation as one blank. Within the following "compound keywords", blanks are *optional* to separate adjacent keywords:

BLOCK DATA	END FUNCTION	END TYPE
DOUBLE PRECISION	END IF	END WHERE
ELSE IF	END INTERFACE	GO TO
END BLOCK DATA	END MODULE	IN OUT
END DO	END PROGRAM	SELECT CASE
END FILE	END SELECT	
END FORALL	END SUBROUTINE	

1.2 Free Source Form

Because blanks must be used as separators or may be used as meaningful characters in character literal constants or in character constant edit descriptors, we say that they are *significant*. There is an exception for blanks in format specifications: blanks may appear within edit descriptors such as BZ, SS, etc.

An "!" character not appearing within a character literal constant or within a character constant edit descriptor indicates the beginning of a **comment**. This comment begins at the "!" and ends at position 132 of the line.

A line with an "!" as the first nonblank character is a **comment line**. A comment line must not be continued.

Summary of free source form

Position(s)	Character	Interpretation
1	!	Indicates a comment line
1 – 132	!	Initiates a comment
1 – 132	Fortran character set	Fortran statement
1 – 132	;	Separates 2 statements
1 – 132	&	Indicates continuation
1 – 132	Only blanks	Comment line

```
!0000000000000000000000000000000000000000000...111111111111111
!000000001111111111222222222233333333334444444...112222222222333
!23456789012345678901234567890123456789012345678...890123456789012
PROGRAM pascal_triangle                         ! initial line
INTEGER basis (13)                              ! initial line
CHARACTER fmt *10                               ! initial line
DATA basis /13*1/, null /23/                    ! initial line
                                                ! comment line
WRITE (*, 1111) &                               ! initial line
    ! writes the first 3 elements                 comment line
    basis(13), &                                ! contin. line
    basis(12), basis(13)                        ! contin. line
1111 FORMAT (5X,'Pascal &
            &triangle',5X,I6/22X,I6,I5)         ! contin. line
                                                ! comment line
DO 20 i=12,2,-1                                 ! initial line
   DO    j=i,12                                 ! initial line
      basis(j) = basis(j) + basis(j+1)          ! initial line
   END DO                                       ! initial line
null = null - 2; 22 FORMAT ('(',I2,'X,13I5)')   ! multipl stmt
```

```
WRITE (fmt, 22) null                     ! initial line
WRITE (*, fmt) (basis(k), k=i-1,13)      ! initial line
20 CONTINUE                              ! initial line
END PROGRAM pascal_triangle              ! initial line
```

1.3   Embedding of Program Lines by INCLUDE

At any line position of a program unit, an INCLUDE line may be used to include additional program lines for processing by the Fortran processor.

An INCLUDE line must be a program line of its own. It must not be labeled, but it may contain an "!" which indicates the beginning of a comment. An INCLUDE line must not be continued.

The forms for an INCLUDE line are:

```
INCLUDE 'C:\COMMON\ARI'
INCLUDE "$USER.PARAM"
INCLUDE 4_'fi.vor_k'
```

The keyword INCLUDE is followed by a character literal constant. This character literal constant may have a leading underline preceded by a *kind type parameter*, which must occur as a sequence of digits. The interpretation of the character constant is processor-dependent; usually, it is the name of a file containing the program line(s) to be included.

The lines to be included may form a complete or uncomplete part of a program unit. Each of these lines must be complete; that is, the first line to be included must not be a continuation line and the last line must not be a continued line. The source form of the lines to be included must be the same as the source form of the program unit containing the INCLUDE line.

The embedding of program lines may be nested; that is, an included line may itself be an INCLUDE line. If INCLUDE lines are nested, an inner INCLUDE line must not reference the same source program part as any of the outer INCLUDE lines. The maximal nesting depth of INCLUDE lines is processor-dependent.

1.4   Classification of Fortran Statements

There are *executable statements, nonexecutable statements, specification statements, execution control statements*, and *input/output statements*.

A Fortran statement is either *executable* or *nonexecutable*.

The following statements are **executable statements**. They cause actions of the program:

ALLOCATE	arithmetic IF	assignment
BACKSPACE	computed GO TO	CALL
CASE	CASE DEFAULT	CLOSE
CONTINUE	CYCLE	DEALLOCATE
DO	ELSE	ELSE IF
ELSEWHERE	END[1]	END DO
END FILE	END FORALL	END FUNCTION
END IF	END PROGRAM	END SELECT
END SUBROUTINE	END WHERE	EXIT
FORALL	FORALL construct stmt.	IF THEN
INQUIRE	logical IF	NULLIFY
OPEN	pointer assignment	PRINT
READ	RETURN	REWIND
SELECT CASE	STOP	unconditional GO TO
WHERE	WHERE construct stmt.	WRITE

All other Fortran statements are **nonexecutable statements**.

Several executable statements may appear only in combination with certain other statements to form executable **constructs**.

CASE construct: SELECT CASE, CASE, END SELECT
DO construct: DO, END DO
IF construct: IF THEN, ELSE IF, ELSE, END IF
FORALL construct: FORALL construct stmt., END FORALL
WHERE construct: WHERE construct stmt., ELSEWHERE, END WHERE

CASE, DO, and IF constructs control the execution of one or more statement blocks. FORALL and WHERE constructs control the selection of array elements for certain assignment operations. Executable statements which are not used to form constructs are **simple executable statements**.

1.5 Statement Ordering

The following diagram gives an overview of the ordering of statements and other language elements within program units and subprograms.

---

[1] END statement of a main program or a subprogram

			PROGRAM, SUBROUTINE, FUNCTION, MODULE or BLOCK DATA statement		
I N C L U D E lines	comment lines and blank lines	FORMAT and ENTRY statements	USE statements		
			IMPLICIT NONE statement		
			PARAMETER stm.s		IMPLICIT statements
					derived type definitions
					interface blocks
					type declaration statements
					specification statements
			DATA statements		statement function def.s
					executable statements
			CONTAINS statement		
			internal subprograms or module subprograms		
			END statement		

The language elements appearing in a particular box may occur in any order. Note that some statements are allowed only in certain program units, in certain subprograms, or in certain constructs. Vertical lines separate groups of language elements which also may occur in any order; for example, FORMAT statements may appear at any position among the specification statements and among the executable statements as well (but not after CONTAINS). Horizontal lines separate groups of language elements which must not be interspersed; for example, the statement function definitions must precede all executable statements and the IMPLICIT statements must precede all DATA statements.

ENTRY statements may appear in a subroutine or function at nearly every position between the USE statement(s) and the CONTAINS statement.

All type declaration statements and the other specification statements must precede all executable statements. The IMPLICIT statement must precede (nearly) all other specification statements. The order of the type declaration statements and the other specification statements may be important if a particular data object is specified or referenced in more than one specification statement.

The END statement is always the last physical line of a program unit or subprogram (except a statement function).

2 TYPE CONCEPT

There are *intrinsic types* and *derived types*. Intrinsic types are provided by the language, whereas derived types are provided by the programmer. All properties of the intrinsic types are known at any point in a Fortran program.

Each intrinsic type is parameterized. By specifying a **kind type parameter** value for a data type, the programmer selects a particular kind of that data type. If the programmer does not explicitly specify a kind type parameter, a default kind type parameter value is assumed, which selects the **default type**, that is to say, the default kind of that data type.

For an intrinsic type, the set of valid values, their internal representation, and their approximation method depend on the value of the selected kind type parameter. Note that the set of supported kind type parameter values is processor-dependent.

In addition to these intrinsic types, the programmer can derive types from intrinsic types and other derived types.

2.1 Intrinsic Types

There are 6 **intrinsic types**. These are the

numeric types   integer, real, double precision real, and complex, and the
nonnumeric types   logical and character.

The double precision real type is a special kind of the real type.

2.1.1 Integer Type

Name: INTEGER.

Sets of values: A set of values of the integer type is a subset of the mathematical integers. Each such set of values has a processor-dependent smallest negative value and a largest positive value.

A Fortran processor must provide at least one internal representation method of integer values. Each of these representation methods is characterized by a processor-dependent kind type parameter value. In addition to the keyword INTEGER, the programmer may specify a kind type parameter. If a kind type parameter is not explicitly specified, the default kind type parameter KIND(0)[1] is assumed. This default kind type parameter selects the **default integer type**.

---

[1] KIND is an intrinsic function. This is a portable way to denote a kind type parameter value. Kind type parameter values even for the default types are processor-dependent (see above).

The value zero is neither positive nor negative. A signed zero and an unsigned zero have the same value.

External representations: An integer value may be represented either as an *integer literal constant* with or without a kind type parameter or as a *BOZ literal constant*.

Operations: Addition, subtraction, multiplication, division, exponentiation, negation, and identity. These operations are defined for all numeric types.

```
INTEGER, PARAMETER :: small = SELECTED_INT_KIND(5)
```

The value of `small` is the kind type parameter value of an integer type that has a minimal range from -10^5 to $+10^5$.

```
INTEGER (KIND=small) x, y      ! is a type declaration
```

If the kind type parameter value for this integer type is 4, we also may write:

```
INTEGER (4) x, y
```

2.1.2 Real Type and Double Precision Real Type

Names: REAL and DOUBLE PRECISION.

Sets of values: A set of values of the real or double precision real type is a subset of the mathematical real numbers. Most real numbers have no exact internal representation. Therefore, a real value has a processor-dependent internal representation, which is an approximation to the exact mathematical number.

Such an approximation has two characteristic properties: the (decimal) *precision* and the (decimal) *exponent range*. A Fortran processor must support at least two different approximation methods of internal representation of real values. Therefore, the real type has at least two different sets of values. These two sets of values must be different with regard to precision, and they also may be different with regard to their range. If the Fortran processor supports exactly two kinds of the real type, the double precision real type is the same as the real type with the higher precision.

The approximation methods are characterized by a processor-dependent kind type parameter value. In addition to the keyword REAL, the programmer may specify a kind type parameter. If a kind type parameter is not explicitly specified, the default kind type parameter KIND(0.0) is assumed. This default kind type parameter selects the **default real type**.

The keyword DOUBLE PRECISION must be written without a kind type parameter. For the double precision real type, the kind type parameter KIND(0.0D0) is assumed. If one needs to use a real type with higher precision but without explicitly specifying a kind type parameter, one must use the double precision real type.

The sets of values contain the real and double precision real zero, respectively. In Fortran processors that distinguish between positive and negative real zeros and between positive and negative double precision real zeros, positive and negative zeros have the same interpretation and designate the same zero value

- in all relational operations,
- as input arguments to intrinsic subprograms other than SIGN, and
- as the scalar numeric expression in an arithmetic IF statement.

External representations: A real value may be represented as a *real literal constant* with or without a kind type parameter. And a double precision real value may be represented as a real literal constant with a kind type parameter value or as a *double precision real literal constant*.

Operations: Addition, subtraction, multiplication, division, exponentiation, negation, and identity. These operations are defined for all numeric types.

```
INTEGER, PARAMETER :: big = SELECTED_REAL_KIND(14, 200)
```

The value of big is the kind type parameter value of a real type with a decimal precision of at least 14 decimals and a minimal exponent range from 10^{-99} to 10^{+99}.

```
REAL (KIND=big) a, b          ! is a type declaration
```

If the kind type parameter value for this real type is 8, we also may write:

```
REAL (8) a, b
```

2.1.3 Complex Type

Name: COMPLEX.

Sets of values: A set of values of the complex type is a subset of the mathematical complex numbers. A complex value has an internal representation consisting of two real values, one for the real part and one for the imaginary part.

The Fortran processor must support at least the same approximation methods of internal representation of the real part and imaginary part as in the case of the real type. Both parts of a complex value must be represented according to the same approximation method.

The approximation methods are characterized by a processor-dependent kind type parameter value. In addition to the keyword COMPLEX, the programmer may specify a kind type parameter. If a kind type parameter is not explicitly specified, the default kind type parameter KIND(0.0) is assumed. This default kind type parameter selects the **default complex type**. The Fortran processor uses the same approximation method of internal representation of the real and the imaginary part as in the case of the default real type.

The sets of values contain the complex zero, which is neither positive nor negative. A signed zero and an unsigned zero have the same value.

External representations: A complex value may be represented as a *complex literal constant* with or without a kind type parameter.

Operations: Addition, subtraction, multiplication, division, exponentiation, negation, and identity. These operations are defined for all numeric types.

2.1.4 Logical Type

Name: LOGICAL.

Sets of values: A set of values of the logical type contains only two values with the interpretation *true* and *false*, respectively.

A Fortran processor must support at least one internal representation method of logical values. Each of these representation methods is characterized by a processor-dependent kind type parameter value. In addition to the keyword LOGICAL, the programmer may specify a kind type parameter. If a kind type parameter is not explicitly specified, the default kind type parameter KIND(.FALSE.) is assumed. This default kind type parameter selects the **default logical type**.

External representations: A logical value may be represented as a *logical literal constant* with or without a kind type parameter.

Operations: Negation, conjunction, inclusive disjunction, logical equivalence, and logical non-equivalence.

2.1.5 Character Type

Name: CHARACTER.

Sets of values: A set of values of the character type contains *character strings*. A **character string** is a sequence of characters. Each character in a string has a position within this string. The positions are numbered from the left to the right beginning with 1, 2, 3, ... The number of the last character position of a string is equal to the **length** of the string. The length may be zero.

A Fortran processor must provide at least one internal representation method of character values. Each of these representation methods is characterized by a processor-dependent kind type parameter value. In addition to the keyword CHARACTER, the programmer may specify a kind type parameter. If a kind type parameter is not explicitly specified, the default kind type parameter KIND('A') is assumed. This default kind type parameter selects the **default character type**.

For a particular kind type parameter value, a character string may contain any *representable characters* of the character set being defined by the representation method for that kind.

External representations: A character value (that is, a character string) may be represented as a *character literal constant* with or without a kind type parameter.

Operation: Concatenation. This operation is defined only for character strings of the same kind.

2.2 Derived Types

Additional data types may be defined by the programmer. For example, a new data type may be derived from intrinsic types; and such a derived type may be used to define another derived type. A derived type has at least one *type component*. Each component of a derived type specifies an intrinsic type or a derived type.

Name: The name of a derived type must be defined in the TYPE definition statement. This name must not be the same as the name of an intrinsic type or the name of another derived type being available in this *scoping unit*.

Set of values: A set of values of a derived type is a combination of the sets of values of the components of this data type. Ultimately, a value of a derived type is a collection of values of intrinsic types.

External representation: The external representation of the values of a derived type is given by a method for the construction of these values; it is called a *structure constructor*.

Operations: There are no intrinsic operations (except the assignment operation). Additional operations may be defined by the programmer with help of *operator functions* and *operator interface blocks*.

2.2.1 Derived Type Definition

A derived type definition is used to define the name of a new derived type and to define the names, attributes, and types of the components of this new derived type. Such a derived type definition begins with a TYPE definition statement, it ends with an END TYPE statement, and contains at least one type component definition. A type component may contain a *default initialization* specification such that each object of the derived type not being explicitly initialized will be implicitly initialized as specified in the type definition.

TYPE [...] **type_name**
 [**PRIVATE**]
 [**SEQUENCE**]
 component_definition
 [**component_definition**]
 ⋮
END TYPE [**type_name**]

If the derived type is defined in the specification part of a module, a PRIVATE statement (without a list of names) may appear anywhere after the TYPE definition statement but before the first component definition. In this case, the component names and thus the internal structure of the derived type are available within the module but unaccessible outside the module.

If the END TYPE statement includes a **type_name**, the corresponding TYPE definition statement must include the same name.

TYPE definition statement

A derived type definition begins with a TYPE definition statement. The form of the TYPE definition statement is:

TYPE [::] **type_name**

TYPE, PRIVATE :: type_name

TYPE, PUBLIC :: type_name

The keyword PRIVATE or PUBLIC may be specified only if the derived type definition appears within the specification part of a module. PRIVATE means that the derived type definition is accessible only within the scoping unit of the module containing this derived type definition; there are no means to make such a derived type definition accessible outside the module. And PUBLIC means that the derived type may become accessible outside the module. A public type also is described as a "visible" type.

SEQUENCE statement, SEQUENCE attribute

Normally, there is no *storage sequence* given by the order of the type components within a derived type definition. But if the derived type definition contains a SEQUENCE statement, the derived type has the **SEQUENCE attribute**. And the sequence of components defines a storage sequence for data entities of this derived type. The SEQUENCE attribute is useful if data objects of this type are specified in COMMON or EQUIVALENCE statements. A SEQUENCE statement may appear before or after the PRIVATE statement, if any, but before the first type component definition.

2.2 Derived Types

A derived type having the SEQUENCE attribute is called a **sequence type**. All components of a sequence type that are of derived type (and the components of such components etc.) also must have the SEQUENCE attribute. And if all ultimate components (after complete resolution) are of default numeric or default logical type and are not pointers, this derived type is called a **numeric sequence type** and the data entities of this type have a *numeric storage sequence*. And if all ultimate components are of default character type and are not pointers, this derived type is called a **character sequence type** and the data entities of this type have a *character storage sequence*.

```
TYPE numeric
  SEQUENCE
  INTEGER in, age, mark
  REAL    cache, tax
  LOGICAL yesorno
END TYPE numeric
```

Data entities of this type have a numeric storage sequence.

Equality of derived data types

Two data entities in a scoping unit have the same derived type if they are declared with regard to the same derived type definition. Two data entities in different scoping units have the same data type if they are declared with regard to the *same* derived type definition, which may be accessed from a module or from the host.

Two data entities in different scoping units are of equal derived type if they are declared with regard to *equal* derived type definitions. Two derived type definitions are **equal** if they have the same name, if they have the SEQUENCE attribute, and if the type components are equal with regard to order, name, rank, shape, type, kind type parameter, and (if applicable) character length.

Two data entities in different scoping units are not of equal derive type if at least one of these entities is private or has a private component.

2.2.1.1 Type Component Definition

The definition of a type component has a form similar to a type declaration statement. A component may be defined to have the DIMENSION attribute and/or the POINTER attribute. In addition, *default initialization* (see below) may be specified for a type component.

type [::] **type_component** [, **type_component**]...

type, attribute [, **attribute**] :: **type_component** [, **type_component**]...

type is a *type specification* of a derived type or of an intrinsic type (optionally with an explicitly specified kind type parameter and/or character length). And **type_component** is a **component_name** or a component name followed by an array specification in parentheses, **component_name (dim [, dim]...)**. If the component is of type character, a character length specification ∗ **own_length** may follow, where **own_length** must be a constant *specification expression*. A *default initialization specification* may follow, which is either an equals symbol followed by an initilization expression for a nonpointer component or a pointer assignment symbol "=>" followed by NULL( ) for a pointer component.

A leading character length specification which is included in the type specification **type** must be a constant specification expression. As an **attribute**, only **DIMENSION (dim [, dim]...)** and/or **POINTER** may be specified. The double colon may be omitted only if neither an **attribute** nor default initialization are specified.

If the **type** is a derived type without the POINTER attribute, this derived type must be defined earlier in the scoping unit or must be accessible there. If it is a derived type with the POINTER attribute, it may be any available derived type or even the derived type being defined.

```
TYPE date
  INTEGER            :: day
  CHARACTER (LEN=3)  :: month
  INTEGER            :: year
END TYPE date
```

The definition of the derived type `date` contains three type components. In a scoping unit where this derived type is available, data objects of type `date` may be declared:

```
TYPE (date) :: birthday, holiday, tax
```

An array object may be declared to be of a derived type:

```
TYPE (date), DIMENSION (25, 13) :: pupil
```

The variable `pupil` is a 2-dimensional array of type `date`. The value of any element of array `pupil` represents a date consisting of three values, as specified in the derived type definition.

Suppose, array `pupil` contains the birthdays of pupils. Then the complete date for the 3rd pupil in classroom no. 12 or a part of that date may be referenced:

```
PRINT *, pupil(3, 12)        ! prints the complete date
PRINT *, pupil(3, 12)%year   ! prints only the year of the birthday
```

Array components

A type component having the DIMENSION attribute is an **array component**. If the array component is not a *pointer component*, the array specification has the same form as for an *explicit-shape array* and the array bounds must be constant *specification expressions*. If the array component is a pointer component, the array specification has the same form as in the array declaration of an *array pointer*.

The array specification may be written as an own array specification following the **component_name** or the array specification may be written immediately after the keyword DIMENSION preceding the "::". In the first case, the array specification applies only to this one component. In the second case, the array specification applies to all components of this component definition having no own array specification.

```
TYPE experiment
  INTEGER :: number
  TYPE (date) :: day
  REAL, DIMENSION (100) :: sensor1, sensor2, hour(24)
END TYPE experiment
```

Where this derived type is available, variables for the storage of measurement results may be declared:

```
TYPE (experiment) :: temperature, density, height
```

Pointer components

A type component having the POINTER attribute is a **pointer component**.

```
TYPE bibliography
  INTEGER                          :: volume, year, pages
  CHARACTER (LEN=72)               :: titel
  CHARACTER, DIMENSION (:), POINTER :: abstract
END TYPE bibliography
```

In the scoping unit where this derived type definition of `bibliography` is available, objects of this type may be declared. Such an object has four components with known memory requirements. These are the default integer components `volume`, `year`, and `pages` and the default character component `titel`. And there is an additional component `abstract`, which is a pointer that may become pointer associated with a 1-dimensional character array.

A pointer component in a derived type definition is allowed to have the same type as the type being defined:

```
TYPE link
  INTEGER                   :: position
  TYPE (link), POINTER      :: left, right
END TYPE link
```

Where this derived type is available, objects of type link may be declared. These objects may be manipulated as elements of a linked list.

Default initialization specification

Type component definitions for a derived type may include **default initializations** such that each object of that derived type not being initialized explicitly by a type declaration statement will have implicitly initialized components as specified in the type definition. Such an object will be initialized by default regardless of whether the type definition is private and inaccessible or not. Default initialization is even applicable to dummy arguments with INTENT(OUT) and to allocatable arrays; in these cases, the component becomes initialized at invocation of the subprogram containing the dummy argument and at execution of the ALLLOCATE statement for the allocatable array, respectively. Unlike explicit initialization by a type declaration statement, default initilization does not imply that the implicitly initialized object has the SAVE attribute.

Nonpointer component: The component becomes initialized by default with the value of the **initialization_expression**. The evaluation of the initialization expression and the assignment of its value to the component obey the rules for "normal" intrinsic assignment statements as if **component_name** were a variable on the left-hand side of an intrinsic assignment. The evaluation is evaluated in the scoping unit of the type definition.

type [, DIMENSION (dim [, dim]...)] ::   &
 component_name [(dim [, dim]...)] [∗ own_length] = initialization_expression

```
TYPE trial
  INTEGER :: number
  TYPE (date) :: next = date(11, 'OCT', 1996)
  REAL, DIMENSION (100) :: result = 0
END TYPE trial
```

Default initialization need not be specified for all components of a type definition; that is, "partial default initialization" of objects is allowed. For the default initialization of an array component, the simplest form of the initialization expression may be a constant array constructor or even a scalar constant.

Pointer component: The pointer association status of the component becomes initialized by default with the pointer association status of the invocation of NULL( ); that is, the status becomes initialized to be disassociated. The

2.2 Derived Types

pointer assignment of NULL( ) to the component obeys the rules for "normal" pointer assignment statements as if **component_name** were a pointer variable on the left-hand side of a pointer assignment statement.

type, POINTER [, **DIMENSION (**: [,:]...**)**] :: &
       component_name [(: [,:]...)] [∗**own_length**] => **NULL( )**

```
TYPE cell
  INTEGER              :: value
  TYPE (cell), POINTER :: left => NULL(), right => NULL()
END TYPE cell
```

The head of a linked list may be declared as:

```
TYPE (cell), TARGET :: head   ! partially initialized by default
```

Multiple initializations: If default initialization is specified in a type component definition of a derived type which is itself a derived type with specified default initialization for its component(s), then the current initialization specification overrides the former one.

```
TYPE date
  INTEGER            :: day = 1
  CHARACTER (LEN=3)  :: month = 'JAN'
  INTEGER            :: year = 2000
END TYPE date

TYPE trial
  INTEGER :: number
  TYPE (date) :: next = date(11, 'OCT', 1996)
  REAL, DIMENSION (100) :: result = 0
END TYPE trial

TYPE (trial) test
```

The default initialization for component `next` of variable `test` is not `date(1,'JAN',2000)` but `date(11,'OCT',1996)`.

Explicit initialization by type declaration statement overrides default initialization:

```
TYPE date
  INTEGER            :: day = 1
  CHARACTER (LEN=3)  :: month = 'JAN'
  INTEGER            :: year = 2000
END TYPE date

TYPE (date) new_years_eve = date(31, 'DEC', 1996)
```

The default initialization for variable `new_years_eve` is overridden by explicit initialization such that the variable becomes initialized with the value `date(31,'DEC',1996)`.

Note that explicit initialization by DATA statement of such a derived type object or subobject is not allowed for which already (partial or complete) default initialization is specified in its type definition.

2.2.1.2 Private and Public Derived Type Definitions

A derived type (definition) or a type component (definition) is **private** if it is accessible only in the module containing the derived type definition. Outside the module, a private type component or a private type are not accessible and there are no means to make them accessible.

A derived type (definition) or a type component (definition) is **public** if it may also be made accessible outside the module containing the derived type definition. Public entities also are described as "visible" entities. Such entities of a module may be made accessible outside the module by a USE statement.

A data type is private if the PRIVATE attribute is specified in its TYPE definition statement, or if the PUBLIC attribute is not specified in its TYPE definition statement and there is either a PRIVATE statement without a list of names or a PRIVATE statement with the name of that derived type in the specification part of the module containing the derived type definition.

A data type is public if the PUBLIC attribute is specified in the TYPE definition statement, or if the PRIVATE attribute is not specified in the TYPE definition statement and there is no PUBLIC statement, a PUBLIC statement without a list of names, or a PUBLIC statement with the name of that derived type in the specification part of the module containing the derived type definition.

If a data type is private, the following characteristics and concepts of this type are available only in the module containing the derived type definition: the name of the type, the names of its components, objects and structure constructors of this type, and subprograms with dummy arguments and function results of this derived type.

A component of a derived type is private if it is a component of a private type or if the derived type definition contains a PRIVATE statement or if the type of the component is another private type. If at least one component of a derived type is private, all components must be private.

```
TYPE point
  PRIVATE
  REAL x, y
END TYPE point
```

This derived type definition may appear only in a module. The type **point** is available within the module and by a USE statement outside the module. The components x and y are only available within the module; the inner structure of the type **point** cannot be made accessible outside the module.

```
TYPE, PRIVATE :: note
  INTEGER number, weight
  LOGICAL ok
  CHARACTER (LEN=72) text
END TYPE note
```

This type **note** is not visible. It cannot be made accessible outside the module.

2.2.2 Structure Objects

Structure objects are *scalar* entities of a derived type such as *structure variables* and *structure constructors*.

The name of a *type component* is needed outside the derived type definition only when a single component of a structure object is referenced by qualifying the name of the parent structure by the name of the corresponding type component.

Structure variables

A structure variable must be declared by a TYPE declaration statement. Note that the type definition of this derived type must be *available* in the same scoping unit before this TYPE declaration statement. "Available" means that the derived type definition must either appear before the declaration or must be accessible by *USE association* or *host association*.

Constructed structure objects: structure constructor

In a scoping unit where the derived type definition and the inner structure of the type are available, values of this derived type may be constructed. A **structure constructor** supplies a series of values.

type_name (expression [, expression]...)

The list of expressions enclosed in parentheses must supply a value for each single component of the derived type **type_name**. And these values must be suitable (in number and order) for the components of this derived type. If the result of such an expression does not agree in type, in kind type parameter, or (if applicable) in character length with those of the corresponding type component, the result is converted according to the rules for intrinsic assignment statements. For a nonpointer component, the shape of the result of the expression also must conform with the shape of the component.

A structure constructor may appear only after the derived type definition.

```
TYPE string
  INTEGER length
  CHARACTER (LEN=max) line
END TYPE string

CHARACTER text *25, mark *8
TYPE (string) colour

READ *, text, mark
colour = string(LEN(text) + LEN(mark), text // mark)
```

An expression is given for each component of the type **string**. After evaluation of these expressions, the constructed value is assigned to the variable **colour**. This variable also is of type **string**.

If the derived type has a pointer component, the **expression** corresponding to this pointer component must evaluate to a result which would be legal on the right-hand side of a *pointer assignment statement*.

```
TYPE (bibliography) :: book
CHARACTER, DIMENSION (1000), TARGET :: source
  ⋮
book = bibliography(1, 1996, 390, 'Language Guide', source)
```

In this example, **source** is the target object for the corresponding component abstract; compare an earlier example on page 2-9.

Constant structure constructors

If all expressions in a structure constructor are constant expressions, it is a **constant structure constructor** or more precisely a *derived type constant expression* or simply a *structure constant*.

```
TYPE date
   INTEGER            :: day
   CHARACTER (LEN=3)  :: month
   INTEGER            :: year
END TYPE date

TYPE (date) :: birthday
  ⋮
birthday = date(11, 'OCT', 1996)
```

A constant is given for each component of the type **date**. This structure constant is assigned to the variable **birthday**, which also is of type **date**.

3 LEXICAL TOKENS

A Fortran statement consists of **lexical tokens**. These are keywords, names, operators, statement labels, literal constants (except complex literal constants), delimiter pairs as (...), /... /, " ... ", ' ... ', ..., and (/... /), and finally, =, =>, &, :, ::, ;, and %.

3.1 Scoping Units

A program unit consists of one or more nonoverlapping *scoping units*. A **scoping unit** consists of all program lines of a derived type definition, of an interface block without any of its embedded interface blocks, or of a program unit or a subprogram without any of its embedded derived type definitions, interface blocks, and subprograms.

The **scope** of a lexical token is that part of a program where the interpretation of the token is unambiguous. If the scope of an entity is the program, the entity has a **global scope**. And if the scope is a scoping unit, the entity has a **local scope**. There are entities with a scope that is only a statement or a part of a statement.

3.2 Keywords

Fortran **keywords** are the names of the statements (that is, the given unmodifiable parts of a statement that identify the statement), the names of the input/output specifiers in input/output statements, and the predefined names of the dummy arguments of the intrinsic subprograms.

There are *no* reserved words in Fortran. A keyword has a predefined interpretation only if it appears at a particular place in a statement. Otherwise it may be used freely at any other place in a statement.

```
      READ (*, 1234, ERR=100) FORMAT
1234 FORMAT (I10)
```

READ and FORMAT are keywords only when they appear at the beginning of a READ statement and FORMAT statement, respectively. And ERR is a keyword only when it appears before an equals in the control information list of an input/output statement.

```
READ = FORMAT + READ * ERR
```

In this case, READ, FORMAT, and ERR are recognized as names of data objects. However, such usage is not recommended because it reduces readability of programs.

3.3 Names

Names identify variables, subprograms, data types, and so on. Normally, names are invented and inserted by the programmer. The scope of a name contains all statements where the name is known and may be used. Names with global scope are *global names*. And names with local scope are *local names*. There are *special names* with a scope consisting only of one statement or consisting only of a part of a statement. Normally, a name is unambiguous in its scope (exception: generic names).

A **name** consists of one through 31 letters, digits, and underscores "_", beginning with a letter. If the **Fortran** processor supports lower-case letters, case is not significant.

```
salary    peter     x12    e605    FortranReferenceManualPages
salary_june_1992            ←— single underscore characters
b_o_l_d__face               ←— consecutive underscore characters
go2heaven_                  ←— trailing underscore character
DINA4     DinA4             ←— identical names
lucky dog    ⎫
luckydog     ⎬ ←— identical names in fixed source form
```

Global names identify the following **global entities:** main programs, external subroutines, entries of external subroutines, external functions, entries of external functions, modules, block data program units, and common blocks.

Local names identify **local entities**, which are classified as follows:

1. Named variables (except variables with *special names*, see below), named constants, named constructs, statement functions, internal subprograms, module subprograms, dummy subprograms, intrinsic subprograms, interface blocks with generic names, derived types, namelist group-names,

2. Type components, and

3. Argument keywords (except of statement functions).

Note that the names of type components form a separate class for each derived data type and that the argument keywords form a separate class for each subprogram.

Except for a common block name, an external subprogram name that is also a generic name, or an external function name within its function definition, a name that identifies a global entity in a scoping unit must not be used to identify a local entity of class 1. (see above) in that scoping unit.

A local name may identify in *other* scoping units another local or global entity. A local name may identify in the *same* scoping unit another local entity of *another* class.

Special names: A *special name* has a scope that consists only of an executable construct, a single statement, or even only a part of a statement. The name of a *dummy argument of a statement function* has a scope that is this statement function definition. The name of a *DO variable* of an implied-DO in a DATA statement or in an array constructor has a scope that is the implied-DO list. The name of an index variable of a FORALL statement or FORALL construct statement has a scope that is the FORALL statement and the FORALL construct, respectively.

If a name of a local or global entity accessible in a scoping unit is the same as a special name in a statement in the scoping unit, the name is interpreted within the construct, the statement, or part of the statement as the dummy argument, as the DO variable of the implied-DO, and as the index variable, respectively, and outside the construct, the statement, or part of the statement as the global or local entity. In any case, the variable (name) has a type, a kind type parameter, and (if applicable) a character length as though it were the name of a variable in the scoping unit containing the statement function definition, the DATA statement, the array constructor, the FORALL statement, and the FORALL construct statement, respectively.

3.4 Operators and Assignment Symbol

Intrinsic operators are global entities. A *defined* operator that is not an extended intrinsic operator is a local entity. Operators may be overloaded. An intrinsic operator may be used to identify a defined operator; in this case, the interpretation of the intrinsic operator is *extended*. Defined operators may also be extended.

The assignment symbol is a global entity. Within a scoping unit, the assignment symbol may identify additional *defined* assignment operations or replace the intrinsic derived type assignment operation. A defined assignment operation may also be extended.

3.5 Statement Labels

The scope of a statement label is a scoping unit; a particular statement label may identify exactly one statement in a scoping unit.

3.6 Literal Constants

There are nine forms of *intrinsic* literal constants: integer, real, double precision real, complex, logical, character, binary, octal, and hexadecimal literal constants. According to the different kinds of the intrinsic types supported by a Fortran processor, literal constants (except binary, octal, and hexadecimal literal constants) may be written with an additional kind type parameter.

The type of a literal constant is not specified explicitly but follows from the form of the constant. Integer, real, double precision real, and complex constants are **numeric constants**. Binary, octal, and hexadecimal constants may be interpreted as special forms of literal constants of type default integer.

3.6.1 Integer Literal Constants

An integer literal constant is a string of digits. It consists of one or more digits, an optional sign, and an optional trailing underscore followed by a kind type parameter. Positive integer literal constants *may* be written with a leading sign "+". Negative integer literal constants *must* be written with a leading sign "−".

[±] **d** [**d**]... [**_ kind**]

Each **d** is a digit. **kind** is a processor-dependent kind type parameter value, which may be written as a string of digits or as a scalar nonnegative integer named constant.

Such an integer constant is interpreted as a decimal value. If there are particular requirements concerning the range of an integer literal constant, a kind type parameter may be written. The value of the kind type parameter must be supported by the **Fortran** processor. If a kind type parameter value is not written, the constant is of type default integer.

```
1    531    +76    9876    -239657    0    ⟵ type default integer
100_2       1024_10        739840_B4       ⟵ with kind type parameter
```

3.6.2 Real Literal Constants

A real literal constant is a string of digits with a point, with an exponent, or with both a point and an exponent. Positive real literal constants and positive exponents *may* be written with a leading sign "+". Negative real literal constants and negative exponents *must* be written with a leading sign "−". A real literal constant may be written with a trailing underscore followed by a kind type parameter.

[±] **dec** [**_ kind**]
[±] **dec E** [±] **exponent** [**_ kind**]
[±] **n E** [±] **exponent** [**_ kind**]

3.6 Literal Constants

dec is a decimal number of the form **n.** or **n.n** or **.n**, where **n** is an unsigned integer literal constant. The **exponent** is an integer literal constant; base is 10. **kind** is a processor-dependent kind type parameter value, which may be written as a string of digits or as a scalar nonnegative integer named constant.

If there are particular requirements concerning the decimal precision and/or the decimal exponent range of the internal representation, a kind type parameter may be written. The value of the kind type parameter must be supported by the Fortran processor. If a kind type parameter value is not written, the constant is of type default real.

Constants of type default real:

12.   -12.34   +.34

12.E-2   value: $12.0 * 10^{-2} = 0.12$
-12.34E+3   value: $-12.34 * 10^3 = -12340.$
+.34E4   value: $0.34 * 10^4 = 3400.$

Real constants with processor-dependent kind type parameter:

-23.4_4   3E-5_10   3.75E7_B4

If the kind type parameter **kind** is equal to the value of KIND(0.0D0), the approximation method of double precision real constants is used; that is, such a real literal constant is equivalent to a double precision real constant.

3.6.3 Double Precision Real Literal Constants

A double precision real literal constant is a string of digits with an exponent or with both a point and an exponent. Positive double precision real literal constants and positive exponents *may* be written with a leading sign "+". Negative double precision real literal constants and negative exponents *must* be written with a leading sign "−".

[±] **dec D** [±] **exponent**

[±] **n D** [±] **exponent**

dec, **exponent**, and **n** are interpreted as for real constants.

A double precision real constant has more precision than a default real constant. The Fortran processor must supply the same approximation method of the internal representation as for a real constant with the kind type parameter KIND(0.0D0).

12.D-2   value: $12.0 * 10^{-2} = 0.12$
-12.34D+3   value: $-12.34 * 10^3 = -12340.$
+.34D4   value: $0.34 * 10^4 = 3400.$

3.6.4 Complex Literal Constants

A complex literal constant is written as a pair of integer, real, and/or double precision real literal constants, which are separated by a comma and enclosed in parentheses.

(real_part , imaginary_part)

The **real_part** and the **imaginary_part** are integer, real, and/or double precision real literal constants.

The parentheses are parts of the constant. The data types of the real part and the imaginary part may be different. If both parts are of type integer, they are internally represented as two values of type default real. If only one part is of type integer, it is converted according to the same approximation method as the other part. If both parts are of type default integer or default real, it is a complex constant of type *default complex*. If both parts are of type double precision real, they are internally represented as two double precision real constants.

If the real part and the imaginary part are of type real but have different kind type parameters, they are represented internally according to one method; the part with less decimal precision will be converted according to the approximation method of the other part. If both parts have the same precision, the kind type parameter and the approximation method of the complex constant are processor-dependent.

In any case, the kind type parameter of a complex constant is the kind type parameter value of the part with the approximation method which is applied to both parts.

```
(7, 3.14)            value: 7. + 3.14 i
(-6.2E-3, 9)         value: −.0062 + 9.0 i
(-6.2E-3_8, 9_4)     value: −.0062 + 9.0 i
(5D-3, 5D-6)         ⟵ double precision complex
```

3.6.5 Logical Literal Constants

There are exactly two logical literal constants. One represents the value *true* and the other represents the value *false*; the form is .TRUE. and .FALSE., respectively. A logical literal constant may be written with a trailing underscore followed by a kind type parameter. The points are parts of the constants.

.TRUE.[_kind]   for the value *true*.

.FALSE.[_kind]   for the value *false*.

kind is a processor-dependent kind type parameter value, which may be written as a string of digits or as a scalar nonnegative integer named constant.

If there are particular requirements concerning the internal representation, a kind type parameter may be written. The value of the kind type parameter must be supported by the **Fortran** processor. If a kind type parameter value is not written, the constant is of type default logical.

.TRUE.
.FALSE.
.FALSE._4   ⟵ with kind type parameter

3.6.6 Character Literal Constants

A character literal constant is a character string enclosed either in quotation marks or in apostrophes. It may be written with a leading underscore preceded by a kind type parameter. The string may be empty, that is, the string may consist of no character at all. Blank characters within character constants are significant; they are parts of the character value and contribute to the length of the character constant. The delimiting quotation marks or apostrophes are not parts of the value of the character literal constant.

[**kind**_]"[**c**]..."

[**kind**_]'[**c**]...'

kind is a processor-dependent kind type parameter value, which may be written as a string of digits or as a scalar nonnegative integer named constant. And each **c** is a character of the **Fortran** character set or another *representable character* (see below) of the processor character set or of another processor-dependent character set that is defined by **kind**.

A quotation mark " " "appearing within the string of characters is represented by two consecutive quotation marks without any intervening blanks if the string is delimited by quotation marks. The included pair of quotation marks is counted as one character. An apostrophe " ' "appearing within the string of the character constant is represented by two consecutive apostrophes without any intervening blanks if the string is delimited by apostrophes. The included pair of apostrophes is counted as one character.

The number of characters between the delimiters is the **length** of the character constant. The length is fixed and may be zero. A zero-length character literal constant is represented by two consecutive quotation marks or two consecutive apostrophes without any intervening blanks.

A **representable character** is

- Any character of the processor-dependent character set if the program unit is written in fixed source form. The **Fortran** processor may exclude some or all control characters.

- Any character of the processor-dependent character set except control characters if the program unit is written in free source form.

If there are particular requirements concerning the internal representation, a kind type parameter may be written. The value of the kind type parameter must be supported by the **Fortran** processor. If a kind type parameter value is not written, the constant is of type default character.

`"American National Standard"`	- length 26
`'o''Henry'`	- length 7
`'Programming Language Fortran 95'`	- length 31
`''''`	- length 1
`'3.1415'`	- length 6
`"2,5"" height"`	- length 11
`''`	- length 0
`gr_"Φωρτραν"`	⟵ with kind type parameter
`de_'Grüße'`	⟵ with kind type parameter

3.6.7 Binary, Octal, and Hexadecimal Literal Constants

A binary, octal, or hexadecimal literal constant is a string of digits (in the binary, octal, and hexadecimal number system, respectively) enclosed either in quotation marks or in apostrophes. The number system is specified by an additional leading B, O, or Z, respectively. The string of digits must contain at least one digit.

Such a **BOZ constant** is interpreted according to its number system. It may appear only in a constant list in a DATA statement, where it may initialize a scalar variable of type integer.

`B"10001"`	`B'111110'`	`B"1111001111001111001111"`	⟵ binary constants
`O"21"`	`O'76'`	`O"17171717"`	⟵ octal constants
`Z"11"`	`Z'3E'`	`Z"3CF3CF"`	⟵ hex. constants

No other special character must appear between the leading B, O, or Z and the last delimiting quotation mark or apostrophe (except blank characters in fixed source form). If the **Fortran** processor supports lower-case letters, the hexadecimal digits A to F may be written as lower-case letters.

4 DATA OBJECTS

Data entities (or simply **data**) are *constants, variables, results of the evaluation of expressions*, and *function results*. **Data objects** (or simply **objects**) are *constants, subobjects of constants*, and *variables*.

Each data object has a data type. The data types of a literal constant and of a *structure constructor* are implicitly given by their respective forms. The types of a named constant and of a named variable may be specified explicitly or implicitly.

A constant has *always* a value. This value may be referenced (that means, used), but a constant can never be redefined during the execution of the program. A variable *may* have a value, may have no value, or may have no value at times. Variables may be defined or redefined during the execution of the program.

The name of an object may be used to specify the type and/or additional *attributes* of the object in a type declaration statement or by other specification statements.

A scalar derived type object is a *structure object* (or simply a *structure*). A structure has components. These structure components are subobjects of the structure object.

Each object has a *rank*; that is, it is either a *scalar* or an *array*. An object is a scalar if it is not an array. A structure is a scalar even if it has an array component.

An *array* is a collection of scalar objects, which have the same type, the same kind type parameter, and (if applicable) the same character length. These *array elements* are ordered (from the point of view of the programmer) such that they form a vector, a matrix, or a cube, and so on.

An *array section* is a subset of the elements of an array. It has (nearly) all properties of an array object, but it has no name. Array sections of a particular array may overlap.

The terms *scalar* and *array* also are used to characterize the *rank* or *shape* of a data entity. For example, the result of the evaluation of an expression or a function result may be scalar or array-valued.

There is a method for the construction of 1-dimensional array values; it is called an *array constructor*.

A *pointer* is a variable which has the POINTER attribute. A pointer may not be referenced or defined until it becomes associated with a *target*.

4.1 Constants

A **constant** has a data type, a kind type parameter, (if applicable) a character length, and a value. If a constant has the PARAMETER attribute, it also has a name. A constant having a name is a **named constant**. *Literal constants* and *constant structure constructors* are constants having no names.

There are six *intrinsic* types of constants: integer, real, double precision real, complex, logical, and character constants. The data type of a literal constant need not be and cannot be specified explicitly, because the type is given by the form of the constant. The data type of a named constant may be specified explicitly or implicitly.

Integer, real, double precision real, and complex constants are called **numeric constants**. *BOZ constants* may be regarded as special purpose integer literal constants.

Subobjects of constants

A subobject of a constant has the same form as a subobject of a variable. Subobjects of constants are: an array element of a named constant, an array section of a named constant, a structure component of a named constant, a character substring of a named character constant or of a character literal constant, and a character substring array section of a named character constant.

A subobject of a constant seems to have similar properties as a variable. It may depend on the values of variables, for example, if the subobject designator references a variable in a subscript expression. Or two identical subobject designators of a constant in a scoping unit may denote different values. Or the actual part of the parent constant which is denoted by a subobject designator may not be known until the execution of the program.

Because subobjects of constants are also constants, they must not be redefined.

```
CHARACTER nix (6) *3, command1, command2
PARAMETER (nix = (/'ls ', 'c  ', 'mcd', 'rm ', 'cat', 'man'/) )
i = 1
command1 = nix(i)(1:1)
⋮
i = 6
command2 = nix(i)(1:1)
```

The first `nix(i)(1:1)` designates the value 'l'. And the second appearance of `nix(i)(1:1)` designates the value 'm'.

4.2 Variables

A **variable** is
- A named scalar variable;
- A named array variable; or
- An unnamed subobject; that is,
  - An array element (= scalar object);
  - An array section (= array object);
  - A structure component (= scalar or array object); or
  - A substring of a character variable (= scalar object).

A subobject is designated by first writing the parent object, of which it is a part, and then writing additional details, which "qualify" the parent object until the subobject is unambiguously specified. These qualifiers are different for array elements, array sections, etc.

A variable has a type, a kind type parameter, (if applicable) a character length, and, sometimes but not always, a value. There are six *intrinsic* types of variables: integer, real, double precision real, complex, logical, and character variables. In a scoping unit where a derived type definition is available, named variables of this derived type may be declared. The type of a named variable may be specified explicitly or implicitly. The integer, real, double precision real, and complex variables are called **numeric variables**.

A variable may be *defined* or *redefined* during the execution of the program; that is, it may be supplied with a *valid* value or with a new valid value. A variable which has a valid value has the definition status "defined". A variable which does not have a valid value has the definition status "undefined".

A variable appearing in an executable statement is either interpreted as the address of the variable (for example on the left-hand side of a numeric assignment statement) or as the value of the variable (for example on the right-hand side of a numeric assignment statement). The form of the variable is independent of its usage as an address or as a value.

Named variables

The type of a named variable is either explicitly specified in a type declaration statement or the type is implicitly specified. The type of an implicitly typed variable depends on the first character of the name of the variable. Implicit typing may be controlled by an IMPLICIT statement. If a variable is affected both by implicit typing and by explicit typing, explicit typing has priority.

Default implicit typing may be in effect for integer and real entities even without writing an IMPLICIT statement. Therefore, variables of type default integer and default real need not be declared by a type declaration statement and there need not be an IMPLICIT statement which specifies implicit typing for them.

Each explicitly or implicitly typed object which does not have the PARAMETER attribute is a variable.

Unnamed variables, subobjects

The type of an unnamed variable depends on the type of the parent object of which it is a subobject. A type declaration is neither needed nor possible.

Array element: an array element is of the same type as its parent array.

Array section: an array section is of the same type as its parent array.

Structure component: a structure component is of the type that is specified for its corresponding type component. The type of the type component is explicitly specified in the derived type definition.

Character substring: a character substring is always of type character.

4.3 Scalars

An object which is no array is a scalar. Its value is a single value in the set of values which characterizes the type of the object. Scalars have *rank* zero.

4.3.1 Character Substrings

Parts of a scalar character object may be referenced. A **character substring** (or simply **substring**) is a contiguous portion of its parent string.

The characters of the scalar parent string are counted from the left to the right, beginning with character position one. For a reference to a substring, its parent string, the starting position of the substring, and the ending position of the substring within the parent string must be specified.

parent_string ([starting_position] : [ending_position])

The **parent_string** must be a scalar named character variable, a character array element, a scalar structure component of type character, or a scalar character constant. The **substring expressions starting_position** and **ending_position** must be scalar expressions. If the **starting_position** is not specified, the default value 1 is assumed. If the **ending_position** is not specified, its default value is equal to the length of the parent string (this corresponds to the last position of the parent string).

If the parent string is a variable, the substring also is a variable.

The results of the substring expressions must be in the range from 1 up to and including the length of the parent string. The **length** of a character substring is the number of characters in the substring, which may be calculated as

MAX(**ending_position** − **starting_position** + 1, 0).

For a *non*zero-length substring, (that is, if the value of the **ending_position** is not less than the value of the **starting_position**), following inequality holds:

$1 \le$ **starting_position** $\le$ **ending_position** $\le$ length of the **parent_string**.

Suppose, the parent string is the character constant 'HANDKERCHIEF', then is

Character string	Value
'HANDKERCHIEF' (:4)	HAND
'HANDKERCHIEF' (8:)	CHIEF
'HANDKERCHIEF' (2:4)	AND

Note, if the parent string is a constant, the substring also is a constant.

```
CHARACTER z *12
DATA z /'KINDERGARTEN'/
```

Suppose, the parent string is the variable z, then is

Character string	Value
z (8:10)	ART
z (:4)	KIND
z (:)	KINDERGARTEN

The reference to z(:) is the same as the reference to z.

```
CHARACTER z_ar (3) *6
DATA z_ar /'LUNEDI', 'SABATO', 'GIORNO'/
```

Suppose, the parent strings are the array elements z_ar(1), z_ar(2), and z_ar(3), then is

Character string	Value
z_ar (1) (:)	LUNEDI
z_ar (2) (5:)	TO
z_ar (3) (3:4)	OR

The reference to z_ar(1)(:) is the same as the reference to z_ar(1).

4.4 Arrays

An **array** is a collection of scalar data, which are rectangularly ordered as a vector, a matrix, or a cube, etc. **Fortran** supports 1-dimensional, 2-dimensional, 3-dimensional, and multi-dimensional arrays up to seven dimensions. An array has a type, a kind type parameter, (if applicable) a character length, certain characteristics concerning its shape, as there are dimensionality and size, possibly a name, and possibly a value.

Corresponding to the six intrinsic types, there are integer, real, double precision real, complex, logical, and character arrays. In a scoping unit where a derived type definition is available, arrays of this derived type may be declared. Integer, real, double precision real, and complex arrays are called **numeric arrays**.

The type of a named array is either explicitly specified in a type declaration statement or the type is implicitly specified. The type of an implicitly typed array depends on the first character of its name. Implicit typing may be controlled by an IMPLICIT statement. If an array is affected by both implicit and explicit typing, explicit typing has priority.

An **array section** is a subset of the scalar data (that is, the elements) of its parent array. It is an array object, but it is unnamed. An array section may be referenced by the qualified name of its parent array. For example, such an array section may be designated by the name of its parent array followed by an *array section subscript list* enclosed in parentheses.

An **array element** is the smallest subobject of an array. It is a scalar and has no name. An array element may be referenced by the qualified name of its parent array. For example, such an array element may be designated by the name of its parent array followed by a *subscript list* enclosed in parentheses.

All elements of an array are of the same type, have the same kind type parameter, and have (if applicable) the same character length. Type, kind type parameter, and length are those of the parent array. An array element may have a value.

Named arrays must be declared by an *array declaration* specifying the characteristics of the array. The specification of the rank and (if applicable) the shape of an array is called an *array specification*.

An array declaration may contain the specification of *array bounds*. Then a *lower bound* and an *upper bound* may be specified for each dimension.

Array declarations and *array specifications* are presented in detail in chapter 6.

```
INTEGER   l (5, -100:0), m (5:9, 101), n (10)
TARGET    l (5, -100:0), m (5:9, 101), n (10)
DIMENSION l (5, -100:0), m (5:9, 101), n (10)
COMMON    l (5, -100:0), m (5:9, 101), n (10)
```

There are also unnamed arrays, for example, the value of an array-valued expression, an array section, or a structure component that is an array.

Fortran supports different kinds of arrays:

Allocatable array: an array with ALLOCATABLE attribute. Its array bounds are not specified in the array declaration but in an ALLOCATE statement.

When this ALLOCATE statement is executed, the array will be allocated, that is to say, created. An existing allocatable array may be deallocated.

Array pointer: an array with POINTER attribute. Its array bounds are not specified in the array declaration but, for example, in an ALLOCATE statement for pointer allocation.

Assumed-shape array: a dummy argument array which receives its shape from its associated actual argument array.

Assumed-size array: a dummy argument array which looks like an explicit-shape array except for the upper limit of the last dimension. Its size must not exceed the size of the associated actual argument.

Automatic array: an array in a subprogram that is no dummy argument array and which has a shape that depends on (at least) one nonconstant specification expression.

Dummy argument array: an array which is a dummy argument in a subprogram interface.

Explicit-shape array: an array whose upper array bounds are specification expressions. The shape and the length of such an array are given.

Variable array: an array which has a shape that depends on at least one *non*constant specification expression.

The **size of an array** is the total number of its array elements. The size is equal to the product of the sizes of the dimensions of the array. The size of a dimension is called the **extent** of the array in that dimension. The extent of an array in a dimension is normally calculated as (**upper_bound − lower_bound** + 1) (see chapter 6). An array may have size zero.

```
DIMENSION vector (15)              ! 1-dim., 15 elements
COMMON tab (-5:5, 1976:1985)       ! 2-dim., 11*10=110  elements
INTEGER temp (-30:40, 1981:1986, 5) ! 3-dim., 71*6*5=2130 elements
```

The number of dimensions of an array is called the **rank** of the array. The **shape** of an array is given by its rank and by the extents of the array in all dimensions. The shape may be described as a 1-dimensional array whose array element values are equal to the extents in the corresponding dimensions. Note that the shape of an array does not say anything about the precise array bounds.

Once an array is declared, its rank remains fixed during the execution of the program. But for a dummy argument array, an automatic array, an array pointer, and an allocatable array, the extents of the array in each dimension and thus its size may vary.

4.4.1 Inner Structure of Arrays

An array is a data object. It consists of a set of scalar array elements, which are ordered rectangularly. This ordering of the array elements as a vector, a square, or a cube, etc. happens only in the imagination of the programmer. In addition to this application-oriented one- or multi-dimensional imaginary structure, there is an *internal* (1-dimensional) imaginary sequence of the array elements. This internal order is called the **array element order**.

The array elements are internally ordered *column-wise*: if a multi-dimensional array is referenced in array element order, the first subscript changes most rapidly and the last subscript changes most slowly.

```
DIMENSION y (4, 3)      !  2-dimensional array with 12 elements
```

	Column 1	Column 2	Column 3
Row 1	y(1,1)	y(1,2)	y(1,3)
Row 2	y(2,1)	y(2,2)	y(2,3)
Row 3	y(3,1)	y(3,2)	y(3,3)
Row 4	y(4,1)	y(4,2)	y(4,3)

Array element	y(1,1)	y(2,1)	y(3,1)	y(4,1)	y(1,2)	y(2,2)	...
Position	1	2	3	4	5	6	...

Array element	y(3,2)	y(4,2)	y(1,3)	y(2,3)	y(3,3)	y(4,3)
Position	7	8	9	10	11	12

```
DIMENSION z (3, 3, 2)   !  3-dimensional array with 18 elements
```

	Col. 1	Col. 2	Col. 3	Col. 1	Col. 2	Col. 3
Row 1	z(1,1,1)	z(1,2,1)	z(1,3,1)	z(1,1,2)	z(1,2,2)	z(1,3,2)
Row 2	z(2,1,1)	z(2,2,1)	z(2,3,1)	z(2,1,2)	z(2,2,2)	z(2,3,2)
Row 3	z(3,1,1)	z(3,2,1)	z(3,3,1)	z(3,1,2)	z(3,2,2)	z(3,3,2)
		Plane 1			Plane 2	

Element	z(1,1,1)	z(2,1,1)	z(3,1,1)	z(1,2,1)	z(2,2,1)	z(3,2,1)	...
Position	1	2	3	4	5	6	...

Element	z(1,3,1)	z(2,3,1)	z(3,3,1)	z(1,1,2)	z(2,1,2)	z(3,1,2)	...
Position	7	8	9	10	11	12	...

Element	z(1,2,2)	z(2,2,2)	z(3,2,2)	z(1,3,2)	z(2,3,2)	z(3,3,2)
Position	13	14	15	16	17	18

Array element order

The position of a single array element $(s_1, s_2, \ldots, s_n)$ within the internal (imaginary) sequence of elements is given by the following formula:

$$1 + (s_1 - j_1) + \sum_{m=1}^{n-1}\left((s_{m+1} - j_{m+1}) * \prod_{i=1}^{m} d_i\right)$$

Where n is the rank of the array,
  s_i is the integer result of the i-th subscript expression,
  j_i is the lower array bound in the i-th dimension,
  k_i is the upper array bound in the i-th dimension, and
  $d_i = max(k_i - j_i + 1, 0)$ is the extent in the i-th dimension.

If the array has *not* size zero, then $j_i \leq s_i \leq k_i$ for $i = 1, 2, \ldots, n$.

```
DIMENSION r (0:15)
  :
r(5) = 94.
```

The array element `r(5)` is at position 6 in the 1-dimensional array `r`, because $(1 + (5 - 0)) = 6$.

```
CHARACTER (LEN=8) s (0:2, -2:3)
  :
s(0, 3) = 'examples'
```

Array element `s(0, 3)` is at position 16 in the 2-dimensional array `s`, because $(1 + (0 - 0) + (3 - (-2)) * (2 - (-0) + 1) = 16$.

If the position of an array element and its character length are given, the first character position of an array element relative to the first character position of the array can be calculated as

  starting_position = 1 + (array_element_position − 1) ∗ character_length .

The array element `s(0, 3)` starts at character position 121, because $1 + (16 - 1) * 8 = 121$.

4.5 Structure Components

A **structure component** is a subobject of a derived type object. It is either one of the components of a (scalar) structure object or it is an array whose elements are themselves components of the corresponding elements of a derived type array.

A structure component does not have a name. It may be referenced by specifying the qualified name of its parent object. Where the name of the type component of the derived type is available, a structure component is identified at least by the designator of the parent object, of which it is a part, and the name of the corresponding type component.

part$_1$ [%**part**$_i$]... %**part**$_n$

part$_1$ is the designator of the parent object of derived type. It is the name of a structure object or of an array object, an array element designator, or an array section designator.

part$_i$ is of derived type. It may be one of the following forms:

- A **component_name** of the preceding **part**$_{i-1}$. This component may have the DIMENSION attribute.
- An array element designator **array (subscript_expr [, subscript_expr]...)**, where **array** is the name of a component (with DIMENSION attribute) of the preceding **part**$_{i-1}$.
- An array section designator **array (section_subscript [, section_subscript]...)**, where **array** is the name of a component (with DIMENSION attribute) of the preceding **part**$_{i-1}$.

And **part**$_n$ is the name of a component of the preceding **part**$_{n-1}$. This component may have the DIMENSION attribute. Not more than one **part**$_k$ may designate a whole array or an array section.

If **part**$_k$ designates a whole array or an array section, the names **component_name** and **array**, appearing in **part**$_{k+1}$ up to **part**$_{n-1}$, and the name **part**$_n$ must not have the POINTER attribute.

Structure components which are array objects, and array sections of structure components also are discussed in chapter 6.

The type of a structure component is given by the type of the component **part**$_n$. If the structure component has a nondefault intrinsic type or a character length other than one, this kind type parameter and this character length, respectively, must be explicitly specified for the type component **part**$_n$ in the derived type definition for **part**$_{n-1}$; that is, the kind type parameter and the character length are constant.

The rank of a structure component is zero if all **part**$_k$ are scalar such as in the case of a structure component whose parent object is an array element. The rank of a structure component is greater than zero if one of the **part**$_k$ designates a whole array or an array section. In this case, the rank of the structure component is equal to the rank of the array or array section.

A structure component has the INTENT or TARGET attribute if the object **part**$_1$ has the respective attribute. A structure component is a pointer if and only if the POINTER attribute is specified in the corresponding type component **part**$_n$ of its derived type definition.

```
TYPE measurement                         ! derived type definitions
  CHARACTER (LEN=10) date
  REAL, DIMENSION (100) :: value
END TYPE measurement

TYPE old_new
  TYPE (measurement) old
  REAL, DIMENSION (100) :: factor
  TYPE (measurement) new
  CHARACTER (LEN=20), DIMENSION (100) :: attribute
END TYPE old_new

TYPE car
  CHARACTER (LEN=10) :: make, type
  INTEGER year_of_manufacture, power
  REAL price
END TYPE car

TYPE (measurement) x                     ! type declarations
TYPE (old_new) z
TYPE (car) priv (10)
```

If the parent object or a type component is an array, then an array element or an array section is written by adding a subscript list or an array section subscript list, respectively, enclosed in parentheses.

Example	Parent object	Type comp.	Struct. comp.
x%date	scalar	scalar	scalar
priv%price	array	scalar	array
priv(5)%price	array element, scalar	scalar	scalar
x%value	scalar	array	array
priv(1:5)%make	array section	scalar	array
z%new%date	structure comp., scalar	scalar	scalar
z%new%value	structure comp., scalar	array	array

4.6 Automatic Variables

The characteristics of a variable may depend on *specification expressions* included in the type declaration or in the array declaration of the variable. These specification expressions may appear in the length specification of a character variable declaration and in the upper and lower bounds of an array declaration.

Such a variable is an **automatic variable** if it is *not* a dummy argument and if, in addition, at least one of these specification expressions is *not* a constant expression. An automatic variable can appear only in a subprogram.

The DATA attribute and the SAVE attribute must not be specified for an automatic variable, neither in a type declaration statement nor by a DATA or SAVE statement.

Automatic variables exist only for the time of the execution of the subprogram. The nonconstant character length of an automatic character variable is calculated once before the execution of the first executable statement of the subprogram. During the execution of the subprogram, the redefinition of any operand in the length specification expression does not change the length of the automatic character variable. The same length remains in effect throughout the current execution of the subprogram.

Automatic arrays are described in chapter 6.

4.7 Association

Association means that an entity is identified in *the same* scoping unit by different names, or that an entity is identified in *different* scoping units by equal names or by different names. If two entities become associated, then corresponding parts of these entities also become associated.

4.7.1 Name Association

There are three ways of name association: *argument association, USE association,* and *host association.*

Argument association

Upon execution of a subprogram reference, that is, upon subprogram invocation, the actual arguments of the subprogram reference become associated with the dummy arguments of the referenced subprogram. These associations established by the subprogram invocation exist only for this subprogram execution. When the subprogram execution terminates, these argument associations also terminate.

USE association

By a USE statement, certain local names in one or more scoping units may become associated with certain names specified within a module. Thus, entities from the module can be accessed in the scoping unit with the USE statement.

Host association

An internal subprogram, a module subprogram, or a derived type definition has access to named entities from its host. These accessible named entities are variables, constants, user-defined subprograms (except statement functions), interface blocks, derived types, generic identifiers, and namelist groups.

If an entity which is accessible in the embedded scoping unit by USE association has the same nongeneric name as an entity in the host scoping unit, the entity in the host is inaccessible by that name within the embedded scoping unit.

```
PROGRAM environment
  TYPE user_def
    :
  END TYPE user_def
  INTEGER i, j
    :
  CONTAINS
    FUNCTION inner
      TYPE (user_def) :: t    ! type user_def of the main program
      INTEGER i               ! local i
        :
      inner = i+j             ! the j of the main program
    END FUNCTION inner
END PROGRAM environment
```

The derived type user_def used in the function inner is the type that is defined in the main program. Because there is an explicit type declaration for the variable i in the function inner, the variable i in the main program is *not* accessible by host association within the function.

A name that appears in the embedded scoping unit as:

- A name declared (by an EXTERNAL statement or interface block) to be an external subprogram name;
- A module name in a USE statement;
- A type name in a TYPE definition statement;
- A data entity name or subprogram name whose characteristics are described by its appearance in a type declaration statement, in a POINTER, SAVE, TARGET, PARAMETER, DATA, DIMENSION, INTRINSIC, COMMON, ALLOCATABLE, or EQUIVALENCE statement;
- A namelist group-name in a NAMELIST statement;
- A generic name in an INTERFACE statement;
- A name of a named construct; or

- A name of a subprogram, an entry point, a result variable, or a dummy argument in a FUNCTION, SUBROUTINE, or ENTRY statement or in a statement function definition

is either the name of a local entity of the scoping unit or the name of a global entity. And any entity of the host that has this as its nongeneric name is inaccessible by that name by host association. Note that entities being local to a subprogram are not accessible to its host.

4.7.2 Pointer Association

A pointer and a target may become associated by **pointer association** such that the target may be referenced or defined by actually referencing or defining the pointer.

A pointer may become associated with another target. But a pointer is associated with not more than one target at the same time. A pointer may become disassociated.

A target may become associated with another pointer. A target may be associated with more than one pointer at the same time. A target may become disassociated from a pointer.

The pointer association status may be *undefined, associated,* or *disassociated*.

4.7.3 Storage Association

Normally, a programmer need not worry about the physical aspects of data, except in connection with the internal representation of data or in connection with *storage association*. **Storage association** means sharing memory by use of COMMON, EQUIVALENCE, or ENTRY statements or sharing memory by function result variables.

There are three different classes of physical **storage units**: *numeric storage units, character storage units,* and *unspecified storage units*.

A **storage sequence** is a sequence of storage units. The size of a storage sequence may be greater than one, one, or even zero; that is, it may contain no storage unit at all. A data object which has a storage sequence occupies an ordered set of storage units. Storage sequences are used to describe relationships between named variables, array elements, character substrings, common blocks, and result variables of functions.

The following rules only apply in context with *storage association*:

A nonpointer scalar object of type default integer, default real, or default logical occupies a single **numeric storage unit**. A nonpointer scalar object of type double precision real or default complex occupies two contiguous numeric storage units. In the complex case, the real part occupies the first storage unit and the imaginary part occupies the second storage unit.

A nonpointer scalar object of a nondefault intrinsic type or of a derived type without the SEQUENCE attribute occupies a single **unspecified storage unit** which is different for each case.

A nonpointer scalar object of type default character and character length one occupies one **character storage unit**. A nonpointer scalar object of type default character and character length *len* occupies *len* character storage units.

A nonpointer array of intrinsic type or of derived type with SEQUENCE attribute occupies a sequence of contiguous storage sequences, one for each array element, in array element order.

A nonpointer scalar object of derived type with SEQUENCE attribute occupies a sequence of contiguous storage sequences corresponding to the sequence of its ultimate components.

A pointer occupies a single unspecified storage unit that is different from that of any nonpointer object and is different for each combination of type, kind type parameters, character length (if applicable), and rank.

A sequence of storage sequences forms a storage sequence. The order of the storage units in such a sequence of storage sequences is that of the individual storage units, taking the storage sequences in succession and ignoring all storage sequences of size zero.

Storage sequence of a common block: Each common block in a scoping unit has a storage sequence. This storage sequence consists of the storage sequences of all variables specified in the scoping unit within COMMON statement(s) for this common block; that is, it consists of all storage sequences of all variables which are "in" the common block. And the order of the single storage sequences corresponds to the order of the variables specified in the COMMON statement(s) in the scoping unit. Variables being accessible in the scoping unit by USE association or host association do not contribute to the storage sequence.

Storage association: Two data objects are storage associated if their storage sequences share at least one storage unit or if the last storage unit of one of these data objects is the immediate neighbour of the first storage unit of the other data object. This may cause further relationships between named variables, array elements, character strings, common blocks, or result variables. If two storage sequences share a particular storage unit, the other corresponding storage units, preceding and following this storage unit, also are associated.

The result variables of the entry points of a function are storage associated if they are all scalar and do not have the POINTER attribute, and in addition to that, if every entry point is of one of the types default integer, default real, double precision real, default complex, or default logical.

An object with TARGET attribute may become storage associated only with another object that has the TARGET attribute.

If two scalar variables occupy the same storage sequence, they are **totally associated**. Two variables are **partially associated** if they occupy at least one storage unit but not the same storage sequence, such as two variables of type default character with different character lengths which are associated by an EQIVALENCE statement.

The *definition status* and the value of a data object affect the definition status and the value of any storage associated object.

A variable with POINTER attribute must be storage associated only with a pointer which has the same data type, the same kind type parameter, (if applicable) the same length, and the same rank as the variable.

4.8 Definition Status

At any time during the execution of a program, the definition status of a variable is either *defined* or *undefined*. A **defined** variable has a (valid) value. The variable remains defined with this value until it becomes either *undefined* or becomes *redefined* with a new value. A variable must be referenced only if it is defined.

If a variable is **undefined**, it has no valid value. In this case, the actual stored value of the variable is normally useless.

A variable is defined if and only if *all* its subobjects are defined. That is, an object is already undefined if only one subobject is undefined.

A pointer which is currently associated with a definable target may be defined or redefined in the same way as a "normal" nonpointer variable.

A variable may be defined already at the beginning of the execution of a program. Such an *initialized* variable has an *initial value*. Such an initial value may be explicitly specified by a DATA statement or type declaration statement. So called "implicit initialization" is given for a nonpointer component of a derived type variable if default initialization is specified for that type component, if the variable is not accessed by USE association or host association, is not dynamically allocated, and either has the SAVE attribute or is declared in the main program, in a module, or in a block data program unit.

Normally, zero-size arrays and zero-length character variables are *always defined*, except allocatable arrays, automatic data objects, and pointers of size zero or length zero, which are not defined until they are allocated or associated.

4.8 Definition Status

All variables which are not initialized, which are not always defined, and which are not associated with initialized objects are *undefined* at the beginning of the execution of a program.

Definition status "defined"

During the execution of a program, variables may become *defined* as a result of certain events, for instance:

The execution of an intrinsic assignment statement other than a masked assignment statement causes the variable on the left-hand side of the equals to become defined.

The execution of a masked assignment statement or FORALL assignment may cause some or all array elements of the array variable(s) on the left-hand side of the equals to become defined.

The execution of a pointer assignment with a defined target causes the pointer on the left-hand side to become defined.

The execution of an input statement causes the input list items to become defined with the values being transferred from the input file. Similar holds for the record(s) written by an internal WRITE statement.

The execution of a DO statement with a DO variable causes this DO variable to become defined.

Normally, the definition of a numeric storage unit, character storage unit, or unspecified storage unit causes all associated numeric, character, or unspecified storage units, respectively, of the same type to become defined.

If both parts of a default complex variable become defined because default real variables become defined which are associated with the real part and the imaginary part, then also the default complex variable becomes defined, and vice versa.

The execution of an input/output statement containing a control information list causes nearly all specified input/output specifiers having the character of output parameters to become defined.

The execution of a statement containing a status variable causes the status variable to become defined.

The execution of an ALLOCATE statement for a variable of a derived type with specified default initialization for an ultimate component causes the component to become defined.

Invocation of a subprogram with INTENT(OUT) dummy argument of a derived type with specified default initialization for an ultimate component causes that component to become defined.

Definition status "undefined"

During the execution of a program, variables may become *undefined* as a result of certain events, for instance:

If a variable of a given type becomes defined, all totally associated variables of different types become undefined.

If a variable of type noncharacter becomes defined, all data objects that are not totally associated become undefined; exceptions: complex variables and their real and imaginary parts.

The execution of a RETURN or END statement in a subprogram causes all local variables of the (instance of the) subprogram to become undefined with the following exceptions:

- Variables in the blank common;
- *Saved* variables, that is, variables with SAVE attribute and certain initialized variables;
- Variables in a named common block that appears in the subprogram and in at least one other scoping unit referencing the subprogram either directly or indirectly;
- Variables accessed from the host scoping unit; and
- Variables accessed from a module that is referenced directly or indirectly by at least one other scoping unit referencing the subprogram directly or indirectly.

If the execution of an input statement causes the occurrence of an error condition or an end-of-file condition, all of the variables specified in the input list or namelist group of the statement become undefined.

The execution of a direct access input statement that specifies a record number for which no record has been written previously causes all of the variables specified in the input list of the statement to become undefined.

The execution of an INQUIRE statement may cause certain specifiers having the character of output parameters to become undefined.

When an allocatable array is deallocated, it becomes undefined.

When a subprogram is invoked, the following variables are undefined:

- An optional dummy argument that is not *present*;
- A dummy argument with INTENT(OUT) attribute except for any ultimate nonpointer component with specified default initialization;
- An actual argument associated with an INTENT(OUT) dummy argument;

- And the result variable of a function except for any nonpointer component with specified default initialization.

5 POINTERS

In Fortran, *pointer* is not a data type, but it is an attribute which may be specified for variables or user-defined functions of any intrinsic or derived type.

A variable or a function that has the POINTER attribute is a **pointer**. This POINTER attribute may be specified in a type declaration statement or by a POINTER statement.

A pointer may be *associated* with a target. If a pointer is associated with a target, then this pointer *points* at this target. And the pointer may be used in place of the target wherever a data entity of the same type, kind type parameter, character length (if applicable), and shape may be used.

A pointer may not be defined and referenced until it is pointer associated. That is, at first the pointer must point at a target, and then this pointer may be used like any other "normal" variable. If the pointer is associated with a target, any reference to the pointer is treated as a reference to the associated target. Unlike other programming languages, the programmer need not distinguish addresses from values. Pointers are implicitly "dereferenced" in Fortran.

There are two different possibilities of pointer association. First: when a *pointer target* is dynamically allocated (that is to say, created) by an ALLOCATE statement for a given pointer, the pointer becomes associated with this target. Second: when a *pointer assignment statement* of the form pointer => target is executed, the pointer becomes associated with a (new) target; we say that the pointer is pointer-assigned to the new target.

An *array pointer* is a named array having the POINTER attribute. The rank but no array bounds are specified in the array declaration for an array pointer by writing only a colon for each dimension. The array bounds may be specified later in an ALLOCATE statement. When this ALLOCATE statement is executed, the array bounds are calculated and a pointer target array of the resulting shape is created and associated with the pointer.

```
INTEGER, POINTER :: p1
LOGICAL, DIMENSION (:, :, :), POINTER :: p2
ALLOCATE (p2 (3, 4, 15))
```

p1 is an integer pointer. And p2 is a 3-dimensional array pointer.

A type component may be specified with the POINTER attribute such that a corresponding structure component is a pointer. This way, a derived type may be defined such that its data objects may be used to form flexible data structures such as linked lists, trees, or graphs.

The ASSOCIATED intrinsic function may be used to determine whether a pointer is currently associated, disassociated, or associated with a given target.

Every target, a pointer points at, must have the TARGET attribute. The TARGET attribute may be specified explicitly in a type declaration statement or by a TARGET statement.

A pointer has a definition status like any other nonpointer variable. Its definition status is that of the currently associated target.

5.1 Pointer Processing

The execution of an ALLOCATE statement for a pointer causes a target object to be created dynamically for the pointer. This target object implicitly has the TARGET attribute. The pointer and this dynamically created **pointer target** become associated and the pointer points at this target. The use of a pointer assignment statement is another possibility to associate a pointer with an(other) target or with a part of a target.

The DEALLOCATE statement may be used to *deallocate* a pointer target created by the execution of an ALLOCATE statement. And the NULLIFY statement may be used to *disassociate* pointers.

5.1.1 Creation of Pointer Targets

The ALLOCATE statement is used to create pointer targets.

ALLOCATE (pointer [, pointer]... [, STAT = status_variable])

Each **pointer** is a name of a scalar variable that has the POINTER attribute, the designator of a scalar structure component that has the POINTER attribute, or the name of an array pointer followed by its array specification (see chapter 6). And the **status_variable** is a scalar integer variable.

The ALLOCATE statement for pointers has the same form as the corresponding statement for array pointers (a special case, see chapter 6) or for allocatable arrays. Precisely: the same ALLOCATE statement may be used to allocate both pointer targets and allocatable arrays at the same time.

If a status variable is specified, it will become defined with the value zero when the ALLOCATE statement has been executed without an error. When an error condition occurs during the execution of the ALLOCATE statement, this status variable will become defined with a processor-dependent positive value.

If the status variable is a pointer (or an array element of an allocatable array), this variable must not be associated (or allocated) by the same ALLOCATE statement. When an error condition occurs during the execution of an ALLOCATE statement without a status variable, the execution of the program is terminated.

It is not an error when an ALLOCATE statement is executed for a pointer which is currently associated with a target. In this case, a new pointer target is created, which has all the attributes that are specified for the pointer. After successful execution of the ALLOCATE statement, the pointer points at this new pointer target and the former association is disassociated. If the former target has been created by an ALLOCATE statement, this target is no longer accessible, except there is an additional association between this pointer target and another pointer. Pointer targets which are no longer accessible are known as "dangling pointers".

5.1.2 Association Status

The association status may be *associated*, *disassociated*, or *undefined*.

Associated: a pointer receives the association status "associated" when

- The pointer is allocated as the result of the successful execution of an ALLOCATE statement referencing the pointer; or
- The pointer is pointer-assigned either to a target that is itself a currently associated pointer or to a target which has an explicitly specified TARGET attribute and which is currently allocated if it is allocatable.

Disassociated: a pointer receives the association status "disassociated" when

- The pointer is nullified by a NULLIFY statement;
- The pointer is deallocated by a DEALLOCATE statement;
- The pointer is pointer-assigned to a currently disassociated pointer; or
- The pointer is an ultimate component of an object for which default initialization specified in its type component definition and
  - a function with this object as its result variable is invoked,
  - a subprogram with this object as an INTENT(OUT) dummy argument is invoked,
  - a subprogram with this object as an automatic data object is invoked,
  - a subprogram with this object as a local object that is not accessed by USE or host association is invoked, or
  - this object is allocated.

Undefined: a pointer receives the association status "undefined"

- When its pointer target is deallocated other than through the pointer (such as through another pointer pointing at the same target);

- When the pointer is pointer-assigned to a currently undefined pointer;
- When the execution of a RETURN or END statement causes the pointer's target to become undefined; or
- During the execution of a RETURN or END statement in the subprogram where the pointer was either declared or accessed unless it is one of the following:
  - A pointer with the SAVE attribute;
  - A pointer in blank common;
  - A pointer in a named common block that appears in at least one other scoping unit that is referencing the subprogram either directly or indirectly;
  - A pointer accessed from a module that is referenced directly or indirectly also by at least one other scoping unit that is referencing the subprogram either directly or indirectly;
  - A pointer accessed by host association; or
  - A pointer that is the result variable of a function with the POINTER attribute.

  In case of these exceptions, the pointer retains its definition status after execution of an END or RETURN statement. When a pointer target becomes undefined because of the execution of an END or RETURN statement, the pointer association status also becomes undefined.

If the association status of a pointer is undefined, the pointer variable (or a subobject of it) must normally not be defined, not be referenced, and not be deallocated. Exception: the pointer may be specified as an actual argument in a reference to any intrinsic inquiry function returning information about the pointer association status, about properties of its data type, about its kind type parameter, about its length, or about argument presence.

If the result variable of a function is a pointer, the pointer association status of this result variable is undefined when the function is invoked. Before returning to the referencing scoping unit, this pointer must be associated with a target or the pointer must receive the pointer association status "disassociated".

If an internal subprogram or a module subprogram accesses a pointer by host association, an association which exists at the time of the subprogram reference remains existent for the time of the execution of the subprogram. The association status may be changed during the excution of the internal subprogram or module subprogram. When the execution of the subprogram is terminated, the pointer association status remains as it is, except the currently associated target becomes undefined by the execution of a RETURN statement or the END statement.

5.1.3 Deallocation of Pointer Targets

A pointer target created by an ALLOCATE statement may be deallocated by the execution of a DEALLOCATE statement. **Deallocation** of a pointer target means that the association between pointer and pointer target is disassociated and that the pointer target object is no longer available.

DEALLOCATE (pointer [, pointer]... [, STAT = status_variable])

Each **pointer** is a name of a scalar variable that has the POINTER attribute, the designator of a scalar structure component that has the POINTER attribute, or the name of an array pointer (*not* followed by an array specification). And the **status_variable** is a scalar integer variable.

The DEALLOCATE statement for pointers has the same form as the corresponding statement for array pointers (a special case, see chapter 6) or for allocatable arrays. Precisely: the same DEALLOCATE statement may be used to deallocate both pointer targets and allocatable arrays at the same time.

If a status variable is specified, it will become defined with the value zero when the DEALLOCATE statement has been executed without an error. When an error condition occurs during the execution of the DALLOCATE statement, this status variable will become defined with a processor-dependent positive value. Such an error condition occurs, for example, when a DEALLOCATE statement tries to deallocate a currently disassociated pointer.

If the status variable is a pointer (or an array element of an allocatable array), this variable must not be deallocated by the same DEALLOCATE statement.

```
INTEGER status_variable
  :
DEALLOCATE (a, b, STAT=status_variable)
```

A pointer must *not* be deallocated by a DEALLOCATE statement if

- The association status is undefined;
- The pointer is currently disassociated;
- The currently associated target is no pointer target which was created by the execution of an ALLOCATE statement;
- The pointer is an array pointer currently associated with an allocatable array; or
- The pointer is currently associated with a portion of a target object that is independent of any other portion of the target object.

When a pointer target is deallocated, the association status of any pointer becomes undefined that is pointing at this pointer target or that is currently associated with a subobject of this pointer target.

When an error condition occurs during the execution of a DEALLOCATE statement without a status variable, the execution of the program is terminated.

```
SUBROUTINE sp
  INTEGER alloc_status, dealloc_status
  INTEGER, POINTER :: ip1, ip2
  ALLOCATE (ip1, ip2, STAT=alloc_status)
  IF (alloc_status > 0) CALL alloc_error
    ⋮
  DEALLOCATE (ip1, ip2, STAT=dealloc_status)
  IF (dealloc_status > 0) CALL dealloc_error
END SUBROUTINE sp
```

5.1.4 Nullification of Pointer Associations

The association between a pointer and a target may be nullified. The association between a pointer and a pointer target which was created dynamically by the execution of an ALLOCATE statement for this pointer is nullified when the pointer target is deallocated by a DEALLOCATE statement. A pointer association may also be nullified by the execution of a NULLIFY statement for the pointer.

NULLIFY (pointer [, pointer]...)

Each **pointer** is a name of a scalar variable that has the POINTER attribute, the designator of a scalar structure component that has the POINTER attribute, or the name of an array pointer (*not* followed by an array specification).

Note that (in contrary to a DEALLOCATE statement) *no* pointer target is deallocated by the execution of a NULLIFY statement.

```
COMPLEX, POINTER :: cp1, cp2, cp3
ALLOCATE (cp1, cp2, cp3)
  ⋮
DEALLOCATE (cp1)
NULLIFY (cp2)
cp3 => cp1
```

After execution of the NULLIFY statement, the pointer cp2 does no longer point at the allocated pointer target. The pointer target continues to exist, but it is inaccessible and cannot be deallocated. The pointer target that was allocated for cp1 is deallocted. Therefore, the pointer cp3 becomes finally disassociated, because cp3 is pointer-assigned to cp1 and cp1 is itself disassociated.

Pointer assignment statements are presented in detail in chapter 8. And array pointers are presented in chapter 6.

6 ARRAY PROCESSING

General aspects of arrays are presented in chapter 4. This chapter deals with array declaration and use of arrays.

6.1 Array Declaration

Named arrays must be explicitly declared to be arrays. Such an array may be declared by a type declaration statement, by an ALLOCATABLE, a COMMON, a DIMENSION, a POINTER, or a TARGET statement. In addition to the name, each array declaration must specify the rank, and, if applicable, the shape of the array. The general form of an **array specification** is:

(dimension [, dimension]...)

Each **dimension** specifies the **bounds** of the array in this dimension. For every dimension, a **lower_bound** and/or an **upper_bound** may be specified, must be specified, may be omitted, or must be omitted, depending on the kind of the array.

The **lower_bound** and the **upper_bound** are particular scalar integer expressions, namely *specification expressions*, which may be negative, positive, or even zero. In the list **dimension [, dimension]...** one to seven dimensions may be specified.

If **lower_bound ≤ upper_bound**, the integer values from **lower_bound** up to and including **upper_bound** determine the valid subscript values (in this dimension) for array element designators and for array section designators. But if **lower_bound > upper_bound** for a particular dimension, there are no valid subscript values in this dimension; that is, the extent of the array in this dimension is zero and, as a result, the size of the array also is zero (see below).

```
SUBROUTINE up (n, p, r, s, t)
REAL, DIMENSION (1:15, 5, 10) :: p, q    ! explicit-shape array
REAL r ( 1:15, n, *)                     ! assumed-size array
REAL, DIMENSION (:, 5:) :: s             ! assumed-shape array
REAL, ALLOCATABLE, DIMENSION (:, :) :: u ! allocatable array
REAL, POINTER, DIMENSION (:) :: v, t     ! array pointer
  :
END SUBROUTINE up
```

Variable array bounds

Array bounds that are *nonconstant* specification expressions may appear in the array declaration only of dummy arguments, of automatic arrays, and of result variables of array-valued functions.

In the first two cases, the array bounds and consequently the shape of the array are determined at entry to the subprogram by evaluating the array bound expressions **lower_bound** and **upper_bound**. Throughout subprogram execution, the bounds of such an array are fixed; that is, they are unaffected by any redefinition or undefinition of operands appearing in the array bounds expressions.

Zero-size arrays

An array has the size zero if its extent in (at least) one dimension is zero. Zero-size arrays are treated as "normal" arrays. Note that arrays may have different shapes even if they have the same size zero.

6.1.1 Explicit-Shape Arrays

An **explicit-shape array** is a named array with explicitly specified upper array bounds.

([**lower_bound** :] **upper_bound** [, [**lower_bound** :] **upper_bound**]...)

The array specification must contain at least the **upper_bound** of every dimension. If the specification of a **lower_bound** is omitted, the default value 1 is assumed.

```
SUBROUTINE ufe (eg, n)
  DIMENSION eg (-5:n+1, 10, n), lofe (55, 14:22)
  ⋮
END SUBROUTINE
```

The dummy argument array `eg` and the local array `lofe` are explicit-shape arrays.

6.1.2 Assumed-Shape Arrays

An **assumed-shape array** is a dummy argument array without the POINTER attribute. It assumes its shape from the associated actual argument array. The array specification must contain at least a colon but no upper bound for each dimension.

([**lower_bound**] : [, [**lower_bound**] :]...)

The size of a dimension of such a dummy argument array is equal to the size of the corresponding dimension of the associated actual argument array. If a **lower_bound** of a dimension is specified, the upper bound of this dimension results from the specified **lower_bound** and the *assumed* extent e of the array in this dimension as (**lower_bound** $+ e - 1$). If the specification of the **lower_bound** of a dimension is omitted, the default value 1 is assumed.

```
DIMENSION af (1950:1989, 2:4)
CALL accident (af)
  ⋮
END
SUBROUTINE accident (ff)
  REAL, DIMENSION (0:, :) :: ff
  ⋮
END SUBROUTINE accident
```

The dummy argument array assumes the shape of the associated actual argument array. Therefore, the first dimension has the size 40 and extends from 0 to 39. And the second dimension has the size 3 and extends from 1 to 3.

6.1.3 Assumed-Size Arrays

An **assumed-size array** is a dummy argument array without the POINTER attribute. Its size is determined by the size of the associated actual argument array. Ranks and sizes of the corresponding actual and dummy arguments may differ. An assumed-size array must not be of a derived type for which default initialization is specified in its type definition.

The array specification of an assumed-size array looks like an array specification of an explicit-shape array, except that the **upper_bound** of the last dimension must be an asterisk *.

([[**lower_bound** :] **upper_bound,**]... [**lower_bound** :] *)

The size of such an assumed-size array is determined as follows:

1. If the specified actual argument is an array which is not of type default character, the dummy argument array has the same size as the actual argument array.

2. If the specified actual argument is an array element which is not of type default character and which has array element position p in an array of size s, then the dummy argument array has the size $(s - p + 1)$.

3. If the specified actual argument is a character array, a character array element, or a character substring of an array element, each of type default character, and if the actual argument has its first character storage unit at position p of an array with t character storage units, the size of the corresponding assumed-size array is $\mathrm{MAX}\,(\mathrm{INT}\,((t-p+1)/l, 0))$, where l is the character length of an array element of the dummy argument array.

If the specification of the **lower_bound** of a dimension is omitted, the default value 1 is assumed. Except for the last dimension, the size of each dimension is given as for an explicit-shape array. But the size of the last dimension is given by the size of the array. That is why for an assumed-size array with rank $r > 1$ the product of the sizes of the first $(r-1)$ dimensions must be less than or equal to the size of the associated actual argument array.

```
DIMENSION aar (3, 4, 5)
CALL cases (aar)
 ⋮
END

SUBROUTINE cases (dar)
  REAL, DIMENSION (6, 0:*) :: dar
   ⋮
END SUBROUTINE cases
```

The assumed-size array `dar` has another rank than the associated actual argument array `aar`, but it has the same size as the actual argument array `aar`. Therefore, the extent of the dummy argument array `dar` is 10 in its second dimension.

6.2 Reference and Use

Whole arrays, *array sections*, and *array elements* may be used. Array sections and array elements do not have names. They are referenced by an *array section designator* and by an *array element designator*, respectively, that is the name of the parent object followed by a particular qualifier.

6.2.1 Whole Arrays

If only the name of an array is written, this is interpreted as the entire array, as all array elements, or as the starting address of the array elements in *array element order*.

A whole array may be a named constant or a named variable. If only the name is written, this does not mean an explicit or implicit order of the references to the single array elements. Only where particularly mentioned in this text, an automatic reference order is in effect; then the array elements are referenced in array element order.

```
   REAL, DIMENSION (100) :: array
   DATA array /100 * 1.0/
   array        = 0.
10 array(33)    = 67.
20 array(50:75) = (/ (-5., i=50,75) /)
```

Statement 10 designates an array element, statement 20 an array section, and all other statements designate the whole array.

6.2.2 Array Elements

An array element is a scalar part of an array entity. It does not have a name. For the reference, the name of the parent array is qualified. A simple case is an array element that is no structure component:

array (subscript_expression [, subscript_expression]...)

where **array** is the name of the parent array. The **subscript list** follows enclosed in parentheses.

Each **subscript_expression** in the subscript list is a scalar integer expression that evaluates to a valid subscript value at the time of the reference. The value of a subscript expression is valid if it is within the interval given by the specified array bounds of the dimension. The number of subscript expressions within the subscript list must correspond to the rank of the array.

```
DIMENSION a (3, 4)
a(3, 2) = 17.      ← valid
a(4, 4) = 29.      ← not valid: the value of the first subscript is too large.
```

A subscript expression may have operands which are array elements or function references. Such function references must not cause **side effects**; that is, the function reference must not redefine any other operand in the subscript list.

For an array element reference, the results of the subscript expressions in the subscript list determine the array element position.

An array element has the INTENT, PARAMETER, or TARGET attribute if its parent object has the respective attribute. But an array element never has the POINTER attribute.

If a complete array is processed in such a way that the program controls the order of the processing of the single elements, then the array elements should be processed *column-wise*. This means that the array elements are processed in canonical *array element order*, which is the most efficient way to determine the position of the elements (relative to the beginning of the array). Any other order of array element references, for example row-wise processing, is allowed but is nearly always extremely inefficient.

Array element of a structure component

The form of an array element of a structure component is:

part$_1$ [%part$_i$]... %part$_n$

where **part$_1$** is the designator of the scalar parent object of derived type. This is either the name of a (scalar) structure object or an array element designator **array (subscript_expression [, subscript_expression]...)** (see above).

Each **part**$_i$ is of derived type. It is either the name of a scalar type component of the derived type of the preceding **part**$_{i-1}$ or it is the array element designator **array (subscript_expression [, subscript_expression]...)** (see above), where **array** is the name of a type component with DIMENSION attribute of the derived type of the preceding **part**$_{i-1}$.

And **part**$_n$ is an array element designator **array (subscript_expression [, subscript_expression]...)** (see above), where **array** is the name of a type component with DIMENSION attribute of the derived type of the preceding **part**$_{n-1}$.

At least **part**$_n$ must have the "normal" form of an array element designator with a subscript list enclosed in parentheses.

```
TYPE calendar
   INTEGER day
   CHARACTER (LEN=10) month
   INTEGER year (5)
END TYPE calendar
TYPE (calendar) new (10), old
```

Then	new(5)%year(2)	is an array element of a structure component,
	old%year(2)	is an array element of a structure component, and
	new(5)%day	is a structure component but no array element.

6.2.3 Array Sections

An array section is an array object. It does not have a name. For the reference, the name of the parent array is qualified. An array section may be a part of a structure component (see below). It also may consist of substrings of array elements of the parent array.

The **extent** of an array section in a particular dimension is given by the number of valid subscript values in that dimension. The **size of an array section** is given by the product of the sizes of all dimensions of the array section.

A simple case is an array section that is no array section of a structure component, no structure component, and no array section of substrings:

array (section-subscript [, section-subscript]...)

where **array** is the name of the parent array. The **section-subscript list** follows enclosed in parentheses.

And every **section-subscript** is a *subscript expression* (see below) that defines a single subscript value, or is a *subscript-triplet* that defines a subscript value sequence, or is a *vector-subscript* that also defines a subscript value sequence. At least one of the section-subscripts must be a subscript-triplet or a vector-subscript, that is, must define a subscript value sequence.

```
DIMENSION f (30, 40), i (5)
PARAMETER (i = (/ 5, 10, 15, 20, 25 /) )
```

`f(1, :)`	is a 1-dimensional array section consisting of all 40 elements of the first row of **f**.
`f(10, 2:40:2)`	is a 1-dimensional array section consisting of every second element of the 10th row of **f**.
`f(1:10, 2:40:2)`	is a 2-dimensional array section of shape $(/\,10,\,20\,/)$ consisting of every second element of the first 10 rows of **f**.
`f(i, 2)`	is a 1-dimensional array section consisting of 5 elements of the second column of **f**.

A **subscript expression** is a scalar integer expression that must have a valid value at the time of the reference, except when the subscript value sequence is empty. If one of the subscript value sequences is empty, the array section has the size zero. A value of a subscript expression is valid if it is within the interval given by the specified array bounds in the corresponding dimension. The number of section-subscripts must correspond to the rank of the **array**. A subscript expression may have operands which are array elements or which are function references without side effects.

An array section consist of all those array elements of the parent array that may be determined by all possible array element subscript lists obtainable from the single subscript value and from the value sequences specified for each section subscript. The **rank of an array section** is equal to the number of subscript value sequences in the section subscript list. The order of the dimensions of an array section is given by the appearances of the subscript value sequences in the section-subscript list from the left to the right. That is, the **shape of an array section** is a 1-dimensional array, whose ith element is the number of integer values in the sequence indicated by the ith subscript triplet or vector subscript.

An array section has the INTENT, PARAMETER, or TARGET attribute if the parent object has the respective attribute. But an array section never has the POINTER attribute.

Array section of a structure component

An array section of a structure component is not a structure component. The form the array section designator is:

part$_1$ [**%part**$_i$]... **%part**$_n$

where **part**$_1$ is the designator of the parent object of derived type. It is either the name of a (scalar) structure object or an array element designator **array (subscript_expression** [**, subscript_expression**]... **)**.

Each **part**$_i$ is of derived type. It is either the name of a type component without the DIMENSION attribute of the preceding **part**$_{i-1}$, or it is an array element designator **array (subscript_expression [, subscript_expression]...)**, where **array** is the name of a type component with DIMENSION attribute of the preceding **part**$_{i-1}$.

And **part**$_n$ is an array section designator **array (section-subscript [, section-subscript]...)** (see above), where **array** is the name of a type component with DIMENSION attribute of the preceding **part**$_{n-1}$. Only **part**$_n$ designates an array object.

With regard to rank and shape of such an array section, the rules of the last section hold.

```
TYPE calendar
  INTEGER day
  CHARACTER (LEN=10) month
  INTEGER year (5)
END TYPE calendar
TYPE (calendar) new (10), old
```

Then   `old%year(:2)`   is an array section of a structure component, and
       `new(7)%year(:)`   is an array section of a structure component.

Array section is structure component

If an array section is a structure component, the form of the array section designator is:

part$_1$ [**%part**$_i$]... **%part**$_n$

where **part**$_1$ is the designator of the parent object of derived type. It is either the **name** of a structure object or of an array object of derived type, or it is an array element designator **array (subscript_expression [, subscript_expression]...)**, where **array** is the name of an array of derived type, or it is an array section designator **array (section-subscript [, section-subscript]...)** (see above), where **array** is the name of an array of derived type.

Each **part**$_i$ is of derived type. It is either the **component_name** of a type component of the type of the preceding **part**$_{i-1}$, or it is an array element designator **array (subscript_expression [, subscript_expression]...)**, where **array** is the name of a type component with DIMENSION attribute of the preceding **part**$_{i-1}$, or it is an array section designator **array (section-subscript [, section-subscript]...)** (see above), where **array** is the name of a type component with DIMENSION attribute of the preceding **part**$_{i-1}$.

And **part**$_n$ is the name of a type component of the type of the preceding **part**$_{n-1}$. Exactly one of these **parts** must designate a whole array or an array section.

With regard to type, kind type parameter, character length (if applicable), and attributes, all the normal rules for structure components hold. If such an array

6.2 Reference and Use

section is written with section-subscripts, the rank and the shape of the array section is given as for the simple form of an array section described earlier. If such an array section is written without a section-subscript, the rank and the shape of the array section is equal to the rank and shape of the part **name**, or **component_name**, or **part**$_n$, respectively, that designates the array object.

```
TYPE calendar
  INTEGER day
  CHARACTER (LEN=10) month
  INTEGER year (5)
END TYPE calendar
TYPE (calendar) new (10), old
```

new(5:10)%day   is an array section that is a structure component,
new(2)%year   is an array section that is a structure component, and
old%year(:)   is an array section that is *no* structure component.

6.2.3.1 Subscript-Triplet

A section-subscript defining a subscript value sequence may be written as a *vector-subscript* or as a *subscript-triplet*. The form of a **subscript-triplet** is:

[**first**] : [**last**] [: **stride**]

where **first** and **last** are *subscript expressions* (see above), specifying the first value and the upper limit, respectively, of the subscript value sequence. If **first** is omitted, the lower bound of the dimension of the parent array is assumed as the default value. The absolute value of **last** is greater than or equal to the last value of the subscript value sequence. If **last** is omitted, the upper bound of the dimension of the parent array is assumed as the default value. The **stride** is a scalar nonzero integer expression specifying the stepsize of the subscript values of the subscript value sequence. If the **stride** is omitted, the value 1 is assumed as the default.

Note that **last** must be written for the last dimension of an assumed-size array.

The **extent of an array section** in a dimension is given by the number of valid subscript values in this dimension, which is calculated as

$$\text{MAX}\,(\text{INT}\,((\textbf{last} - \textbf{first} + \textbf{stride})\,/\,\textbf{stride},\, 0))\,.$$

If **stride** > 0, the subscript value sequence contains the values **first** + $n*$**stride** for $n = 0, 1, 2, \ldots$ up to the upper limit **last**. The subscript value sequence is *empty* if **first** > **last**.

If **stride** < 0, the subscript value sequence contains the values **first** − $n*$**stride** for $n = 0, 1, 2, \ldots$ down to lower limit **last**. The subscript value sequence is *empty* if **first** < **last**.

Note that the values of the subscript expressions of a subscript-triplet need not be valid subscript values. But the selected array elements must be within the specified array bounds.

Subscript-triplet	Subscript values
5 : 12	5, 6, 7, 8, 9, 10, 11, and 12.
5 : 12 : 3	5, 8, and 11.
23 : 3 : 2	the subscript value sequence is empty.
4 : -3 : -1	4, 3, 2, 1, 0, -1, -2, and -3.
4 : -4 : -3	4, 1, and -2.
10 : 990 : -2	the subscript value sequence is empty.

Suppose, array f is declared as:

```
DIMENSION f (-20 : 1000)
```

Then the following array sections consist of the following array elements of the parent array f:

Array section	Array elements in array element order
f(5 : 12)	f(5), f(6), f(7), f(8), f(9), f(10), f(11), and f(12).
f(5 : 12 : 3)	f(5), f(8), and f(11).
f(23 : 3 : 2)	array section of size zero.
f(4 : -3 : -1)	f(4), f(3), f(2), f(1), f(0), f(-1), f(-2), and f(-3).
f(4 : -3 : -2)	f(4), f(2), f(0), and f(-2).
f(10 : 990 : -2)	array section of size zero.

```
DIMENSION a (4, 3, 5)
```

a(3, 2, :)	is a 1-dimensional array section that has the shape $(/\,5\,/)$. Its size is 5.
a(:, 3, :)	is a 2-dimensional array section that has the shape $(/\,4, 5\,/)$. It consists of all array elements of the second columns of all 5 planes of array a. Its size is 20.
a(2:4, :, 3:5)	is a 3-dimensional array section that has the shape $(/\,3, 3, 3\,/)$. Its size is 27.
a(3, 3, 5:)	is a 1-dimensional array section that has the shape $(/\,1\,/)$.

The selected array elements of an array section are chained in array element order. In this case, the array element order is the order defined for the subobject. The array element order of an array section has nothing to do with the array element order of its parent array.

6.2 Reference and Use

```
DIMENSION r (4, 3)    ! <-- parent array
```

The following array section of the parent array **r** has the same shape as the parent array. It consists of *all* array elements of array **r**, but the array element order of the new array object is different from that of its parent array **r**.

Array section: $r(4:1:-1, 3:1:-1)$

	Column 1	Column 2	Column 3
Row 1	r(4,3)	r(4,2)	r(4,1)
Row 2	r(3,3)	r(3,2)	r(3,1)
Row 3	r(2,3)	r(2,2)	r(2,1)
Row 4	r(1,3)	r(1,2)	r(1,1)

The order of the array elements of the array section is as follows:

Array element	r(4,3)	r(3,3)	r(2,3)	r(1,3)	r(4,2)	r(3,2)	...
Position	1	2	3	4	5	6	...

Array element	r(2,2)	r(1,2)	r(4,1)	r(3,1)	r(2,1)	r(1,1)
Position	7	8	9	10	11	12

```
DIMENSION s (3, 3, 2)    ! <-- parent array
```
Array section: $s(3:1:-2, 1:3:2, 1:2)$

The array section has the following form:

	Column 1	Column 2
Row 1	s(3,1,1)	s(3,3,1)
Row 2	s(1,1,1)	s(1,3,1)

Plane 1

	Column 1	Column 2
Row 1	s(3,1,2)	s(3,3,2)
Row 2	s(1,1,2)	s(1,3,2)

Plane 2

The order of the array elements of the array section is as follows:

Array elem.	s(3,1,1)	s(1,1,1)	s(3,3,1)	s(1,3,1)	s(3,1,2)	s(1,1,2)	s(3,3,2)	s(1,3,2)
Position	1	2	3	4	5	6	7	8

6.2.3.2 Vector-Subscript

A section-subscript defining a subscript value sequence may be written as a *subscript-triplet* or as a *vector-subscript*. A **vector-subscript** is a 1-dimensional integer array expression. Every array element of such a (vector-subscript) array expression must be defined with a valid subscript value.

```
INTEGER   u (3),  v (4)
DIMENSION z (5, 7)
```

Suppose, arrays u and v are defined as follows:

```
u = (/1, 3, 2/)
v = (/2, 1, 4, 3/)
```

Then the array section $z(3, v)$ is a 1-dimensional array section consisting of the following elements (in this order):

    z(3, 2)   z(3, 1)   z(3, 4)   z(3, 3)

And the array section $z(u, 2)$ is a 1-dimensional array section consisting of the following array elements (in this order):

    z(1, 2)   z(3, 2)   z(2, 2)

And the array section $z(u, v)$ is a 2-dimensional array section consisting of the following array elements:

	Column 1	Column 2	Column 3	Column 4
Row 1	z(1,2)	z(1,1)	z(1,4)	z(1,3)
Row 2	z(3,2)	z(3,1)	z(3,4)	z(3,3)
Row 3	z(2,2)	z(2,1)	z(2,4)	z(2,3)

Array element	z(1,2)	z(3,2)	z(2,2)	z(1,1)	z(3,1)	z(2,1)	...
Position	1	2	3	4	5	6	...

Array element	z(1,4)	z(3,4)	z(2,4)	z(1,3)	z(3,3)	z(2,3)
Position	7	8	9	10	11	12

Array sections *with* vector-subscripts are subject to certain restrictions which are normally not effective for array sections *without* vector-subscripts. There are restrictions on their use as internal files, as actual arguments, as targets in pointer-assignment statements, and on their use on the left-hand side of intrinsic assignment statements.

6.2.3.3   Array Sections of Substrings

If an array section is of type character, a character string notation may be written immediately after the "normal" array section designator. This designates a subobject of an array section that has the same shape as the array section; and each element of the subobject consists only of the specified character positions of the array elements of the array section. Such a subobject of an array (section) also is an array section, but it is no character string.

CHARACTER (LEN=10) z (5, 10, 10)

Then z(:, :, 4)(6:10) is a 2-dimensional array section that has the shape (/ 5, 10 /). Its array elements are character substrings of the corresponding elements of its parent array z and have the character length 5.

6.3 Memory Management and Dynamic Control

If a subprogram needs an array which is no dummy argument array and which has a shape that depends on one or more variables, then the programmer may use an *automatic array*. And if, for some time, a local array is needed that has a shape which must be dynamically calculated before the array is allocated, then the programmer may use an *allocatable array* or an *array pointer*.

6.3.1 Automatic Arrays

An **automatic array** may be declared only in the scoping unit of a subprogram. It is an explicit-shape array which is no dummy argument array and which has at least one array bound that is a *nonconstant* specification expression.

The lifetime of an automatic array is limited. It is automatically created (allocated) on entry to the subprogram and is automatically deallocated on return to the referencing scoping unit.

```
SUBROUTINE exchange (n, f1, f2)
  COMMON m
  REAL, DIMENSION (n+2, m*m-1) :: f1, f2, hf
  hf = f1;  f1 = f2;  f2 = hf
END SUBROUTINE
```

The automatic array hf is allocated when the subroutine **exchange** is invoked, and it is deallocated when the END statement of the subroutine is executed. The programmer need not take care of allocation and deallocation operations.

An automatic array *exists* only for the execution of the (instance of the) subprogram. Therefore, an automatic array must not be specified to have the SAVE attribute and it must not be initialized by a DATA statement or in a type declaration statement. As long as an automatic array exists, that is, throughout the execution of the (instance of the) subprogram, a redefinition of any operand in one of the nonconstant bounds expressions has no effect on the current shape of the automatic array.

6.3.2 Allocatable Arrays

An **allocatable array** is a named array (but *no* dummy argument array and *no* function result) for which the ALLOCATABLE attribute is specified in a type declaration statement or by an ALLOCATABLE statement. The array declaration contains only a colon for each dimension but no array bounds. The form of such an array specification is:

(: [, :]...)

The explicit shape for an allocatable array is specified in an ALLOCATE statement. And the (storage for the) array is allocated when this ALLOCATE statement is executed; that is, the array becomes dynamically allocated. The allocatable array is available until it is automatically deallocated or until a DEALLOCATE statement is executed for this array.

Type, kind type parameter, character length (if applicable), name, rank, and the ALLOCATABLE attribute of an allocatable array must be specified in the specification part of the scoping unit with containing ALLOCATE statement or must be accessible from the host or from a module. Only the array bounds of all dimensions remain to be determined during the execution of the program.

Note that the TARGET attribute is allowed to be specified for an allocatable array; that is, a pointer may point at an allocatable array.

The rank of an allocatable array is given by the number of colons in its array declaration. An allocatable array may be declared by a type declaration statement, or by an ALLOCATABLE, DIMENSION, or TARGET statement.

```
REAL, DIMENSION (:, :), ALLOCATABLE :: a
INTEGER, ALLOCATABLE :: b (:, :, :)
```

a is a 2-dimensional and b is a 3-dimensional allocatable array.

As long as no ALLOCATE statement is executed for an allocatable array, that is, as long as it is not allocated, its array bounds, its shape, and its size are undefined.

Only the lower and upper bounds of all dimensions are dynamically determined by the execution of the ALLOCATE statement for the array. Thus the extents in all dimensions, the shape, and the size of the array become defined.

ALLOCATE (array (dim [, dim]...) [, array (dim [, dim]...)]... &
[, STAT = status_variable])

Each **array** is the name of an allocatable array available in the scoping unit containing the ALLOCATE statement. **(dim [, dim]...)** is an array specification, which has the form as for an explicit-shape array. And the **status_variable** is a scalar integer variable. The array bounds expressions must be scalar integer expressions. Note that they need not be specification expressions.

6.3 Memory Management and Dynamic Control

The ALLOCATE statement for allocatable arrays has the same form as the corresponding statement for pointers. Precisely: one ALLOCATE statement may be used to allocate allocatable arrays and pointer targets at the same time. See chapter 5 for the use of the status variable.

An allocatable array is a "normal" local array but never a structure component. Its size may be zero. And it may be a zero-length character array.

When an ALLOCATE statement is executed, the array bounds are determined for each dimension, and then the (storage for the) array is allocated. As long as the allocatable array remains allocated, the redefinition of any operand in its array bounds expressions has no effect on the current array bounds of the allocatable array. Its array bounds expressions may become even undefined.

```
INTEGER alloc_error
REAL, ALLOCATABLE :: a (:, :), b (:, :, :)
  :
ALLOCATE (a (-n:-1, n), b (25:75, 2, n), STAT=alloc_error)
```

Allocation status: An allocatable array becomes **allocated** after successful execution of an ALLOCATE statement until it becomes *deallocated*.

A **deallocated** allocatable array is no longer allocated. If an allocatable array is currently not allocated, either no ALLOCATE statement has been (successfully) executed or the array is already deallocated. Only a currently allocated allocatable array may be defined and referenced.

By execution of a RETURN or END statement in a subprogram, the following currently allocated allocatable arrays in the subprogram retain their allocation status and their definition status:

- An allocatable array with SAVE attribute;
- An allocatable array which is accessible by host association; or
- An allocatable array in the scoping unit of a module which also is accessed by another scoping unit that is currently in execution.

If none of these three cases applies and the allocatable array is accessible in the subprogram by USE association, the association status is processor-dependent when RETURN or END is executed. And if none of the above four cases applies for the allocatable array in the subprogram, it is **automatically deallocated** when RETURN or END is executed.

Deallocation: A currently allocated allocatable array may be deallocated by the execution of a DEALLOCATE statement. **Deallocation** of an allocatable array means that the storage allocated for the array is no more available.

DEALLOCATE (array [, array]... [, STAT = status_variable])

The DEALLOCATE statement for allocatable arrays has the same form as the corresponding statement for pointers. Precisely: one DEALLOCATE statement may be used to deallocate allocatable arrays and pointer targets at the same time. See chapter 5 for the use of the status variable.

```
INTEGER dealloc_stat
DEALLOCATE (a, b, STAT=dealloc_stat)
```

An allocatable array with TARGET attribute must not be deallocated by using an associated pointer. If such an allocatable array is deallocated, all currently associated pointers become undefined. An allocatable array with undefined allocation status must not be deallocated.

```
SUBROUTINE ff
  REAL, ALLOCATABLE, DIMENSION (:, :) :: df1, df2
  READ (*, *) ja, je
  ALLOCATE (df1 (ja:je, 7), df2 (1950:je, 7), STAT=istat)
  IF (istat > 0) CALL al_error
  CALL compare(1950, ja, je, df1, df2)
  DEALLOCATE (df1, df2, STAT=istat)
  IF (istat > 0) CALL deal_error
END SUBROUTINE ff
```

6.3.3 Array Pointers

An **array pointer** is a named array for which the POINTER attribute is specified in a type declaration statement or by a POINTER statement. The array declaration contains only a colon for each dimension but no array bounds. The form of such an array specification is:

(: [, :]...)

Type, kind type parameter, character length (if applicable), name, rank, and the POINTER attribute of an array pointer must be specified in the specification part of the program unit or subprogram or must be accessible from the host or from a module. Only the array bounds of all dimensions remain to be determined during the execution of the program.

The rank of an array pointer is given by the number of colons in the array declaration. An array pointer may be declared in a type declaration statement, or by a DIMENSION, or POINTER statement.

During the execution of the program, the array bounds and the shape are determined either by the execution of an ALLOCATE statement, which specifies the explicit shape of the pointer target, or by the execution of a pointer assignment statement, which associates a target with the array pointer.

6.3 Memory Management and Dynamic Control

When an ALLOCATE statement is executed for the array pointer, the array bounds are determined for each dimension, and then the pointer target array is allocated (created). This dynamically created pointer target (implicitly) has the TARGET attribute. The array pointer and this pointer target array become pointer associated, and the array pointer may be used, as any other pointer, to reference and to define this pointer target. By the execution of pointer assignment statements, additional pointers may be associated with this pointer target array or with parts of it. In this case, two or more pointers are pointing at the same target.

```
REAL, DIMENSION (:, :), POINTER :: c
INTEGER, POINTER :: d (:, :, :)
```

c ist a 2-dimensional and d is a 3-dimensional array pointer.

The lower bound and the upper bound of all dimensions are dynamically determined by the execution of the ALLOCATE statement for the array pointer. Thus the extents of the array in all dimensions, the shape of the pointer target array, and its size become defined.

The ALLOCATE statement for array pointers has the same form as the corresponding statement for allocatable arrays (see above).

The ALLOCATE statement must specify the array bounds of *all* dimensions of an array pointer. The array bounds expressions must be scalar integer expressions. Note that they need not be specification expressions.

The size of the array pointer may be zero. And it may be a zero-length character array.

```
INTEGER alloc_error
REAL, POINTER :: a (:, :), b (:, :, :)
ALLOCATE (a (-n:-1, n), b (25:75, 2, n), STAT=alloc_error)
```

If the array bounds of an array pointer are determined such that the array pointer ap becomes associated with a target t by pointer-assignment, such as $ap => t$, then the lower bounds are given by the function result of $LBOUND(t)$ and the upper bounds by the result of $UBOUND(t)$.

Deallocation of pointer targets: The pointer target of an array pointer allocated with an ALLOCATE statement may be deallocated (like any other target of this kind) by the execution of an DEALLOCATE statement; see chapter 5 for exceptions.

```
SUBROUTINE pp
  REAL, POINTER, DIMENSION (:, :) :: ap1, ap2
  READ (*, *) ja, je
  ALLOCATE (ap1 (ja:je, 7), ap2 (1950:je, 7), STAT=istat)
  IF (istat > 0) CALL al_err
  CALL compare(1950, ja, je, ap1, ap2)
  DEALLOCATE (ap1, ap2, STAT=istat)
  IF (istat > 0) CALL deal_err
END SUBROUTINE pp
```

Nullification: The association between an array pointer and a target array may be nullified like any other pointer association (see chapter 5).

6.4 Construction of Array Values

An **array constructor** is a sequence of scalar values which is interpreted as a 1-dimensional array value. The scalar values in this sequence are given by a value-list consisting of single scalar values, of array values, and/or of implied-DOs.

```
INTEGER x (31), f (3, 4)
⋮
x = (/ 0, 1, x(11) - 2, 3, 4, (i+5, 0, i=5,10), 11, 12, f /)
```

An array constructor has a type, a kind type parameter, and (if applicable) a character length, but it does not have a name (except when it is a named array constant). Type, kind type parameter, and length of the array constructor correspond to those of its elements. Therefore, all values in the value-list must have the same type, kind type parameter, and (if applicable) character length.

(/ value-list /)

The **value-list** may be empty. Each value of the list may be a scalar expression, an array expression, or the following *implied-DO*:

(value-list, do_variable = first, last [, stride])

The **do_variable** is the name of a scalar integer variable. And **first**, **last**, and **stride** are scalar integer expressions. If the stride is omitted, the value 1 is assumed as the default stride.

Array expression: An array expression in the value-list is interpreted as a sequence of scalar values in array element order.

Implied-DO: An implied-DO in the value-list is interpreted as a sequence of expressions which is expanded under control of the DO variable. Loop control is the same as for a "normal" DO construct. The value-list in an implied-DO may contain another implied-DO; that is, implied-DOs may be nested. Note that each of the implied-DOs in such a nest must have its own unambiguous DO variable.

```
REAL, DIMENSION (28) :: x, y
LOGICAL l (7:5)
INTEGER v (5:7), z (77)
l = (/ /)
v = (/ 5+1, 6, 27-3 /)
x = (/ y(10:14) + y(20:24), SQRT(r), (r/i, i=5,25), 99. /)
z = (/ (((i, i=1,5), j=1,4), k=1,3), 0, ((i, i=9,6,-1), j=1,4) /)
```

Empty value list: If the value list of an array constructor is empty, this is interpreted as a (1-dimensional) array of size zero.

Constant array constructor

The value-list of a **constant array constructor** contains only *constant expressions*. A constant array expression may be named.

```
INTEGER x (5), y (7), a (6)
CHARACTER name (3) *5
PARAMETER (a = (/ 0, 0, 0, 1, 1, 1 /) )
  :
x    = (/ 1, 3, 5, 7, 9 /)             ! or  x = (/ (1, l=1,10,2) /)
y    = (/ 1, 1, 1, 1, 1, 1, 1 /)        ! or  y = (/ (1, k=1,7) /)
name = (/'Peter', 'Jerry', 'Brian'/)   ! all the same length
```

6.5 Operations on Arrays

The numeric, relational, logical, and character intrinsic operators, the intrinsic assignments, and most intrinsic functions (namely the elemental functions) are defined for scalar operands and arguments, respectively, but they may also be applied element-by-element to array operands and arguments, respectively.

6.5.1 Array Expressions

An **array operand** may be a whole array, an array section, an array constructor, an array function reference, or an array expression enclosed in parentheses.

Two arrays are **conformable** if they have the same shape, that is, if they have the same rank and if the corresponding dimensions have the same size. A scalar entity is per definition conformable with every array. Such a scalar may be interpreted as a conformable array with all array elements having the same value as the scalar.

```
REAL, DIMENSION (5, 3) :: a, b (8, 5), c (5, 3, 4), d
```

Conformable arrays or array sections are:

a	and	d
a(:, 3)	and	b(4:8:1, 1)
a(3:5, 2)	and	b(2, 3:5)
a(1:3, 1)	and	c(2, :, 4)
a	and	b(1:5, 1:3)
a	and	b(4:8, 3:5)
a(1:4, :)	and	c(2:5, 2, 2:4)

If the binary *intrinsic* operators +, −, *, ... are applied to array operands, the operands must be conformable. Such operations are performed element-by-element and produce a resultant array conformable with the array operand(s). In this case, corresponding elements of the operands are involved as in a scalar-like operation and produce the corresponding element in the conformable array-valued result. These elemental operations can be performed in any order or even simultaneously. Such an operator is called an **elemental operator**.

```
REAL a (3, 3), b (9, 3), c (3, 9)
```

Array expressions are:

```
    a + b(3:9:3, :)
    a(:, 1) * a(:, 2) / ( a(:, 3) - a(:, 1) )
    c(1, 7:9) ** a(3, 3)
    ( b(2:3, 1:2) + 4.7E11 ) * ( c(1:2, 5:6) - 0.8E15 )

    INTEGER, DIMENSION (3, 4, 5) :: a, b
    REAL, DIMENSION (3, 4, 5)    :: c, d
    ⋮
5   c = a * b
8   d = d + 5
```

Statement 5 is no matrix multiplication, but it is an elemental multiplication of two arrays. The resultant array has the same shape as the operand arrays. In statement 8, the value 5 is added to *each* element in array d.

6.5.2 Array Subprograms

The programmer may use elemental intrinsic subprograms which are defined for scalar arguments but which may also be referenced with array arguments. He may write his own array-valued functions and he may use a lot of array intrinsic functions.

There are intrinsic functions for

- Matrix multiplication, vector-matrix multiplication (MATMUL) and dot-product (DOT_PRODUCT);
- Numeric and logical computations with reduction of the rank of the result (SUM, PRODUCT, MAXVAL, MINVAL, COUNT, ANY, and ALL);
- Inquiry of certain characteristics of arrays (ALLOCATED, SIZE, SHAPE, LBOUND, and UBOUND);
- Construction of arrays (MERGE, SPREAD, RESHAPE, PACK, and UNPACK);
- Manipulation (such as transposition) of arrays (TRANSPOSE, EOSHIFT, and CSHIFT); and
- Determination of the position of certain array elements (MAXLOC and MINLOC).

6.5.3 Array Assignments

If the left-hand side of an intrinsic assignment statement is an array, this statement is an **array assignment statement**.

If the left-hand side is an array and the right-hand side is scalar, all elements of the array on the left-hand side are (re)defined with the same value of the right-hand side. If the right-hand side also is an array entity, both sides of the equals must be conformable array entities.

An arr assignment operation is *elemental*. That is, the left-hand side of an array assignment statement is defined element-wise for corresponding array elements; the first element of the left-hand side becomes defined with the first element of the right-hand side, the second element of the left-hand side becomes defined with the second element of the right-hand side, and so on. There is no order given for the processing of these elemental assignments; they may be even processed in parallel.

```
LOGICAL x (9, 5, 3), y (9, 3), z (3, 3)
z = x(1:5:2, 2, :) .OR. y(9:5:-2, :)
```

is equivalent with the following scalar assignment statements in any order:

```
z(1, 1) = x(1, 2, 1) .OR. y(9, 1)
z(2, 1) = x(3, 2, 1) .OR. y(7, 1)
z(3, 1) = x(5, 2, 1) .OR. y(5, 1)
z(1, 2) = x(1, 2, 2) .OR. y(9, 2)
z(2, 2) = x(3, 2, 2) .OR. y(7, 2)
z(3, 2) = x(5, 2, 2) .OR. y(5, 2)
z(1, 3) = x(1, 2, 3) .OR. y(9, 3)
z(2, 3) = x(3, 2, 3) .OR. y(7, 3)
z(3, 3) = x(5, 2, 3) .OR. y(5, 3)
```

The actual assignment (i. e., storing) of the result of the right-hand side is done not before all computations on the right-hand side are done and not before all address computations for array elements or other subobjects on the left-hand side are done.

The evaluation of the right-hand side must neither affect nor be affected by the computations for the reference to the left-hand side; side effects are not allowed.

```
  REAL, DIMENSION (n) :: a, b, c
  ⋮
5 a(2:n) = a(1:n-1) * b(2:n) + c(2:n)
  ⋮
  DO 8, i=2,n
     a(i) = a(i-1) * b(i) + c(i)
8 END DO
```

Statement 5 and the DO construct 8 are not equivalent because the DO construct contains data dependencies.

When array assignments are to be evaluated not for all but only for particular array elements, *WHERE assignments* or *FORALL assignments* may be used (see chapter 8).

7 EXPRESSIONS

An **expression** is a formula for the computation of a value. It consists of operands and operators. **Operands** are constants, variables, array constructors, structure constructors, function references, and (sub)expressions enclosed in parentheses. An operand is a scalar or an array.

3.1415	.TRUE.	'Berlin'	← scalar (literal) constants
"garden"(1:4)	'parking'(:4)		← subobjects of constants
(/7,13,24,6/)	(/1,0,0,1/)		← array constructors
x	dor	week name	← scalar named variables, whole arrays
dor(4)	t(10)	week(14)	← array elements
dor(1:3)		week(2:19)	← array sections
address%po		address%cip	← structure components
name(:5)		name(11:21)	← character substrings
SIN(x)	LOG(X=10)	f(dor,2)	← function references
(r*SIN(x/y) + pi/2 + 55.571)			← subexpression

Operators designate the computations (operations). There are *binary* operators processing two operands, and *unary* operators processing one operand. Operators are either *intrinsic* or *defined*. Intrinsic operators are predefined and may be used in all parts of the program. Defined operators must be defined by the programmer and can be used only in the scoping unit where the operator is defined or where its definition is accessible.

Intrinsic operators:

Numeric operators:	+	−	*	/	**	
Relational operators:	.GT.	.GE.	.LT.	.LE.	.EQ.	.NE.
	>	>=	<	<=	==	/=
Logical operators:	.NOT.	.AND.	.OR.	.EQV.	.NEQV.	
Character operator:	//					

An expression evaluates to a value, which has a type, (if applicable) a kind type parameter, a character length, and a shape. It is scalar or array-valued.

Two arrays are **conformable** if they have the same shape. A scalar is by definition conformable with every array.

Elemental operators are defined for scalar operands but may also be applied to conformable operands. When an array operand is being processed by such an operator, the operation is performed element-by-element without any given order thus that each array element of the result has the same array element position as the corresponding processed array element of the operand. For an unary elemental operator, the result of the element-wise operations is an array with the same shape as the operand. For a binary elemental operator with two array operands, the operands must be in shape conformance und the result of the element-wise operations is an array with the same shape as the operands. For a binary elemental operator with one scalar operand and one (non-zero size) array operand, the scalar operand is treated as if it were expanded into an array whose shape is that of the array operand and whose array element values are equal to that of the scalar operand.

The intrinsic operators are predefined as elemental operators. A defined operator is an elemental operator if its defining *operator function* is an *elemental function*.

An expression is not a statement, but it is a part of a statement. There are different kinds of expressions:

- Predefined expressions:
  - numeric intrinsic expressions, such as   `(a + b) * SIN(c)`
  - relational intrinsic expressions, such as `d <= e`
  - logical intrinsic expressions, such as   `f .AND. (g .OR. h)`
  - character intrinsic expressions, such as   `i // j // k`
- Defined expressions, such as   `l .plus. (m .rest. line)`

Numeric intrinsic expressions, relational intrinsic expressions, logical intrinsic expressions, and character intrinsic expressions process operands of intrinsic types by intrinsic operators. Thus, a numeric intrinsic expression evaluates to a numeric result, a logical intrinsic expression evaluates to a result of type logical, and a character intrinsic expression evaluates to a result of type character. A relational intrinsic expression evaluates to a logical result and may be used as an operand only in a logical expression.

Defined expressions also may contain operands of derived types which may be processed by defined operators or even by extended intrinsic operators. A defined expression also may be a numeric, logical, or character expression if it evaluates to a result of such an intrinsic type.

An expression is either a *scalar expression* or an *array expression*. A **scalar expression** contains only scalar (primary) operands. An **array expression** has at least one (primary) *array operand*. An **array operand** is a whole array, an array section, an array constructor, a reference to an array-valued function, or an array expression enclosed in parentheses.

Constant expressions, initialization expressions and *specification expressions* are special purpose expressions mainly appearing in specification statements. They are presented at the end of this chapter.

Interpretation

At first, a **Fortran** processor must establish the interpretation of an expression on the basis of the *interpretation rules*. Then the **Fortran** processor may actually evaluate an *equivalent* expression instead of the given expression. A parenthesized expression is treated as a data entity by the processor; that is, subexpressions enclosed in parentheses are invariant against internal equivalence transformations.

Two numeric intrinsic expressions are (mathematically) **equivalent** if their mathematical results — in accordance with the **Fortran** rules — are equal for all possible operands. Two relational or logical intrinsic expressions are (logically) **equivalent** if their logical results are equal for all possible operands.

7.1 Numeric Intrinsic Expressions

A *numeric expression* is used to express numeric computations. A numeric expression evaluates to a scalar numeric value or to a numeric array value. A **numeric value** is an integer, real, double precision real, or complex value. A double precision real value is a special case of a real value. A **numeric intrinsic expression** consists of *numeric operands* and *numeric operators*.

```
a + b * c - (3 + x) ** 2 / y + SIN(z)
```

Numeric operands are numeric constants, numeric variables, numeric array constructors, high precedence numeric defined subexpressions (see 7.6.1), numeric function references, and (intrinsic and defined) (sub)expressions enclosed in parentheses.

Simple numeric expressions may consist only of one operand without an operator:

```
x     10    0    ijk_1    str7    1    COS(y)    ff(1:3, 1985·1988)
```

More complicated numeric expressions consist of one or more operands which are processed by *numeric intrinsic operators*. They may contain numeric subexpressions enclosed in parentheses. These parentheses must appear in pairs.

Numeric intrinsic operators are

Operator	Operation	Use	Interpretation
**	Exponentiation	x ** y	Raise x to the power y
*	Multiplication	x * y	Multiply x by y
/	Division	x / y	Divide x by y
+	Addition	x + y	Add x and y
+	Identity	+ x	Same as x (without a sign)
−	Subtraction	x - y	Subtract y from x
−	Numeric Negation	- x	Negate x

Except for the division, the interpretation of the numeric intrinsic operators is the same as the interpretation of the corresponding algebraic operators for scalar operands. The interpretation of the division depends on the data types of the operands (see below).

Precedence of operators: The precedence of the operators is the same as in the algebra: the operator ** takes precedence over *, /, +, and −. And the operators * and / take precedence over + and −.

Interpretation rules: If a numeric intrinsic expression consists of several operands and operators, the expression is interpreted in accordance with the following rules:
- A numeric intrinsic expression is interpreted from the left to the right. That is, if there is an operand between two operators of the same precedence (except exponentiation), the left operator is combined with the operand.
- A parenthesized subexpression is treated as a data entity.
- If an expression includes operators of different precedence, the precedence of the operators controls the order of the combination of operators and operands.
- A sequence of exponentiations is combined from the right to the left; for example, 2**3**4 is interpreted as $2^{(3^4)}$.

5 - a ** 2 + b * 3   is interpreted as $(5 - (a^2)) + (b * 3)$

because exponentiation has precedence over subtraction and because multiplication has precedence over addition.

A numeric expression must not contain consecutive numeric operators:

```
a + - b         a ** - 2       ! not allowed
a + (-b)        a ** (-2)      ! allowed
```

Type[1], kind type parameter, and shape

An integer, real, or complex expression evaluates to a result that is an integer, real, or complex value, respectively. Type, kind type parameter, shape, and value of a numeric intrinsic expression result from those of the operand(s) and from the interpretation of the expression.

Unary numeric operator: If the operator + (plus) or − (minus) is applied to one numeric operand, the type of the expression (and thus the type of the result) is the same as the type of the operand. There are similar rules for the kind type parameters and for the shapes of the operand and of the result.

---

[1] Note: DOUBLE PRECISION is the same as REAL (KIND=KIND(1.0D0))

7.1 Numeric Intrinsic Expressions

Binary numeric operator: If a numeric operator is applied to two operands of the same numeric type, the type of the expression (and thus the type of the result) is the same as the type of the operands. That is, if all operands of a numeric expression are of the same type, the result of the expression also is of this type. And if all its operands also have the same kind type parameter and the same shape, the result of the expression also has this kind type parameter and this shape.

A binary numeric operator may be applied to two operands of different shapes (if one operand is scalar), two operands of the same numeric type but with different kind type parameters, or two numeric operands of different numeric types.

Precedence of numeric types: If at least two operands of a numeric intrinsic expression are of different types, the precedence of the numeric types determines the data type of the (result of the) expression:

Type	Precedence
complex	high
real	middle
integer	low

`7.3 + 17 - 2.E-4`   `! is a real expression.`

Type conversion: When a binary numeric operator is applied to two operands of different numeric types, first the value of the operand with the lower type precedence is internally converted to the type of the operand with the higher type precedence, then the operator is applied to these two values of the same type. The result of the operation has the type of the operand with the higher type precedence. The only exception to this rule is the exponentiation where a real or complex value is raised to an integer power.

Kind type parameter conversion: When a binary numeric operator is applied to two integer operands with different kind type parameters, first the value of the operand with the smaller decimal range is internally converted to the kind type parameter of the other operand, then the operator is applied to these two integer values having the same kind type parameter. The result of the operation has the same kind type parameter as the operand with the greater range. The only exception to this rule is the exponentiation where an integer value is raised to an integer power.

When a binary numeric operator is applied to two real operands, two complex operands, or one real and one complex operand with different kind type parameters, first the value of the operand with the smaller decimal precision is internally converted to the kind type parameter of the other operand, then the operator is applied to these two real or two complex values having the same kind type parameter. The result of the operation has the same kind type parameter as the operand with the higher decimal precision.

If a binary numeric operator is applied to an integer operand and a real or complex operand, the result of the operation has the same kind type parameter as the real or complex operand, respectively.

If both operands are of type *default* real or *default* complex, the result also is of type default real or default complex, respectively. If both operands are of type double precision real, or if one operand is of type default real and the other operand is of type double precision real, the result also is of type double precision real.

Shape: If a binary numeric operator is applied to two operands of the same shape, the result of the operation also has this shape. If the shapes of the operands are different, the result has the same shape as the array operand.

Type, kind type parameter, and interpretation of a+b, a−b, a*b, and a/b are presented in the following table:

Operand types: i = integer, r = real, c = complex.
Operator $\otimes$ stands for + or − or * or /.

Operands			Result of a $\otimes$ b		
a	b	Kind	Type	Kind	Interpretation
i	i	$ki(a) = ki(b)$	i	$ki(a)$	$a \otimes b$
i	i	$ra(a) < ra(b)$	i	$ki(b)$	$INT(a, ki(b)) \otimes b$
i	i	$ra(a) > ra(b)$	i	$ki(a)$	$a \otimes INT(b, ki(a))$
i	r		r	$ki(b)$	$REAL(a, ki(b)) \otimes b$
i	z		z	$ki(b)$	$CMPLX(a, ki(b)) \otimes b$
r	i		r	$ki(a)$	$a \otimes REAL(b, ki(a))$
r	r	$ki(a) = ki(b)$	r	$ki(a)$	$a \otimes b$
r	r	$prec(a) < prec(b)$	r	$ki(b)$	$REAL(a, ki(b)) \otimes b$
r	r	$prec(a) > prec(b)$	r	$ki(a)$	$a \otimes REAL(b, ki(a))$
r	c	$ki(a) = ki(b)$	c	$ki(a)$	$CMPLX(a, ki(a)) \otimes b$
r	c	$prec(a) < prec(b)$	c	$ki(b)$	$CMPLX(a, ki(b)) \otimes b$
r	c	$prec(a) > prec(b)$	c	$ki(a)$	$CMPLX(a, ki(a)) \otimes CMPLX(b, ki(a))$
c	i		c	$ki(a)$	$a \otimes CMPLX(b, ki(a))$
c	r	$ki(a) = ki(b)$	c	$ki(a)$	$a \otimes CMPLX(b, ki(a))$
c	r	$prec(a) < prec(b)$	c	$ki(b)$	$CMPLX(a, ki(b)) \otimes CMPLX(b, ki(b))$
c	r	$prec(a) > prec(b)$	c	$ki(a)$	$a \otimes CMPLX(b, ki(a))$
c	c	$ki(a) = ki(b)$	c	$ki(a)$	$a \otimes b$
c	c	$prec(a) < prec(b)$	c	$ki(b)$	$CMPLX(a, ki(b)) \otimes b$
c	c	$prec(a) > prec(b)$	c	$ki(a)$	$a \otimes CMPLX(b, ki(a))$

7.1 Numeric Intrinsic Expressions

Type, kind type parameter, and interpretation of `a ** b` correspond to the last table with some exceptions:

Operand types: i = integer, r = real, c = complex.

Operands			Result of `a ** b`		
a	b	Kind	Type	Kind	Interpretation
i	i	$ra(a) > ra(b)$	i	$ki(a)$	`a ** b`
r	i		r	$ki(a)$	`a ** b`
c	i		r	$ki(a)$	`a ** b`
other	other	other	corresponding to the last table		

In the last two tables, the type conversion functions INT, REAL, and CMPLX are the generic intrinsic functions defined in chapter 14; and the functions ki, $prec$, and ra have the same interpretation as the intrinsic functions KIND, PRECISION, and RANGE, respectively.

If the operands of the exponentiation are of type integer and if the exponent is negative, the interpretation of `a ** b` is the same as the interpretation of `1 / (a ** ABS(b))`; ABS has the same interpretation as the intrinsic function ABS. This equivalent numeric expression is subject to the rules of *integer division* (see below).

`2 ** (-3)` is interpreted as `1 / (2 ** 3)` and evaluates to 0 .

Invalid operations: Any numeric operation is prohibited during program execution whose result is not defined by the arithmetic currently used by the Fortran processor. For instance, the following operations may be allowed or prohibited depending on the Fortran processor: division by zero, zero raised to zero power, and zero raised to a negative power. Note that raising a negative real value to a real power is always prohibited.

Integer division

If the result of a division of two integer operands is *mathematically* not a whole number, the mathematical result is converted to integer type, that is to say, it is truncated towards zero. Thus the noninteger part of the mathematical result is lost.

`8 / 3`   evaluates to 2
`(-8) / 3`   evaluates to -2
`- 100/99`   evaluates to -1

7.2 Relational Intrinsic Expressions

A **relational intrinsic expression** compares the results of two numeric expressions or of two character expressions. A relational intrinsic expression must not compare a numeric expression with a logical or character expression, a logical expression with any intrinsic expression, or a character expression with any noncharacter expression.

    17+4 .LE. nn   klm == airline + i3   'Jean' .NE. 'Joan'

A relational expression may appear only as an operand in a logical expression. It evaluates either to a scalar value (that is, the value *true* or the value *false*) or to an array value of type default logical.

There are the following **relational intrinsic operators** for numeric or character data entities:

Operator			Operation: comparison	Use		Interpretation
.LT.	or	<	Less Than	x .LT. y	x < y	$x < y$
.LE.	or	<=	Less Than Or Equal To	x .LE. y	x <= y	$x \leq y$
.EQ.	or	==	Equal To	x .EQ. y	x == y	$x = y$
.NE.	or	/=	Not Equal To	x .NE. y	x /= y	$x \neq y$
.GT.	or	>	Greater Than	x .GT. y	x > y	$x > y$
.GE.	or	>=	Greater Than Or Equal To	x .GE. y	x >= y	$x \geq y$

7.2.1 Numeric Relational Intrinsic Expressions

A **numeric relational intrinsic expression** compares the results of two numeric intrinsic expressions. If one operand is of type complex and the other one of any numeric type, only the operators == and /= (or .EQ. and .NE., respectively) are allowed.

A scalar numeric relational intrinsic expression evaluates to the default logical value *true* if and only if the operands satisfy the relation specified by the operator; otherwise the expression evaluates to the default logical value *false*. If the operands are conformable arrays, the result of the expression is produced element-wise for corresponding array elements.

When the types or kind type parameters of the operands in the numeric relational expression

    numeric_expression$_1$ **relational_operator numeric_expression**$_2$

are different, the type and/or the kind type parameter of the result of one operand is internally converted in accordance to the rules for the evaluation of the expression (**numeric_expression**$_1$ + **numeric_expression**$_2$).

17 .LE. 13	evaluates to *false*.
ABS(x-2.) < 1E-75	tests a real value.

7.2.2 Character Relational Intrinsic Expressions

A **character relational intrinsic expression** compares the results of two character expressions lexically according to the processor-dependent collating sequence. The two operands must have the same kind type parameter, but their character lengths may be different.

A scalar character relational intrinsic expression evaluates to the default logical value *true* if and only if the operands satisfy the lexical relation specified by the operator; otherwise the expression evaluates to the default logical value *false*. If the operands are conformable arrays, the result of the expression is produced element-wise for corresponding array elements.

The operands are compared one character at a time in order, beginning with the first character of each operand. If the operands have different lengths, the shorter one is treated as though it were extended on the right with *padding characters* to the length of the longer operand. If the operands are of type default character, these **padding characters** are blanks; otherwise the padding characters are processor-dependent.

The result of the character relational intrinsic expression

> **character_expression**$_1$ **relational_operator character_expression**$_2$

is given by the first character c_1 of the value of **character_expression**$_1$ that is different from the corresponding character c_2 of the value of **character_expression**$_2$. The **character_expression**$_1$ is (lexically) *less than* the **character_expression**$_2$ if and only if the character position of c_1 precedes the character position of c_2 in the collating sequence of the processor. There are similar rules for the other relational operators. Two character expressions of length zero are always lexically equal when being compared in such relational expressions.

Collating sequence

The collating sequence is largely processor-dependent. The standard document demands only that the digits are in numeric order (with 0 before 1), that the upper-case letters are in alphabetic order, and that the blank character precedes both 0 and A. If the Fortran processor supports lower-case letters, there are similar rules for these lower-case letters.

Either *all* upper-case letters precede *all* digits or the other way round. There is an equivalent rule for lower-case letters, but the order is independent of the actual relation between upper-case letters and digits. There is no rule for the relation between upper-case letters and lower-case letters.

```
'books' .LT. 'text'      is true.
'Text' > 'Texts'         is false.
```

Only the application of the operators .EQ. (or ==) and .NE. (or /=) is independent of the collating sequence. If the programmer needs other processor-*in*dependent (lexical) comparisons of character intrinsic expressions, then intrinsic functions such as LGE, LGT, LLE, and LLT may be used. These intrinsic functions use the *ASCII collating sequence*. There are additional intrinsic functions, for example, ACHAR returns for a given integer value the corresponding ASCII character, and IACHAR returns for a given character the integer value which is the position of this character in the ASCII collating sequence.

7.3 Logical Intrinsic Expressions

Logical intrinsic expressions are used to express logical computations. A logical expression evaluates either to a scalar logical value (that is, the value *true* or the value *false*) or to a logical array value. A **logical intrinsic expression** consists of *logical operands* and *logical operators*.

```
phi .OR. lambda .AND. psi .OR. (.NOT. alpha(b))
```

Logical operands are logical constants, logical variables, logical array constructors, high precedence logical defined subexpressions (see 7.6.1), logical function references, relational intrinsic expressions, and logical (intrinsic and defined) (sub)expressions enclosed in parentheses.

Simple logical expressions may consist only of one operand without an operator. More complicated logical expressions consist of one or more operands which are processed by *logical intrinsic operators*. They may include logical subexpressions enclosed in parentheses. These parentheses must appear in pairs.

Logical intrinsic operators are

Operator	Operation	Use	Interpretation
.NOT.	Logical Negation	.NOT. x	*true* if x is *false*
.AND.	Conjunction	x .AND. y	*true* if both x and y are *true*
.OR.	Inclusive Disjunction	x .OR. y	*true* if x and/or y are *true*
.NEQV.	Non-equivalence	x .NEQV. y	*true* if x is *true* and y is *false* or if y is *true* and x is *false*
.EQV.	Equivalence	x .EQV. y	*true* if both x and y are *true* or if both x and y are *false*

7.3 Logical Intrinsic Expressions

The precise interpretations of the logical intrinsic operators are given in the following truth tables:

x	y	.NOT. y	x .AND. y	x .OR. y	x .EQV. y	x .NEQV. y
true	true	false	true	true	true	false
true	false	true	false	true	false	true
false	true	false	false	true	false	true
false	false	true	false	false	true	false

Interpretation rules: If a logical intrinsic expression consists of several operands and operators, the expression is interpreted in accordance with the following rules:

- A logical intrinsic expression is interpreted from the left to the right. That is, if there is an operand between two operators of the same precedence, the left operator is combined with the operand.
- A parenthesized subexpression is treated as a data entity.
- If an expression includes operators of different precedence, the precedence of the operators controls the order of the combination of operands and operators.

Precedence of operators:

Operator	Precedence
.NOT.	highest
.AND.	high
.OR.	low
.EQV. or .NEQV.	lowest

.NOT.a .OR. b.AND.c   is interpreted as   (.NOT.a) .OR. (b.AND.c)

Logical expressions must not include two or more consecutive logical operators. But there are two exceptions which are demonstrated in the next example.

l .OR. m .AND. 2+4*8 .LE. 40 .NEQV. p   is (grammatically) correct.

.TRUE. .OR. .EQV. .TRUE.   is not valid.

r .AND. .NOT. s   is allowed and is interpreted as   r .AND. (.NOT. s).

r .OR. .NOT. s   is allowed and is interpreted as   r .OR. (.NOT. s).

Once the interpretation of a logical intrinsic expression has been fixed, not all operands need be evaluated if the result of the expression may be determined in another way.

```
x > y .OR. (23-x*6 > x*y)
```

If x is actually greater than y, the parenthesized subexpression need not be evaluated.

Kind type parameter and shape

Kind type parameter, shape, and value of a logical intrinsic expression result from those of the operand(s) and from the interpretation of the expression.

Unary logical operator: If the operator .NOT. is applied to a logical operand, the kind type parameter of the result of the operation is the same as the kind type parameter of the operand. There is a similar rule for the shape of the operand and of the result.

Binary logical operator: If a logical operator is applied to two logical operands with the same kind type parameter, the kind type parameter of the result of the operation is the same as the kind type parameters of the operands.

A binary logical operator may be applied to two operands having different shapes (if one is a scalar) or to two logical operands with different kind type parameters.

Kind type parameter conversion: If a binary logical operator is applied to two logical operands with different kind type parameters, the kind type parameter of the result is processor-dependent.

Shape: If a binary logical operator is applied to two operands of the same shape, the result of the operation also has this shape. If the operands have different shapes, the result has the same shape as the array operand.

7.4 Character Intrinsic Expressions

A *character expression* evaluates to a character value, that is, a value of type character. The result is either a scalar value (a character string) or an array value. A **character intrinsic expression** consists of *character operands* and *character operators*.

```
a // 'B' // (c(d) // 'EFG')
```

Character operands are character constants, character variables, character array constructors, high precedence character defined subexpressions (see 7.6.1), character function references, and character (intrinsic or defined) (sub)expressions enclosed in parentheses.

Simple character expressions may consist only of one operand without an operator:

```
'Example'       'EXAMPLE'       '1'       t10       line(s)
```

7.4 Character Intrinsic Expressions

More complicated character intrinsic expressions consist of two or more operands which are processed by *character intrinsic operators*. They may include character (sub)expressions enclosed in parentheses. These parentheses must appear in pairs.

There is only one **character intrinsic operator**:

Operator	Operation	Use	Interpretation
//	Concatenation	x // y	Concatenate x with y

The character operands may have different character lengths but must have the same kind type parameter.

A character expression has a character length which is the character length of the result of the expression. The result of the concatenation of two scalar character operands $c1$ and $c2$ is a character string with a length which is the sum of the lengths of the operands, $LEN(c1) + LEN(c2)$. The first part of the resulting string consists of the characters of the first operand $c1$ and the second part consists of the characters of the second operand $c2$.

'Kinder' // 'Garten'   evaluates to the string   KinderGarten

A Fortran processor may evaluate only as much of a character intrinsic expression as is needed at this point to determine the result.

```
CHARACTER (LEN=2) a, b, c, cf
  :
a = b // cf(c)
```

The character function `cf` need not be evaluated, because `a` has character length 2 and the first operand `b` defines already `a` on the left-hand side of the equals.

Except in character assignment statements, variables with *assumed character length* $*(*)$ or $LEN = *$ must not be concatenated in character intrinsic expressions. Named character constants with assumed character length may appear in all character expressions.

Interpretation rules: If a character intrinsic expression consists of several operands and operators, the expression is interpreted from the left to the right. That is, if there is an operand between two operators, the left operator is combined with the operand. A parenthesized subexpression is treated as a data entity; but in this case, parentheses have no effect on the result of the evaluation.

'BOOK' // 'WORM' // 'LOVER'   is interpreted as
('BOOK' // 'WORM') // 'LOVER'

The result of the evaluation is the character string BOOKWORMLOVER.

Kind type parameter, character length, and shape

Kind type parameter, character length, shape, and value of a character intrinsic expression result from those of its operands and from the interpretation of the expression.

Kind type parameter: The two operands of a concatenation must have the same kind type parameter. The kind type parameter of the result is the same as the kind type parameter of the operands.

Character length: The character lengths of the operands may be different. The length of the result of the concatenation is equal to the sum of the lengths of the operands.

Shape: If the concatenation operator is applied to two operands of the same shape, the result of the concatenation also has this shape. If the shapes of the operands are different, the result has the same shape as the array operand.

7.5 Defined Expressions

A defined expression may consist of operands of derived and intrinsic types and of *defined operators*, intrinsic operators, and *extended* intrinsic operators. Note that a defined expression may also be a numeric, logical, or character expression if it evaluates to a result of such an intrinsic type.

A **defined expression** and an intrinsic expression differ in that a defined expression has at least one operand of derived type, or in that a defined expression includes a defined operator or an extended intrinsic operator.

Operands are (as in the case of an intrinsic expression) constants, variables, array constructors, function references, and (sub)expressions enclosed in parentheses. Defined expressions also may include structure constructors as operands.

New *operators* may be defined for operands of any data types. And the interpretation of such a defined operator may be extended. In a similar way, the interpretation of an intrinsic operator may also be extended. The extension of operators is sometimes described as "overloading".

7.5.1 Defined Operators and Extended Operators

For the definition of a new unary or binary *defined operator*, an *operator interface block* and at least one *operator function* must be supplied. The operator function defines the operation to be performed by the application of the operator. And the operator interface block specifies for a given operator which operator functions

7.5 Defined Expressions

are responsible for the interpretation of the operator. Similar measures are necessary for the extension of an intrinsic operator.

A defined operator is an **elemental operator**, if its defining operator function is an *elemental function*.

Operator function

An **operator function** is a "normal" external function or module function with one or two dummy arguments. It may also be a dummy argument function.

The dummy arguments of an operator function must be nonoptional arguments representing data objects. They must be specified with the INTENT(IN) attribute.

The function, that is, the result variable of the operator function, must not be a character entity with assumed length.

An operator function may be referenced like any other function as an operand in an expression. And in a scoping unit where an operator interface block is available, this function may be invoked implicitly (i.e. automatically) when the associated operator is encountered during expression evaluation.

Operator interface block

An **operator interface block** is a special *generic interface block* for a defined operator, for an extended defined operator, or for an extended intrinsic operator.

INTERFACE OPERATOR (operator) ⟵ INTERFACE statement
⋮   ⟵ *interface definition(s)* of operator function(s); may be absent
   MODULE PROCEDURE ... ⟵ MODULE PROCEDURE stmt(s).; may be absent or must be absent
⋮
END INTERFACE [OPERATOR (operator)] ⟵ END INTERFACE statement

where **operator** is the operator to be defined or extended. It is either an intrinsic operator or it is a defined operator written as **.op.**, where **op** is a sequence of 1 to maximal 31 letters. Such a designator of a defined operator must be different from the logical literal constants .TRUE. and .FALSE..

Interface definition: If the operator function is an *external function* or a *dummy argument function*, an operator interface block must be specified containing an interface definition which begins with a FUNCTION statement, ends with an END statement, and may include specification statements which characterize the interface, that is, the function result and the dummy arguments.

The *specific name* of an operator function must appear only once in all interface blocks being available in a scoping unit.

If the operator function is a *module function*, no interface definition must be specified in the operator interface block, but the name of the operator function must be specified in a MODULE PROCEDURE statement in the interface block:

MODULE PROCEDURE operator_function [, operator_function]...

A MODULE PROCEDURE statement may appear in an operator interface block only if the interface block is embedded in a module or if it references a module. And a module function specified in a MODULE PROCEDURE statement must be accessible in the operator interface block.

Unary and binary defined operators may be used in the same way as intrinsic operators. That means, a unary defined operator appears immediately before its operand and a binary defined operator appears between its two operands.

A unary defined operator has the highest operator precedence, and a binary defined operator has the lowest operator precedence. An extended intrinsic operator has the same precedence as the intrinsic operator.

Note that two or more interface blocks for the same operator being available in a scoping unit are interpreted as a single generic interface.

Implicit reference

An operator in an expression is interpreted in a scoping unit as a defined operator or as an extended intrinsic operator (that is, as an implicit reference to an operator function),

- If an operator interface block for this operator is available (that is to say, specified or accessible) in the scoping unit with the operator;
- If the operator function specified in the operator interface block can be referenced in the scoping unit;
- If the operand(s) agree with the corresponding dummy argument(s) of the operator function with regard to data type, kind type parameter, and, if applicable, character length;
- and if
  - either the operator function is an elemental function and the operands for a binary operator are conformable,
  - or the operand(s) and the corresponding dummy argument(s) agree with regard to rank and, if applicable, shape.

If this is true for a unary operator, the operand is used as the actual argument in the implicit reference to the operator function. And in the case of a binary operator, the left operand is used as the first actual argument in the implicit reference to the operator function and the right operand is used as the second actual argument.

7.5.1.1 Nonextended Defined Operator

If the operator function is an *external function* or a *dummy argument function*, the **operator interface block** for a nonextended defined operator must include an interface definition of the operator function. This interface definition begins with a FUNCTION statement and ends with an END statement:

INTERFACE OPERATOR (operator)   ⟵ INTERFACE statement
   [...] **FUNCTION operator_function (...)** [...]   ⟵ FUNCTION statement
   ⋮   ⟵ spec. part; may be absent
   END [**FUNCTION** [**operator_function**]]   ⟵ END statement
END INTERFACE [**OPERATOR (operator)**]   ⟵ END INTERFACE stmt.

And if the operator function is a *module function*, the form of the **operator interface block** for a nonextended defined operator is:

INTERFACE OPERATOR (operator)   ⟵ INTERFACE statement
   MODULE PROCEDURE operator_function   ⟵ MODULE PROCEDURE st.
END INTERFACE [**OPERATOR (operator)**]   ⟵ END INTERFACE stmt.

```
INTERFACE OPERATOR (.dist.)
  FUNCTION distance (a, e)
    REAL, DIMENSION (2) :: a, e
  END FUNCTION distance
END INTERFACE OPERATOR (.dist.)
    ⋮
IF ((destin .dist. station1) .GT. (destin .dist. station2)) THEN
  CALL departure(station2)
    ⋮
ENDIF
END

FUNCTION distance (a, e)
  REAL, DIMENSION (2) :: a, e
  distance = SQRT( ABS( (a(1) - e(1))**2 - (a(2) - e(2))**2))
END FUNCTION distance
```

The defined operator .dist. is a binary operator. Its operator function `distance` is an external function.

An operator may be defined for operands of intrinsic types, but it may also be defined for operands of derived types.

7.5.1.2 Extended Defined Operator

A defined operator may be extended by specifying more than one operator function in the operator interface block or by accessing two or more interface blocks for the operator in the scoping unit. It makes no difference whether these additional operator functions are module functions specified in a MODULE PROCEDURE statement or whether they are external functions or dummy argument functions with corresponding interface definitions within the interface block.

All these functions specified in the operator interface block for a particular operator must be valid operator functions for this operator. They must have the same number of dummy arguments. And they must satisfy similar unambiguity rules as those specific subprograms which have a common generic name: within a scoping unit, any two operator functions for the same operator must have a dummy argument at the same position such that these two arguments have a different data type, kind type parameter, or rank.

When such an extended defined operator is encountered, the characteristics of the operands determine which of these operator functions will actually be invoked and executed.

7.5.1.3 Extended Intrinsic Operator

If the operator specified in the INTERFACE statement of an operator interface block is an intrinsic operator, this defines an extension of the intrinsic operator. An operator function for such an extended intrinsic operator may extend the operator only for those data types of its operands that do not belong to the data types for which this operator is predefined.

The number of arguments of an operator function for an extended intrinsic operator must agree with the number of operands needed for the intrinsic operator. That is, a unary intrinsic operator may be extended only unary, and a binary intrinsic operator may be extended only binary.

There are similar unambiguity rules as for extended defined operators. In the case of an extended intrinsic operator, the unambiguity rules are applied both to the operator functions which are specified in the interface block(s) for this extended intrinsic operator and to an imaginary collection of additional operator functions defining the predefined meaning of the intrinsic operator. For any valid combination of types, kind type parameters, and ranks of the operands, a corresponding operator function could be thought.

When the extended intrinsic operator is encountered, the characteristics of the operands determine which of these operator functions will actually be invoked and executed.

7.6 Common Rules for Expressions

7.6.1 Precedence of Operators

If there are numeric, logical, relational, or character intrinsic operators, and/or defined operators in the same expression, the combination of operands with operators is determined by the *precedence of operators*.

Expression	Operator	Operation	Precedence Level	
High Precedence Defined Expressions	*Unary* Def. Op.	(user defined)	1.	highest
Numeric Intrinsic Expressions	**	Exponentiation	2.	
	*	Multiplication	3.	
	/	Division	3.	
	+	Addition	4.	
	+	Identity	4.	
	−	Subtraction	4.	
	−	Numeric Negation	4.	
Character Intrinsic Expressions	//	Concatenation	5.	
Relational Intrinsic Expressions	.EQ.	Equal To	6.	
	==	Equal To	6.	
	.GE.	Greater Than Or Equal To	6.	
	>=	Greater Than Or Equal To	6.	
	.GT.	Greater Than	6.	
	>	Greater Than	6.	
	.LE.	Less Than Or Equal To	6.	
	<=	Less Than Or Equal To	6.	
	.LT.	Less Than	6.	
	<	Less Than	6.	
	.NE.	Not Equal To	6.	
	/=	Not Equal To	6.	
Logical Intrinsic Expressions	.NOT.	Logical Negation	7.	
	.AND.	Conjunction	8.	
	.OR.	Inclusive Disjunction	9.	
	.EQV.	Equivalence	10.	
	.NEQV.	Non-equivalence	10.	
Low Precedence Defined Expressions	*Binary* Def. Op.	(user defined)	11.	lowest

If an intrinsic operator is extended, this extended operator has the same precedence as the intrinsic operator, even when the operator is used with its extended interpretation. If operators and operands are to be combined without regard to the rules of operator precedence, parentheses may be used.

```
r .AND. .NOT. s        5 + .fahrtocel. 13       - .celtofahr. n
```

7.6.2 Interpretation of Expressions

Operators are combined with operands according to the following interpretation rules in this order:

1. A parenthesized (sub)expression is treated as a data entity.
2. The precedence of the operators determines how operands are combined with operators in expressions with several operators.
3. Successive exponentiations are combined from the right to the left.
4. Successive multiplications and/or divisions are combined from the left to the right.
5. Successive additions and/or subtractions are combined from the left to the right.
6. Successive concatenations are combined from the left to the right.
7. Successive logical conjunctions (logical products) are combined from the left to the right.
8. Successive logical inclusive disjunctions (logical sums) are combined from the left to the right.
9. Successive logical equivalences and/or non-equivalences are combined from the left to the right.
10. Successive binary defined operators are combined from the left to the right.

```
a + b ** 2 .LT. 7 .OR. 1 .AND. c .EQ. 'letter' // 'head'
```

is interpreted as

```
((a + (b ** 2)) .LT. 7) .OR. (1 .AND. (c .EQ. ('letter' // 'head')))
```

Once the **Fortran** processor has established the interpretation of an expression, it may evaluate an equivalent expression (see page 7-3 *Interpretation*).

```
3 / 4 * 8       evaluates to    0
3 * 8 / 4       evaluates to    6
```

The last two expressions are not equivalent (according to **Fortran** rules). The result of the evaluation depends on the order of the evaluation.

7.6.3 Evaluation of Expressions

Operands

A whole assumed-size array must not be used as an operand in an expression.

A scalar variable must not be used as an operand in an expression until it is defined with a valid value. Accordingly, all elements of an array must be defined when this array is used as an operand.

When a variable of derived type is used as an operand, all components of the variable must be defined. All characters of a character operand must be defined when it is used as an operand.

If an operand is a pointer or a pointer enclosed in parentheses, the associated target is used in the expression. Type, kind type parameter, length (if applicable), shape, and value of the operand are those of the associated target.

Elemental operations

Elemental operations on array operands are performed element-wise as a set of scalar operations which may be executed in any order. The operand(s) being involved in such a scalar-like operation and the result are corresponding array elements, that is to say, elements in the same array element position within their respective parent array. When a binary elemental operator has a scalar operand and an array operand (with at least one array element), the scalar operand is treated as if it were expanded into an array whose shape is that of the array operand and whose array element values are equal to that of the scalar operand.

Note that whether or not an overloaded defined operator or an extended intrinsic operator is interpreted as an elemental operation may depend on the encountered operand(s) during program execution.

Side effects

Certain *side effects* are not allowed. The evaluation of a function must not change any data entity appearing in the expression with the function reference except the function result. This rule applies to *direct* side effects such as the definition of actual arguments or equivalenced data objects. This rule also applies to *indirect* side effects such as the definition of common block variables affecting other functions referenced in the same statement. Note that function references in the logical expression in a logical IF statement, in the mask expression in a WHERE statement, and function references in the scalar integer expressions defining the value sets for index variables in a FORALL statement are allowed to affect data entities appearing in the statement which is conditionally executed.

```
IF ( f(x)) a = x
WHERE ( g(x)) b = x
```

The functions f and g may redefine x.

```
a(i) = f(i)
y = g(x) + x
```

If the function reference f(i) modifies the actual argument i, or if the function reference g(x) modifies the actual argument x, the respective assignment statement is not valid.

The evaluation of an actual argument must not affect the type of the expression with the function reference and the type of the expression must not affect the evaluation of the actual argument; exception: generic functions. Note that the type of the result of a generic function may depend on the types of the arguments in the function reference.

If array elements or array sections appear as operands, subscript expressions are to be evaluated. The evaluation of a subscript expression does neither affect nor is affected by the type of the expression with the array element or array section. There is a similar rule for the evaluation of character substring expressions. The appearance of an array constructor in an expression may require the evaluation of the parameters of an included implied-DO. The type of the expression with this array constructor does neither affect nor is affected by the evaluation of the implied-DO parameters.

Zero-size arrays and zero-length character strings

A Fortran processor needs to evaluate neither a subscript expression of a zero-size array nor a character substring expression of a zero-length substring if such an array or substring, respectively, appears as an operand in an expression.

Evaluation of single operands in expressions with several operands

Not all operands in an expression need be evaluated and an operand need not be completely evaluated if the result of the expression can be determined in a different way. Such a situation may arise, for example, during the evaluation of a logical expression or if an operand is a zero-size array or a zero-length character string.

```
x > y .OR. a+b*c <= 17 .OR. l(z)
```

If the Fortran processor knows that x is actually greater than y, the rest of the logical expression on the right-hand side of the first .OR. need not be evaluated.

```
x + w(z)
```

If `x` is a zero-size array, `w(z)` need not be evaluated.

If a statement includes an expression with a function reference which need not be evaluated for the execution of the statement, then all those data objects become undefined after the evaluation of the expression which might have become defined by the evaluation of the function reference.

```
xx + ww(z1)
x > y .OR. a+b*c <= 17 .OR. ll(z2)
```

Suppose, `xx` is a zero-size array and `x > y` is true: If the function `ww` would define the actual argument `z1` or if function `ll` would define the actual argument `z2`, then `z1` and `z2`, respectively, are undefined after the evaluation of the respective expression.

```
CHARACTER (LEN=2) a, b, c, f
a = b // f(c)
```

The character function `f` need not be evaluated because the value of `b` determines the value for the left-hand side `a` completely.

Order of evaluation of function references

A Fortran processor may evaluate function references appearing in one statement in any order. And the result of each function reference must be independent of the order of the evaluations of the functions; exceptions: function references in logical IF statements, in the actual argument list of function references, in WHERE statements, and in FORALL statements.

```
IF ( l(z)) f(x)
y = f( l(z))
WHERE ( lf(x)) b = ff(x)
```

The functions `l` and `lf` are evaluated first and then (possibly) the functions `f` and `ff`, respectively.

7.7 Special Expressions

For the specification of constant values, sometimes not only (literal or named) constants but even more complicated *constant expressions* may be written. For the initialization of variables and for the definition of named constants within type declaration statements and other specification statements, *initialization expressions* may be written. These two classes of special expressions can be evaluated already during program compilation. *Specification expressions* form a third class of special expressions which may be used to specify certain properties of data objects in the specification part of subprograms. These last forms of special expressions can be evaluated during program execution on invocation of the subprogram.

7.7.1 Constant Expressions

A **constant expression** is an expression in which each operator is intrinsic and each of its operands is one of the following constant data entities:

- A constant or a subobject of a constant,

- An array constructor where each element and the loop parameters of each implied-DO are expressions whose operands are constant expressions,

- A structure constructor where each component and the loop parameters of each implied-DO are constant expressions,

- An elemental intrinsic function reference where each actual argument is a constant expression,

- A transformational intrinsic function reference except NULL where each actual argument is a constant expression,

- A reference to the transformational intrinsic function NULL,

- A reference to one of the intrinsic inquiry functions LBOUND, SHAPE, SIZE, UBOUND, LEN, BIT_SIZE, KIND, DIGITS, EPSILON, HUGE, MAXEXPONENT, MINEXPONENT, PRECISION, RADIX, RANGE, and TINY, where each actual argument is either a constant expression or a variable whose kind type parameters, character lengths, and bounds inquired about are not assumed, are not defined by an expression that is not a constant expression, and are not definable by an ALLOCATE statement or a pointer assignment statement,

- An implied-DO variable within an array constructor where the loop parameters of the implied-DO are constant expressions, or

- A constant expression enclosed in parentheses,

where each subscript, section subscript, substring starting character position, and substring ending character position is a constant expression. Any other named or unnamed variables, any non-intrinsic function reference, and any defined operator are prohibited within constant expressions.

Depending on the data type of the expression, there are **integer constant expressions**, **numeric constant expressions**, **logical constant expressions**, and **character constant expressions**.

7.7 Special Expressions

```
2                                          ← integer constant expression
RANGE(i) / 10                              ← integer constant expression
COUNT( (/ .TRUE., .FALSE., .FALSE. /) )    ← integer constant expression
- 23.5 + (4 * 47) ** 2                     ← real constant expression
1. - ABS(-.4)                              ← real constant expression
.TRUE.                                     ← logical constant expression
SQRT(63.) - 8. < 0.                        ← logical constant expression
'timetable'                                ← character constant expr.
'timetable'(1:4) // ' sheet'               ← character constant expr.
```

7.7.2 Initialization Expressions

An **initialization expression** is a constant expression in which the exponentiation is permitted only with an integer power and each of its operands is one of the following:

- A constant or a subobject of a constant,

- An array constructor where each element and the parameters of its implied-DOs are expressions whose operands are initialization expressions,

- A structure constructor where each component is an initialization expression,

- An elemental intrinsic function reference where each actual argument is an initialization expression of type integer or character,

- A reference to one of the transformational intrinsic functions REPEAT, RESHAPE, SELECTED_INT_KIND, SELECTED_REAL_KIND, TRANSFER, and TRIM, where each actual argument is an initialization expression,

- A reference to the transformational intrinsic function NULL,

- A reference to one of the intrinsic inquiry functions LBOUND, SHAPE, SIZE, UBOUND, LEN, BIT_SIZE, KIND, DIGITS, EPSILON, HUGE, MAXEXPONENT, MINEXPONENT, PRECISION, RADIX, RANGE, and TINY, where each actual argument is either an initialization expression or a variable whose properties inquired about are not assumed, are not defined by an expression that is not an initialization expression, and are not definable by an ALLOCATE statement or a pointer assignment statement,

- An implied-DO variable within an array constructor where the loop parameters of the implied-DO are initialization expressions, or

- An initialization expression enclosed in parentheses,

where each subscript, section subscript, substring starting character position, and substring ending character position is an initialization expression. Any other named or unnamed variables, any non-intrinsic function references, and any defined operators are prohibited within initialization expressions.

```
2                                    ←— integer initialization expr.
RANGE(i) / 10                        ←— integer initialization expr.
COUNT( (/ .TRUE., .FALSE., .FALSE. /) )←— no initialization expression,
                                         prohibited transformational
                                         intrinsic function
- 23.5 + (4 * 47) ** 2               ←— real initialization expression
1. - ABS(.-4)                        ←— no initialization expression,
                                         real actual argument
.TRUE.                               ←— logical initialization expr.
SQRT(63.) - 8. < 0.                  ←— logical initialization expr.
'timetable'                          ←— character initialization expr.
'timetable'(1:4) // ' sheet'         ←— character initialization expr.
```

If an initialization expression includes a reference to an inquiry function for a kind type parameter, a character length, or an array bound of an object specified in the same specification part, the kind type parameter, length, or bound must be specified lexically earlier in the specification part. This prior specification may appear to the left of the inquiry function in the same statement.

7.7.3 Specification Expressions

In type declaration statements and in other specification statements within the specification part of a subprogram, *specification expressions* may be used to specify array bounds and character lengths. These specification expressions may contain not only "normal" operands but also operands which are composed of other data entities, for instance array constructors, structure constructors, subprogramm references, and subexpressions in parentheses. The constituent parts of these compound operands may be written as *restricted expressions*. Specification expressions are special forms of restricted expressions. Both are subject to particular restrictions because they are evaluated during program execution after invocation and before execution of the subprogram.

A **restricted expression** is an expression in which each operator is intrinsic and each of its operands is one of the following:

- A constant or a subobject of a constant,

- A variable which is a dummy argument that has neither the OPTIONAL nor the INTENT(OUT) attribute, or a variable that is a subobject of such a dummy argument,

7.7 Special Expressions

- A variable that is in a common block or a variable that is a subobject of a variable in a common block,
- A variable that is made available by USE association or host association or a variable that is a subobject of such a variable,
- An array constructor where each element and the parameters of its implied-DOs are expressions whose operands are restricted expressions,
- A structure constructor where each component is a restricted expression,
- A reference to one of the intrinsic inquiry functions LBOUND, SHAPE, SIZE, UBOUND, LEN, BIT_SIZE, KIND, DIGITS, EPSILON, HUGE, MAXEXPONENT, MINEXPONENT, PRECISION, RADIX, RANGE, and TINY, where each actual argument is either a restricted expression or a variable whose properties inquired about are not dependent on the upper bound of the last dimension of an assumed-size array, are not defined by an expression that is not a restricted expression, and are not definable by an ALLOCATE statement or a pointer assignment statement,
- A reference to any other standard intrinsic function where each argument is a restricted expression,
- A reference to a pure function which is not an intrinsic function, is not an internal function, is not a statement function, and is not a recursive function, where each actual argument is a restricted expression,
- An implied-DO variable within an array constructor where the loop parameters of the implied-DO are restricted expressions, or
- A restricted expression enclosed in parentheses,

where each array subscript, section subscript, substring starting character position, and substring ending character position is a restricted expression. Any other named or unnamed variables, any references to another class of functions, and any defined operators are prohibited within restricted expressions.

A **specification expression** is a scalar restricted expression of type integer.

```
LEN(line) - 2       ! line is a character variable
KIND(x) - KIND(y)   ! x and y are real variables
UBOUND(f, 2)        ! f is a 2-dimensional local array
```

Type, kind type parameter, and (if applicable) character length of a variable in a specification expression must be explicitly or implicitly declared lexically earlier in this scoping unit, or the declaration must be available by USE association or host association. If type, kind type parameter, or length are declared implicitly, a lexically later type declaration statement for such a variable may confirm the implicit declaration, but must not change it.

If a specification expression includes a reference to a function inquiring about a kind type parameter, a character length, or an array bound of an entity in the same specification part, this kind type parameter, this character length, and this array bound, respectively, must be specified lexically earlier in this specification part. If a specification expression includes a reference to the value of an element of an array specified in the same specification part, the array bounds must be specified lexically earlier in this specification part. The prior specification may be on the left of the inquiry function reference in the same statement.

8 ASSIGNMENTS

Assignments are executable statements. There are *intrinsic assignment statements*, *defined assignment statements*, *pointer assignment statements*, *masked array assignment statements*, and *indexed assignment statements*.

During execution of an intrinsic assignment statement, a variable becomes defined (with a value) or redefined (with a new value). During execution of a defined assignment statement, a variable may become defined or redefined; other actions may also be possible. During execution of a pointer assignment statement, a pointer may become associated with a target such that afterwards the pointer points to that target. During execution of a masked array assignment (WHERE), one or more array assignment statements are evaluated element-for-element for those array element positions that are selected by a given logical array expression. During execution of an indexed assignment (FORALL), one or more assignment statements and/or pointer assignment statements are evaluated for those left-hand side variables and right-hand side expressions that are selected by given index variables and by an optionally given logical scalar expression.

8.1 Intrinsic Assignment Statements

Intrinsic assignment statements cannot be identified by characteristic keywords. The form of an intrinsic assignment statement is:

variable = expression

The **expression** is evaluated first. Then its result is assigned (possibly after certain conversions) to the **variable**. The two parts of the assignment statement are called the *left-hand side* and the *right-hand side*.

The *numeric assignment statement*, the *logical assignment statement*, the *character assignment statement*, and the *assignment statement for derived types* are predefined; that is, the interpretation of the assignment symbol is given by the language. For these intrinsic assignment statements, the left-hand side and the right-hand side must be conformable and both sides must be of numeric type, of type logical, of type character (with the same kind type parameter), or of the same derived type.

Intrinsic array assignment statement: If the left-hand side of an assignment statement is an array, the statement is an **array assignment statement**. The left-hand side must not be an array section with a vector subscript with two or more array elements having the same value. If the left-hand side is a zero-size array, no value is assigned to the left-hand side.

If the left-hand side is an array and the right-hand side is scalar, all elements of the left-hand side are defined with the same scalar value of the right-hand side. If the right-hand side is an array value, the left-hand side must also be an array and both sides must have the same shape.

The left-hand side of an array assignment statement is defined element-wise for corresponding array elements; that is, the first element of the left-hand side is defined with the value of the first element of the right-hand side, the second element of the left-hand side is defined with the value of the second element of the right-hand side, and so on. The order of these element-wise assignments is processor-dependent.

```
REAL, DIMENSION (10) :: x = (/ (i, i=1,10) /)
x(1:10) = x(10:1:-1)
```

After execution of the array assignment statement, array x is defined with the values 10, 9, 8, ... , 2, 1.

Interpretation: The assignment (i. e., the storing of the value of the right-hand side at the address of the left-hand side) is not done until all evaluations of the right-hand side are completed and all address computations for subobjects (for instance array elements or character substrings) of the left-hand side are completed. Note that the same data objects may appear on both sides and that even the same subobjects may be affected on both sides. But side effects are not allowed; that is, the evaluation of expressions included in the left-hand side must neither affect nor be affected by the evaluation of the right-hand side.

The equals in an intrinsic assignment statement does not have the normal mathematical interpretation.

```
n = n + 1
```

is not a contradiction in the Fortran language. The interpretation of this statement is: add 1 and the (current) value of n first; then store the result of this addition in the address of the variable n. Thus n is redefined with a new value and its former value is lost.

Pointers: If the left-hand side is a pointer, this pointer must be pointer associated with a definable target at the time of the execution of the assignment statement. Type, kind type parameter, character length (if applicable), and shape of the target must conform with those of the right-hand side. And the result of the evaluation of the right-hand side is assigned to the target the left-hand side is pointing at. If the right-hand side consists only of a pointer, the associated target is referenced.

8.1 Intrinsic Assignment Statements

8.1.1 Numeric Assignment Statement

If both the left-hand side and the right-hand side are of numeric type, it is a **numeric assignment statement**.

numeric_variable = numeric_expression

The left-hand side and the right-hand side may be of different numeric types and may have different kind type parameters.

```
REAL x, var, f, y
x = 17.5D-1 + var * 2.5 - f(y)
```

When a numeric assignment statement is executed, the following steps are performed in sequence:

1. Evaluation of the numeric expression on the right-hand side and all expressions used to determine the variable on the left-hand side.
2. Conversion of the result of the right-hand side according to the data type and/or the kind type parameter of the left-hand side if types and/or the kind type parameters of both sides are different.
3. Assignment (i.e., storing) of the (possibly converted) result of the right-hand side to the numeric variable (that is, in the address) of the left-hand side.

```
DOUBLE PRECISION dup
REAL xray, pi, r
dup = 4.7E3 * xray - 2.5 * r * pi
r   = 3.1415 - dup
```

These numeric assignment statements are interpreted as

```
dup = DBLE(4.7E3 * xray - 2.5 * r * pi)
r   = REAL(3.1415 - dup, KIND(r))
```

If a right-hand side of type default real is assigned to a left-hand side of type double precision real, only the internal representation of the result of the right-hand side is converted. Note that the precision of the internal representation of the right-hand side is not improved.

8.1.2 Logical Assignment Statement

If both the left-hand side and the right-hand side are of type logical, it is a **logical assignment statement**.

logical_variable = logical_expression

The left-hand side and the right-hand side may have different kind type parameters.

```
LOGICAL a, b, c, d
a = (b .OR. c) .AND. d
```

When a logical assignment statement is executed, the following steps are performed in sequence:

1. Evaluation of the logical expression on the right-hand side and all expressions used to determine the variable on the left-hand side.

2. Conversion of the result of the right-hand side according to the kind type parameter of the left-hand side if the kind type parameters of both sides are different.

3. Assignment (i. e., storing) of the (possibly converted) result of the right-hand side to the logical variable (that is, in the address) of the left-hand side.

8.1.3 Character Assignment Statement

If both the left-hand side and the right-hand side are of type character, it is a **character assignment statement**.

character_variable = character_expression

The left-hand side and the right-hand side may have different character lengths but they must have the same kind type parameter. If the left-hand side has character length zero, no value is assigned.

```
CHARACTER file *12
file = file(1:8) // '.f90'
```

The left-hand side and the right-hand side may overlap; that is, character positions may appear on the left-hand side which are used on the right-hand side.

When a character assignment statement is executed, the following steps are performed in sequence:

1. Evaluation of the character expression on the right-hand side and all expressions used to determine the variable on the left-hand side.

2. Conversion of the result of the right-hand side if both side have different character lengths: if the length of the **character_variable** is greater than the length of the **character_expression**, the result of the character expression is filled with trailing *padding characters* until it has the same length as the left-hand side; if the length of the **character_variable** is less than the

8.1 Intrinsic Assignment Statements

length of the **character_expression**, the result of the character expression is truncated from the right until it has the same length as the left-hand side.

Within a character assignment statement for the default character type, these **padding characters** are blanks; otherwise, the padding characters are processor-dependent.

3. Assignment (i.e., storing) of the (possibly converted) result of the right-hand side to the character variable (that is, in the address) of the left-hand side.

The value of the character expression of the right-hand side need be defined only for character positions which are used to define the left-hand side.

```
CHARACTER a *2, b *4
a = b
```

For this character assignment statement, only positions b(1:1) and b(2:2) need be defined. It is not necessary that b(3:4) is defined.

8.1.4 Assignment Statement for Derived Types

When both the left-hand side and the right-hand side are of the same derived type, the statement is an intrinsic **assignment statement for derived types** if no assignment interface block and no assignment subroutine are available for this derived type in the scoping unit with this assignment statement.

derived_type_variable = defined_derived_type_expression

```
TYPE employee
  CHARACTER name *15, first_name *10
  INTEGER number
  LOGICAL married
  REAL salary
END TYPE employee

TYPE (employee), DIMENSION (1000) :: region_north

region_north(78) = employee('Smith', 'Pit', 78, .TRUE., 840.64)
```

When an assignment statement for derived types is executed, the following steps are performed in sequence:

1. Evaluation of the defined expression on the right-hand side and all expressions used to determine the variable on the left-hand side.

2. Assignment (i.e., storing) of the result of the right-hand side to the variable (that is, in the address) of the left-hand side.

An intrinsic assignment statement for derived types is interpreted such as if the assignment is executed component-wise (for corresponding components). A nonpointer component is treated according to the rules of intrinsic assignments. And a component with POINTER attribute is treated according to the rules for pointer assignment statements.

The last assignment statement is interpreted as the following set of component-wise assignments:

```
region_north(78)%name       = 'Smith'   ! character assignment
region_north78)%first_name  = 'Peter'   ! character assignment
region_north(78)%number     = 78        ! numeric assignment
region_north(78)%married    = .TRUE.    ! logical assignment
region_north(78)%salary     = 840.64    ! numeric assignment
```

8.2 Defined Assignment Statements

The defined assignment statement has the same form as an intrinsic assignment statement:

variable = expression

If the left-hand side and the right-hand side are *not both* of numeric type, *not both* of type logical, *not both* of type character with the same kind type parameter, and *not both* of the same derived type, or if both sides are of the same derived type but an *assignment interface block* and an *assignment subroutine* for this type (see below) are available, then the statement is a **defined assignment statement**.

In this case, the equals does not have the "normal" intrinsic interpretation but its interpretation (as an assignment operator for this left-hand side and this right-hand side) is determined by an *assignment subroutine* and an *assignment interface block*, which must be available in the scoping unit with this assignment statement. The assignment subroutine specifies which operations are to be performed for the execution of the defined assignment statement. For example, the assignment subroutine may specify how the result of the right-hand side should be treated before it is assigned (or not) to the left-hand side. And the assignment interface block specifies the assignment subroutine(s) belonging to the defined assignment statement.

The interpretation of *intrinsic* assignment statements cannot be overridden by a defined assignment statement. Defined assignments are extensions of the intrinsic interpretation of the assignment symbol.

A defined assignment is an **elemental assignment** if the assignment subroutine has the ELEMENTAL property, that is, if it is an elemental subroutine.

8.2 Defined Assignment Statements

Assignment subroutine

An **assignment subroutine** is a "normal" external subroutine, module subroutine, or dummy argument subroutine with exactly two dummy arguments.

The two dummy arguments must be nonoptional dummy arguments representing data objects. The first dummy argument must have the INTENT(OUT) or INTENT(INOUT) attribute and the second dummy argument must have the INTENT(IN) attribute.

An assignment subroutine may be referenced like any other subroutine by the execution of a CALL statement. But in a scoping unit where an appropriate assignment interface block for a defined assignment statement is available, the assignment subroutine may be invoked automatically during execution of the assignment statement.

Assignment interface block

An **assignment interface block** is a special *generic interface block* for a defined assignment.

INTERFACE ASSIGNMENT (=) ⟵ INTERFACE statement
⋮ ⟵ *interface definition(s)* of assignment subroutine(s); may be absent
MODULE PROCEDURE ... ⟵ MODULE PROCEDURE statement(s); may be absent or must be absent
⋮
END INTERFACE [ASSIGNMENT (=)] ⟵ END INTERFACE statement

Interface definition: If the assignment subroutine is an *external subroutine* or a *dummy argument subroutine*, an assignment interface block must be specified containing an interface definition which begins with a SUBROUTINE statement, ends with an END statement, and may include specification statements, which characterize the interface, that is, the dummy arguments.

If the assignment subroutine is a *module subroutine*, no interface definition must be specified in the interface block, but the name of the assignment subroutine must be specified in a MODULE PROCEDURE statement within the interface block:

MODULE PROCEDURE assign_subr [, assign_subr]...

A MODULE PROCEDURE statement must appear in an assignment interface block only if the interface block is contained within a module or if it references a module. And the module subroutine specified in the MODULE PROCEDURE statement must be available in the interface block by USE association.

Note that two or more assignment interface blocks being available in a scoping unit are interpreted as a single generic interface.

Implicit reference

An assignment statement is interpreted as a defined assignment statement (that is, as an implicit reference to an assignment subroutine)

- If an assignment interface block is available (i. e., specified or accessible) in the same scoping unit;
- If the assignment subroutine is specified in the assignment interface block and can be referenced in the scoping unit;
- If the left-hand side and the right-hand side of the assignment statement agree with the corresponding dummy arguments of the assignment subroutine with regard to data type, kind type parameter, and, if applicable, character length;
- and if
  - either the assignment subroutine is an elemental subroutine, and either the left-hand side and the right-hand side having the same shape or with the right-hand side being scalar,
  - or the left-hand side and the right-hand side and the corresponding first and second dummy argument agree with regard to rank and, if applicable, shape.

If this is true, the left-hand side is used as the first actual argument and the result of the evaluation of the right-hand side is used as the second actual argument for the implicit reference to the assignment subroutine.

8.2.1 Nonextended Defined Assignment

If the assignment subroutine is an *external subroutine* or a *dummy argument subroutine*, the **assignment interface block** for a nonextended defined assignment must include an interface definition of the assignment subroutine. This interface definition begins with a SUBROUTINE statement and ends with an END statement:

INTERFACE ASSIGNMENT (=)	←— INTERFACE statement
[...] **SUBROUTINE assign_subr (...)**	←— SUBROUTINE statement
⋮	←— specification part; may be absent
END [SUBROUTINE [assign_subr]]	←— END stmt. of the assignm. subr.
END INTERFACE [ASSIGNMENT (=)]	←— END INTERFACE statement

And if the assignment subroutine is a *module subroutine*, the form of the **assignment interface block** for a nonextended defined assignment is:

8.2 Defined Assignment Statements

INTERFACE ASSIGNMENT (=)	⟵ INTERFACE statement
MODULE PROCEDURE assign_subr	⟵ MODULE PROCEDURE stmt.
END INTERFACE [ASSIGNMENT (=)]	⟵ END INTERFACE statement

```
INTERFACE ASSIGNMENT (=)
  SUBROUTINE integer_to_logical (left, right)
    LOGICAL left
    INTEGER right
  END SUBROUTINE integer_to_logical
END INTERFACE
LOGICAL ls
INTEGER julian
  :
ls = julian - 15    ! defined assignment
IF (ls) CALL sub
  :
END
SUBROUTINE integer_to_logical (left, right)
  LOGICAL left
  INTEGER right
  IF (right /= right/2*2) THEN
    left = .FALSE.
  ELSE
    left = .TRUE.
  END IF
END SUBROUTINE integer_to_logical
```

The assignment subroutine is an external subroutine. The statement `ls = ...` is an assignment statement with a logical left-hand side and with an integer right-hand side.

8.2.2 Extended Defined Assignment

A defined assignment statement may be extended by specifying more than one assignment subroutine in the assignment interface block or by accessing two or more assignment interface blocks in a scoping unit. It makes no difference whether these additional assignment subroutines are module subroutines, external subroutines, or dummy argument subroutines.

All these subroutines specified in the assignment interface block must be valid assignment subroutines. And they must satisfy similar unambiguity rules as specific subprograms having a common generic name: Within a scoping unit, any two assignment subroutines must have a dummy argument at the same position such that these two arguments have a different data type, kind type parameter, or rank.

When such an extended assignment is encountered, the characteristics of the left-hand side and of the right-hand side determine which of the assignment subroutines will be actually invoked and executed.

8.3 Pointer Assignment Statement

A **pointer assignment statement** may be used to associate a pointer with a (new) target. Afterwards the pointer points at this new target and not at the former target. The form of a pointer assignment statement is

pointer => target

The **target** may be one of the following:
- a variable having the TARGET attribute,
- a pointer, that is, a variable having the POINTER attribute, or
- an expression which evaluates to a pointer.

The **pointer** is called the *left-hand side* and the **target** is called the *right-hand side* of the pointer assignment statement.

The right-hand side must have either the POINTER attribute or the TARGET attribute. It may be a subobject of a variable with the TARGET attribute. But it must not be an array section with a vector subscript. If it is a structure component of a derived type variable without the TARGET attribute, the corresponding type component must have the POINTER attribute.

The right-hand side must have the same type, the same kind type parameter, the same character length (if applicable), and the same rank as the pointer on the left-hand side.

If the right-hand side is *not* a pointer, the pointer on the left-hand side becomes associated with the target on the right-hand side; that is, the pointer points at the target on the right-hand side. In this case, the right-hand side may be even an allocatable array (with TARGET attribute).

If the right-hand side is a pointer, the pointer on the left-hand side receives the same pointer association status as the pointer on the right-hand side. If the right-hand side is a pointer which is currently associated, the pointer on the left-hand side becomes associated with the same target as the pointer on the right-hand side. If the right-hand side is a pointer which is currently not associated with a target, or if the right-hand side is a reference to the intrinsic NULL function, the pointer on the left-hand side also becomes "disassociated". And if the right-hand side is a pointer with undefined association status, the left-hand side also receives the pointer association status "undefined".

```
REAL, POINTER :: line (:), short;   REAL, TARGET   :: page (45, 10)
TYPE struct
  INTEGER i
  REAL, POINTER :: comp
  CHARACTER (LEN=78) text
END TYPE struct
TYPE (struct) :: structure
  :
line          => page(7, 1:10)          ! pointer => no pointer
short         => structure%comp         ! pointer => pointer
structure%comp => f(aa1, aa2)           ! pointer => function result
  :
CONTAINS
  FUNCTION f (da1, da2)
    REAL, POINTER :: f
    :
  END FUNCTION f
END
```

8.4 Masked Array Assignments

Sometimes, array assignment statements must be executed only for selected elements, for example:

```
REAL, DIMENSION (100) :: denominator, reci
READ (14) denominator
reci = 1/denominator
```

Suppose, the used Fortran processor allows the execution of the above array assignment statement only for those array elements for which the denominator(i) is nonzero. If the array expression and the array assignment is to be preserved, a WHERE statement or a WHERE construct may be used containing an appropriate logical mask to avoid the illegal element-wise assignment operations. Such a statement is called a *masked array assignment*.

8.4.1 WHERE Statement

The WHERE statement may be used to control the execution of a single array assignment statement by a logical array mask.

WHERE (mask_expression) array_variable = expression

The **mask_expression** is a logical array expression which must have the same shape as the **array_variable**.

During execution of a WHERE statement, the mask expression is evaluated first. The result of the evaluation is called the **control mask**. Then the expression on the right-hand side of the array assignment statement is evaluated only for those array element positions corresponding to *true* elements of the mask expression. Finally, only these selected results are assigned to the array variable on the left-hand side of the array assignment statement. The other array elements of the array variable are not (re)defined. The elemental assignments are executed according to the rules for (scalar) assignment statements.

```
REAL, DIMENSION (100) :: denominator, reci
READ (14) denominator
WHERE (denominator /= o.) reci = 1/denominator
```

8.4.2   WHERE Construct

The WHERE construct statement, the END WHERE statement, and optional (masked and nonmasked) ELSEWHERE statements may be used to control the execution of one or more sequences of array assignment statements by one logical array mask. Such sequence of array assignment statements is called an **assignment block**.

[where-name :] WHERE (mask_expression)   is a WHERE construct stmt.
ELSEWHERE [where-name]   is a (nonmasked) ELSEWHERE statement.
ELSEWHERE (mask_expression) [where-name]
                             is a masked ELSEWHERE statement.
END WHERE [where-name]   is an END WHERE statement.

where-name is a name identifying the WHERE construct. And **mask_expression** is a logical array expression.

The WHERE construct statement is the first statement of a WHERE construct. The END WHERE statement is the last statement of a WHERE construct. And one nonmasked ELSEWHERE statement and one or more masked ELSEWHERE statements may be used if several assignment blocks are to be executed such that each of these assignment blocks is executed for another selection of array element positions.

An assignment block within the body of a WHERE construct is allowed to contain

- WHERE constructs,
- WHERE statements, and
- intrinsic and defined array assignment statements.

Thus, WHERE constructs may be nested.

8.4 Masked Array Assignments

In a WHERE construct, the mask expression(s) and the left-hand side of each array assignment statement within the body of the WHERE construct must have the same shape.

Each statement belonging to and contained within a WHERE construct may be labeled, but only the outmost WHERE construct statement may be used as a branch target statement.

If the WHERE construct statement includes a **where-name**, the corresponding END WHERE statement *must* include the same name and the corresponding (masked or nonmasked) ELSEWHERE statement *may* include this name. If the WHERE construct statement does not include a **where-name**, neither the END WHERE statement nor the ELSEWHERE statements must include a **where-name**.

Note that the name of a WHERE construct statement is no statement label. Therefore, this identification must appear in position 7 to 72 if the program unit is written in fixed source form.

The most simple form of a WHERE construct begins with a WHERE construct statement, ends with an END WHERE statement, and includes an assignment block with a sequence of (intrinsic and/or defined) array assignment statements.

[**where-name :**] **WHERE (mask_expression)**
   array_variable$_1$ **= expression**$_1$
   array_variable$_2$ **= expression**$_2$
   ⋮
END WHERE [**where-name**]

The where-block may be empty.

During execution of this WHERE construct, the mask expression is evaluated first. Its result is the **control mask** which is internally stored. Then, the array assignment statements within the body of the WHERE construct are evaluated in order as though they were single WHERE statements.

```
REAL xarray (100)
INTEGER sparse (100)
LOGICAL, DIMENSION (100) :: mask = .FALSE.
mask(1:100:2) = .TRUE.
WHERE (mask)
  xarray = xarray - 1.0
  sparse = 1
END WHERE
```

Every second element of array **xarray** is reduced by 1.0 beginning with the first array element. And the corresponding elements of array **sparse** are marked (i. e., defined) with the value 1. The other elements of arrays are not redefined.

A (nonmasked) ELSEWHERE statement may be used to separate two assignment blocks within the body of a WHERE construct. After execution of the first assignment block for those array element positions corresponding to *true* values in the mask expression, the second assignment block is executed *exactly* for those array element positions corresponding to *false* values in the mask expression.

A masked ELSEWHERE statement may be used if the body of the WHERE construct includes (at least) two assignment blocks such that the first assignment block is executed for those array element positions corresponding to *true* values in the mask expression of the WHERE construct statement and the second assignment block is executed *at most* for those array element positions corresponding to *false* values in the mask expression of the WHERE construct statement. The selection of these remaining array element positions depends on the mask expression of the masked ELSEWHERE statement. Thus, the selection of array element positions to be used during the execution of the second assignment block depends on two mask expressions.

[**where-name :**] **WHERE (mask_expression**$_1$ **)**
   array_variable$_1$ = **expression**$_1$
   array_variable$_2$ = **expression**$_2$
   $\vdots$

ELSEWHERE [**(mask_expression**$_2$ **)**] [**where-name**]
   array_variable$_k$ = **expression**$_k$
   array_variable$_{k+1}$ = **expression**$_{k+1}$
   $\vdots$

END WHERE [**where-name**]

Both assignment blocks may be empty.

During execution of this WHERE construct, the first mask expression is evaluated. Its result is the **control mask** which is internally stored. In addition, the value (.NOT. control mask) is internally stored which is the **pending control mask**. Then the array assignment statements within the where-block are evaluated in order as though they were single WHERE statements with the established control mask. During execution of the ELSEWHERE statement, the second mask expression is evaluated, if present; otherwise, an array having only *true* elements is taken as default control mask. Then the current control mask for the elsewhere-block is internally stored as the result of (pending control mask .AND. second mask expression). An finally, the array assignment statements within the (masked or nonmasked) elsewhere-block are evaluated in order as though they were single WHERE statements with the established control mask ((.NOT. **mask_expression**$_1$) .AND. **mask_expression**$_2$).

8.4 Masked Array Assignments

```
REAL, DIMENSION (12, 31)    :: pressure, temperature, voltage
INTEGER, DIMENSION (12, 31) :: rpm, minutes
:
WHERE (temperature <= 100.)
  pressure  = .99*pressure;    rpm = rpm - 100
ELSEWHERE         ! ELSEWHERE (temperature > 100.)
  pressure  = 1.01*pressure
  rpm       = rpm + 100
  minutes   = minutes + 1
END WHERE
```

The array assignment statements within the where-block are executed for those array element positions where the temperature is not greater than 100. And the array assignment statements within the elsewhere-block are executed only for those array element positions where the temperature is greater than 100.

WHERE in WHERE: More complicated WHERE constructs may include WHERE statements and even WHERE constructs. If a WHERE construct contains a WHERE statement, the current control mask for this WHERE statement is established as the value of (control mask .AND. mask expression) by taking the control mask of the host assignment block and the mask expression in the WHERE statement. If a WHERE construct contains another WHERE construct, the current control mask for this enclosed WHERE construct is established as the value of (control mask .AND. mask expression) by taking the control mask of the host assignment block and the mask expression of the inner WHERE construct statement. Similarly, the pending control mask for the enclosed WHERE construct is established as the value of (control mask .AND. (.NOT. mask expression)) by taking the control mask of the host assignment block and the result of the mask expression of the enclosed WHERE construct statement. After execution of the contained WHERE statement and WHERE construct, respectively, the current control mask of the host assignment block is active again.

8.4.3 Common Rules for Masked Array Assignments

If a WHERE construct contains a WHERE statement, a masked ELSEWHERE statement, or another WHERE construct, each mask expression and each left-hand side of an array assignment statement within the outmost WHERE construct must have the same shape.

Each statements within a WHERE construct is executed in sequence. Defined array assignments must be elemental assignments.

Function reference: If the expression in the right-hand side or the array variable in the left-hand side of an array assignment statement within a WHERE statement or within a WHERE construct includes a reference to a *nonelemental* function, then the actual argument expressions in this function reference and the function itself are evaluated completely without consideration of the mask expression. And then, if the result of the function is an array and the reference is not within the argument list of a nonelemental function, only those array elements are selected for use in evaluating the expression which correspond to *true* values in current control mask.

If the expression in the right-hand side or the array variable in the left-hand side of one of the array assignment statements includes a reference to an elemental function as an *operand* (but not within the actual argument list of a reference to a nonelemental function), then this function is processed only for those elements corresponding to *true* elements of the current control mask.

If an array constructor appears in an array assignment statement or in a mask expression within a WHERE statement or WHERE construct, the array constructor is evaluated without consideration of any masked control and then the array assignment statement is executed and the mask expression is evaluated, respectively.

```
REAL, DIMENSION (10, 10) :: x, y
WHERE (x > 0.)
   y = LOG(x)
ELSEWHERE
   x = -x + 1.E-3
   y = PRODUCT( LOG(x)) - 1
END WHERE
```

The reference to the elemental function LOG(x) in the where-block is evaluated element-wise for all positive array elements of x. But in the elsewhere-block, LOG(x) is evaluated for the complete array x, because, in this case, LOG(x) is not an operand on the right-hand side of the assignment statement.

Side effects: The assignment to an array variable in the left-hand side of an array assignment statement which also appears as a variable within the mask expression has no effect on the current control mask. A function reference within the mask expression is allowed to (re)define variables appearing also in an array assignment statement within the WHERE statement or within the WHERE construct.

Note that an array assignment statement within the body of a WHERE construct may affect subsequent array assignment statements, WHERE statements, WHERE constructs, and masked ELSEWHERE statements.

8.5 Indexed Assignments

The FORALL statement and the FORALL construct may be used to control multiple executions of pointer assignment statements and (scalar or array) assignment statements by sets of scalar *index variable values*. Mostly, these index variables are used to determine the array element position of the variable on the left-hand side of each contained assignment statement and pointer assignment statement. The selection of values from these sets of index variable values may be limited by an optional scalar logical mask.

8.5.1 FORALL Statement

The FORALL statement supports multiple executions of a single contained assignment statement or pointer assignment statement. Execution depends on given index variable values and on the value of an optional scalar logical mask expression.

FORALL (triplet [, triplet]... [, mask]) variable = expression
FORALL (triplet [, triplet]... [, mask]) pointer => target

Each **triplet** specifies a value set for an **index variable** and has the following form: **index_variable = first : last [: stride]**

index_variable is a named scalar variable of type integer. And **first**, **last**, and **stride** are scalar integer expressions. **mask** is a scalar logical expression.

The index variables may appear within **variable**, **expression**, **pointer**, and **target**. They also may appear as operands within the scalar logical mask expression. The scalar integer expressions **first**, **last**, and **stride** must not include index variables being defined within the same WHERE statement. A contained assignment statement must not (re)define an index variable of the WHERE statement.

8.5.1.1 Execution of the FORALL Statement

At first, the value sets of the index variables are determined. Then, the optional scalar mask expression is evaluated. And finally, the assignment statement or pointer assignment statement is executed.

Determination of the value sets of the index variables

The scalar integer expressions **first**, **last**, and **stride** are evaluated in any order, and their results $m1$, $m2$ and $m3$, respectively, are converted, if necessary, to the kind type parameter of the **index_variable**. $m3$ must not be zero; if a **stride** is not present, $m3$ has the default value 1.

The value set of an index variable contains at most $\mathcal{M} = (m2 - m1 + m3)/m3$ elements. If $\mathcal{M} \leq 0$ for a particular index variable, the value set is empty and the execution of the FORALL statement is complete; otherwise, the index variable has the

$$\text{value set} = \{m1 + (k-1)*m3 \mid k = 1, 2, \ldots, \mathcal{M}\}$$

If n index variables are given and the optional scalar mask expression is absent, the Cartesian product of the value sets of the index variables defines the set of n-tupels of index variable values controlling the execution of the contained assignment statement or pointer assignment statement. If a mask expression is present, only a subset of these n-tupels of index variable values may be activated for execution control of the FORALL.

Evaluation of the scalar mask expression

If present, the scalar logical mask expression is evaluated for all n-tupels in the Cartesian product of the value sets of all index variables; otherwise, .TRUE. is used as default mask expression. Those n-tupels of index variable values for which the result of the mask expression is *true* are determined to form the set of **active** n-tupels of index variable values.

Execution of the assignment or pointer assignment statement

The contained assignment statement or pointer assignment statement is executed for all active n-tupels of index variable values:

For an intrinsic or defined assignment statement, the **expression** on the right-hand side and all expressions on the left-hand side used for the determination of the **variable** are evaluated for all active n-tupels of index variable values. Then for all active n-tupels, each result of the evaluation of the **expression** is assigned to its corresponding **variable**. Both the evaluations and the assignments may be executed in any order.

For a pointer assignment statement, all expressions on the right-hand side used for the determination of the **target** and all expressions on the left-hand side used for the determination of the **pointer** are evaluated for all active n-tupels of index variable values. Then for all active n-tupels, the **target** is evaluated. And finally for all active n-tupels, each **target** and its corresponding **pointer** are associated. Both the evaluations and the associations may be executed in any order.

```
REAL, DIMENSION (100, 100)    :: x, y
LOGICAL, DIMENSION (100, 100) :: m
  :
FORALL (i=1:100) x(i,i) = 1.
```

The value 1.0 is assigned to the elements of the main diagonal of matrix x.

8.5 Indexed Assignments 8-19

```
FORALL (i=1:100:2, j=1,100, i < j .AND. m(i,j)) x(i,j) = y(i,j)
```

Each second row of the upper triangle of matrix x is redefined at those array element positions where the corresponding element of matrix m is *true*.

8.5.2 FORALL Construct

The FORALL construct statement and the END FORALL stement may be used to control multiple executions of one or more assignment statements and/or pointer assignment statements by a single list of index variables and by a single optional scalar logical mask.

[**forall-name :**] **FORALL (tripel** [**, tripel**]... [**, mask**] **)**
  is a FORALL construct statement.
END FORALL [**forall-name**] is an END FORALL statement.

forall-name is a name identifying the FORALL construct. **tripel** und **mask** are defined as for FORALL statements (see last section).

The FORALL construct statement is the first statement of a FORALL construct. The END FORALL statement is the last statement of the FORALL construct.

The body of a FORALL construct is allowed to contain

- FORALL constructs,
- FORALL statements,
- WHERE constructs,
- WHERE statements,
- intrinsic and defined assignment statements, and
- pointer assignment statements.

Thus, FORALL constructs may be nested.

Each statement belonging to and contained within a FORALL construct may be labeled, but only the outmost FORALL construct statement may be used as a branch target statement.

If the FORALL construct statement includes a **forall-name**, the corresponding END FORALL statement *must* include the same name. If the FORALL construct statement does not include a **forall-name**, neither the END FORALL statement must include a **forall-name**.

Note that the name of a FORALL construct statement is no statement label. Therefore, this identification must appear in position 7 to 72 if the program unit is written in fixed source form.

A FORALL construct is interpreted as though it were a FORALL statement that contains several statements and/or constructs (see above) whose execution is controlled by the same list of index variables and, if present, by the same scalar logical mask.

The most simple form of a FORALL construct begins with a FORALL construct statement, ends with an END WHERE statement, and includes within its body a sequence of assignment statements and/or pointer assignment statements.

[forall-name :] FORALL (...)
   variable$_1$ = expression$_1$
   pointer$_1$ => target$_1$
   variable$_2$ = expression$_2$
   pointer$_2$ => target$_2$
   ⋮
END FORALL [forall-name]

The body of the forall construct may be empty.

```
FORALL (i=1:365, j=1991:1996)
  diff(i,j)   = max_x(i,j) - min_x(i,j)
  signal(i,j) = diff(i,j) - normal(i,j)
END FORALL
```

Firstly, the elemental assignments to `diff` are performed. Then, the computed results for `diff` are used for the evaluation of the right-hand side of the second assignment statement.

More complicated FORALL constructs may include WHERE statements, WHERE constructs, FORALL statements, and other FORALL constructs.

For the list of index variables, for the scalar logical mask expression, for the assignment statements and pointer assignment statements within the body of a FORALL construct including any nested statements and constructs, and for the execution of the FORALL construct, the same rules apply as are given for the FORALL statement (see last section).

WHERE within FORALL: If a FORALL construct contains a WHERE construct, the statements within the body of the WHERE construct are executed in sequence. The nonscalar mask expression of a contained WHERE statement, WHERE construct statement, or masked ELSEWHERE statement is evaluated only for all active n-tupels of index variable values, possibly controlled by control masks of outer WHERE constructs.

Each array assignment statement in a WHERE statement or WHERE construct within the body of a FORALL construct is executed for all active n-tupels of index variable values under control of the current control mask for the array assignment statement.

8.5 Indexed Assignments

```
REAL, DIMENSION (25, 100) :: x, y
LOGICAL, DIMENSION (25, 100) :: sparse
  ⋮
FORALL (i=1:10)
  WHERE (.NOT. sparse(i,:))
    x(i,:) = y(i,:)
    sparse(i,:) = .TRUE.
  END WHERE
  x(i,:) = x(i,:) * 1000.
END FORALL
```

The first 10 rows of matrix **x** are redefined element-wise at those array element positions corresponding to *false* values of matrix **sparse**. Then, the affected elements of **sparse** are set to .TRUE. Finally, the elements of the first 10 rows of **x** are scaled by 10^3.

FORALL within FORALL: If a FORALL construct contains a FORALL statement or FORALL construct, the expressions **first**, **last**, and possibly **stride** for any index variable of the inner FORALL statement and FORALL construct statement, respectively, are evaluated for all active n-tupels of index variable values of the outer FORALL construct. Then, the m value sets of the (suppose) m index variables of the inner FORALL are determined for each single active n-tupel. Thus, the total set of all $(m+n)$-tupels of index variable values for the inner FORALL is determined. If present, the scalar mask expression of the inner FORALL statement and FORALL construct statement, respectively, is evaluated for these $(m+n)$-tupels of index variable values thus determining the set of all active $(m+n)$-tupels for the inner FORALL; otherwise, the value .TRUE. is used as default mask expression. And finally, all statements contained within the inner FORALL are executed for each active $(m+n)$-tupel of index variable values.

```
FORALL (i=1:9)
  FORALL (j=1:i, ABS(x(i,j)) < 1.E-7)
    x(i,j) = y(i,j)
  END FORALL
END FORALL
```

Matrix **x** is redefined at those array element positions of the lower diagonal matrix and of the main diagonal where the magnitude of **x** is less than 10^{-7}.

8.5.3 Common Rules for Indexed Assignments

The scope of an index variable is the FORALL statement or FORALL construct defining the index variable. Data type and kind type parameter are given as though the variable were declared within the host scoping unit. An index variable cannot have any other attribute.

Any explicitly or implicitly (by a defined operation or assignment) referenced subprogram within the scalar mask expression or within any contained assignment statement must be a pure subprogram.

For a contained assigment being a defined assignment, the belonging assignment subroutine must not reference (the value) of the variable on the left-hand side of the defined assignment statement.

A single assignment statement or pointer assignment statement contained in a FORALL statement or within the body of a FORALL statement must not be used such that multiple assignments to the same variable and multiple associations with the same pointer, respectively, occur. This rule is also valid for indirect redefinitions of the left-hand side variable through a pointer. Note that more than one assignment statement or pointer assignment statement are allowed to appear in the body of FORALL construct redefining the same variable and associating with the same pointer, respectively.

```
INTEGER, DIMENSION (100) :: a, b, indirect
  :
FORALL (i=1:100) a(indirect(i)) = b(i)
```

This FORALL –Anweisung is allowed only if array **indirect** contains are permutation of the value sequence (1, 2,... , 100).

```
FORALL (i=1:101:50) radius = a(i)
```

This FORALL statement is illegal because it would cause three assignments to the scalar variable **radius**.

```
FORALL (i=1:101:50, i > 75) radius = a(i)
```

The set of active index variable values includes only the value 101. Thus, no multiple assignment occurs and the FORALL statement is legal.

Normally, the left-hand side of an assignment statement and of a pointer assignment statement within a FORALL statement or FORALL construct designates an array element or an array section with *all* index variables being used for the identification of the subobject.

Side effects: A function reference in the logical expression within a FORALL statement or FORALL construct statement, and a function reference in one of the scalar integer expressions defining the value set for an index variable in a FORALL statement or FORALL construct statement is allowed to (re)define variables appearing also in a statement being executed under control of that FORALL statement and FORALL construct statement, respectively.

9 DECLARATIONS AND SPECIFICATIONS

Specification statements are nonexecutable statements. They are used, for example, to specify properties of data entities and subprograms or to specify the accessibility of certain entities in modules. Every data entity has a type, a rank and, if necessary, other properties. Most of these properties are called *attributes*. If a data entity has a name, that is, if it is a named variable, a named constant, or a function result, then these attributes may be specified in a type declaration statement, by separate specification statements, or they are given implicitly.

A named entity may appear in the same scoping unit in different specification statements. And different named entities which have the same attribute(s) may be distributed over several specification statements; for example, a specification part may include several INTEGER statements.

```
INTEGER ug, og, mine
POINTER mine
CHARACTER z, line (72)
```

These specifications may also be written as follows:

```
INTEGER ug
INTEGER og
INTEGER mine
POINTER mine
CHARACTER z
CHARACTER line
DIMENSION line (72)
```

A particular attribute may be specified only once for a named entity in the scoping unit. And the name of the entity may appear only once in a particular specification statement. There are exceptions for the DIMENSION attribute, for the length specification in CHARACTER statements, and for names which are affected by IMPLICIT statements.

The declarations of data entities and the specifications of their attributes may appear in any order, except when these entities are used afterwards in other specification statements in the same specification part.

```
REAL (KIND=2) cc
⋮
REAL (KIND(cc)) dd
```

The function reference KIND(cc) appears in the kind type parameter specification for dd. Therefore, the declaration of cc must appear before the declaration of dd.

There are the following *type declaration statements*:

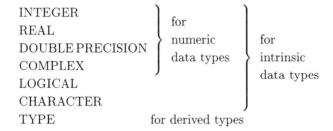

INTEGER
REAL
DOUBLE PRECISION } for numeric data types } for intrinsic data types
COMPLEX
LOGICAL
CHARACTER
TYPE   for derived types

In addition, there are the following *specification statements*:

ALLOCATABLE	COMMON	**DATA**	**DIMENSION**
EQUIVALENCE	EXTERNAL	IMPLICIT	INTENT
INTRINSIC	NAMELIST	**OPTIONAL**	**PARAMETER**
POINTER	PRIVATE	PUBLIC	SAVE
SEQUENCE	TARGET		

The boldface statements are *attribute specification statements*.

Except for the IMPLICIT statement and the SEQUENCE statement, each specification statement consists of a keyword (that is, the name of the statement) and at least an additional list of names of those entities for which certain properties are to be specified. Except for the PRIVATE, PUBLIC, and SAVE statements, these lists of names must not be empty.

9.1 Attributes

Each data entity has properties such as a data type, a kind type parameter, a length (if applicable), a rank, and mostly a shape. In addition, a data entity may have other properties.

Except for the type, the kind type parameter, the length, and the name, these properties are **attributes**. Some attributes may be specified for entities that are no data entities.

Attributes may be specified either explicitly or implicitly. Attributes for named data entities may be specified explicitly in type declaration statements, in attribute specification statements, and in COMMON statements.

Except for the DATA attribute, the name of an attribute is equal to the key word which is used to specify the attribute in a type declaration statement.

9.1 Attributes

Attribute	Use
ALLOCATABLE	allocatable array
DATA	initialization of a variable
DIMENSION	rank and shape of an array
EXTERNAL	external subprogram
INTENT	intended use of dummy argument
INTRINSIC	intrinsic subprogram
OPTIONAL	optional dummy argument
PARAMETER	named constant expression
POINTER	pointer
PRIVATE	never accessible outside a module
PUBLIC	may be accessible outside a module
SAVE	retains certain values in a subprogram
TARGET	a pointer may point at a target

9.1.1 ALLOCATABLE Attribute

The ALLOCATABLE attribute is used to specify that an array is an *allocatable array*.

```
REAL, ALLOCATABLE, DIMENSION (:, :) :: aa2, aa22, aa3 (:, :, :)
```

or

```
DIMENSION aa2 (:, :), aa22 (:, :), aa3 (:, :, :)    ! array specif.
REAL, ALLOCATABLE :: aa2, aa22, aa3                 ! type and attr.
```

or

```
REAL :: df2, df22, df3                              ! type
DIMENSION df2 (:, :), df22 (:, :), df3 (:, :, :)    ! array specif.
ALLOCATABLE df2, df22, df3                          ! attribute
```

When the ALLOCATABLE attribute is specified, the following rules apply:

The ALLOCATABLE attribute may be specified only for arrays. The array specification must not include array bounds.

The ALLOCATABLE attribute must not be specified for dummy arguments or function results.

An array that has the ALLOCATABLE attribute must not have an additional POINTER or PARAMETER attribute.

9.1.2 DATA Attribute

The DATA attribute is used to specify *initial values* for nonpointer variables and an initial pointer association status for pointer variables. This kind of initialization is described as "explicit initialization". Initialized nonpointer variables are already defined with a valid value when the execution of the program begins. Initialized variables may be redefined during the execution of the program. The DATA attribute may be specified by a DATA statement. Explicit initialization may also be specified for a variable in a type declaration statement.

```
INTEGER :: start = 1, border = 0, max = 50 ! type and init. value
```

or

```
INTEGER start, border, max              ! type
DATA start, border, max /1, 0, 50/      ! initial value
```

When a variable is initialized, the following rules apply:

The assignment of the initial value to a variable is done according to the rules for intrinsic assignment statements.

A variable or a part of it may be explicitly initialized only once in a program.

The DATA attribute cannot be specified within the scoping unit of an interface block. Therefore, a DATA statement must not appear in an interface block; an initial value in a type declaration statement may be specified in an interface block, but it is ineffective.

If a variable is initialized, all associated variables also become initialized.

The following entities must not be explicitly initialized: a dummy argument, a function result, a variable in a named common block outside a block data program unit, a variable in the blank common, an allocatable array, and an automatic variable.

A variable that has the DATA attribute and is not in a common block automatically has the SAVE attribute.

9.1.3 DIMENSION Attribute

The DIMENSION attribute is used to specify that a data entity is an array.

```
REAL, DIMENSION (8) :: market, fruit (2:8) ! type and array spec.
```

or

```
REAL market, fruit                      ! type
DIMENSION market (8), fruit (2:8)       ! array specification
```

or

```
REAL market, fruit                  ! type
COMMON market (8), fruit (2:8)      ! array specification
```

9.1.4 EXTERNAL Attribute

The EXTERNAL attribute is used to specify that a name is an external subprogram, a dummy argument subprogram, or a block data program unit. The specification is needed, for example, if the name of an external subprogram is supplied as an actual argument or if a user-defined subprogram is used that has the same name as an intrinsic subprogram. The EXTERNAL attribute may be specified (processor-dependent) for the name of a block data program unit which is to be loaded from a program library.

```
COMPLEX, EXTERNAL :: plane, inside, outside  ! type and attribute
```

or

```
COMPLEX plane, inside, outside      ! type
EXTERNAL plane, inside, outside     ! attribute
```

When the EXTERNAL attribute is specified, the following rules apply:

An entity that has the EXTERNAL attribute must not have an additional POINTER, TARGET, or INTRINSIC attribute.

A dummy argument that has the EXTERNAL attribute is a dummy argument subprogram.

If the EXTERNAL attribute is specified for the name of an intrinsic subprogram, the intrinsic subprogram cannot be used in this scoping unit.

A statement function, an internal subprogram, and a module subprogram must not have the EXTERNAL attribute.

9.1.5 INTENT Attribute

The INTENT attribute is used to specify the intended use of dummy arguments. There are the following INTENT attributes:

INTENT(IN): The INTENT(IN) attribute specifies that the dummy argument is intended for use as an *input argument*, that is, to receive values from the invoking (sub)program. Such a dummy argument may be used (referenced) but must neither be redefined nor become undefined during the execution of the program.

INTENT(OUT): The INTENT(OUT) attribute specifies that the dummy argument is intended for use as an *output argument*, that is, to return values to the invoking (sub)program. Such a dummy argument may be used (referenced) only if it has (gotten) a valid value. An associated actual argument must be a variable. If such a dummy argument is of a derived type, it may have components for which *default initialization* is specified in its type definition.

INTENT(INOUT): The INTENT(INOUT) attribute specifies that the dummy argument is intended for use as an *input/output argument*. Such a dummy argument may receive values from the invoking (sub)program and may return values to the invoking (sub)program. An associated actual argument must be a variable.

```
INTEGER, INTENT (IN)  :: in1, in2      ! type and attribute
REAL, INTENT (OUT)    :: out           ! type and attribute
```

or

```
INTEGER        in1, in2                ! type
REAL           out                     ! type
INTENT (IN)    in1, in2                ! attribute
INTENT (OUT)   out                     ! attribute
```

When an INTENT attribute is explicitly specified, the following rules apply:

An INTENT attribute may be specified only for dummy arguments that are variables.

A dummy argument which has an INTENT attribute must not have an additional POINTER, or SAVE attribute.

If a dummy argument of a derived type has an INTENT attribute, its components have (implicitly) the same INTENT attribute.

9.1.6 INTRINSIC Attribute

The INTRINSIC attribute is used to specify that a name is (the specific or generic name of) an intrinsic function. If it is a specific name, it may be supplied as an actual argument.

```
INTEGER, INTRINSIC :: IABS, IDIM, MAX0    ! type and attribute
```

or

```
INTEGER    IABS, IDIM, MAX0               ! type
INTRINSIC  IABS, IDIM, MAX0               ! attribute
```

An entity that has the INTRINSIC attribute must not have an additional POINTER, TARGET, or EXTERNAL attribute.

9.1.7 OPTIONAL Attribute

The OPTIONAL attribute is used to specify for a dummy argument that the subprogram reference need not contain an actual argument corresponding to the dummy argument. A dummy argument with OPTIONAL attribute is an "optional" dummy argument.

```
REAL, OPTIONAL :: da3          ! type and attribute
```
or
```
REAL      da3                  ! type
OPTIONAL da3                   ! attribute
```

9.1.8 PARAMETER Attribute

The PARAMETER attribute is used to specify a name that identifies a **named constant**. A named constant is a name for the result of an *initialization expression*. It may be used everywhere a literal constant of the same type may be used (exceptions see below).

```
REAL, PARAMETER :: r=2.5, pi=3.1415, d=22./(2*pi) ! type and attr.
```
or
```
REAL r, pi, d                                     ! type
PARAMETER (r = 2.5, pi = 3.1415, d = 22./(2*pi) ) ! attribute
```

When the PARAMETER attribute is specified, that is, if a named constant is defined, the following rules apply:

The assignment of the initialization expression is done according to the rules for intrinsic assignment statements.

The PARAMETER attribute may be specified only for data entities.

The PARAMETER attribute must not be specified for dummy arguments, function results, allocatable arrays, pointers, and not for variables in a common block.

If the PARAMETER attribute is specified for a data object of a derived type, the components of the object also (implicitly) have the PARAMETER attribute.

A named constant must not have an additional POINTER, TARGET, or SAVE attribute.

A variable must not have the PARAMETER attribute.

Another named constant may appear in the initialization expression for the definition of a named constant:

- If the other named constant is already defined lexically earlier in the same type declaration statement or PARAMETER statement;
- If the other named constant is already defined lexically earlier in another type declaration statement or PARAMETER statement; or
- If the other named constant is already accessible by USE association or host association.

A named constant must be used neither in a format specification nor as a part of another numeric constant.

9.1.9 POINTER Attribute

The POINTER attribute is used to specify that a data entity is a *pointer*.

```
REAL, POINTER :: coeff, inter, result    ! type and attribute
```

or

```
REAL    coeff, inter, result             ! type
POINTER coeff, inter, result             ! attribute
```

When the POINTER attribute is specified, the following rules apply:

The pointer attribute may be specified only for variables and for function results.

An entity that has the POINTER attribute must not have an additional ALLOCATABLE, EXTERNAL, INTENT, INTRINSIC, PARAMETER, or TARGET attribute.

A pointer must not be specified to have an initial value, but it may be initialized with the pointer association status "disassociated".

9.1.10 PRIVATE Attribute

The PRIVATE attribute is used to specify that certain entities in a module cannot be made accessible outside the module. The PRIVATE attribute may be specified for variables, module subprograms, derived data types, named constants, namelist groups, type components, and generic interface blocks. Entities with PRIVATE attribute are described as "private" entities.

```
INTEGER, PRIVATE :: code, mark, mask     ! type and attribute
```

or

```
INTEGER code, mark, mask                 ! type
PRIVATE code, mark, mask                 ! attribute
```

9.1 Attributes

When the PRIVATE attribute is specified, the following rules apply:

The PRIVATE attribute may be specified only in the scoping unit of a module.

If there is no PRIVATE attribute effective for an entity that may have an accessibility attribute, this entity (implicitly or explicitly) has the PUBLIC attribute.

An entity that has the PRIVATE attribute must not have an explicitly specified PUBLIC attribute.

If a group-name (for namelist input/output) has the PUBLIC attribute, no variable in the namelist group must have the PRIVATE attribute.

9.1.11 PUBLIC Attribute

The PUBLIC attribute is used to specify that certain entities in a module can be made accessible outside the module. The PUBLIC attribute may be specified for variables, module subprograms, derived data types, named constants, namelist groups, type components, and generic interface blocks. Entities with PUBLIC attribute are described as "public" or "visible" entities.

```
REAL, PUBLIC :: pr_1, pr_2, pr_10      ! type and attribute
```
or
```
REAL     pr_1, pr_2, pr_10             ! type
PUBLIC   pr_1, pr_2, pr_10             ! attribute
```

When the PUBLIC attribute is specified, the following rules apply:

The PUBLIC attribute may be specified only in the scoping unit of a module.

The PUBLIC attribute is the default attribute; that is, if no accessibility attribute (PUBLIC or PRIVATE) is specified, all module entities which may have an accessibility attribute have implicitly the PUBLIC attribute.

An entity that has an explicitly specified PUBLIC attribute must not have an additional PRIVATE attribute.

If a group-name (for namelist input/output) has the PUBLIC attribute, no variable in the namelist group must have the PRIVATE attribute.

If a derived data type has the PRIVATE attribute, no data entity of this type must have the PUBLIC attribute.

9.1.12 SAVE Attribute

The SAVE attribute is used if certain properties of the specified variables are to be preserved when the END statement or a RETURN statement of a subprogram is executed. Variables with SAVE attribute are described as "saved" entities.

A SAVE attribute which is specified in the scoping unit of a main program is ineffective. If it is specified in the scoping unit of a subprogram, the following properties of an affected variable are preserved (for the next execution of the subprogram): its definition status, its value, its pointer association status (if it is a pointer), and its allocation status (if it is an allocatable array).

```
REAL, SAVE :: x, temp, mean    ! type and attribute
```

or

```
REAL x, temp, mean             ! type
SAVE x, temp, mean             ! attribute
```

When the SAVE attribute is explicitly specified, the following rules apply:

The SAVE attribute may be specified only for variables.

If the name of a common block enclosed in slashes is specified in a SAVE statement, the SAVE attribute is (implicitly) specified for *all* variables in this common block.

The SAVE attribute must not be specified for dummy arguments, function results, subprogram names, automatic variables, namelist groups, named constants and not for single variables in a common block.

A variable that has an explicitly specified SAVE attribute must not have an additional INTENT, OPTIONAL, EXTERNAL, or INTRINSIC attribute.

The DATA attribute implies (except for variables in named common blocks) the SAVE attribute; that is, a variable that has the DATA attribute also (implicitly) has the SAVE attribute. This SAVE attribute may be confirmed in a type declaration statement or by a SAVE statement.

If a SAVE statement appears without a list, the SAVE attribute must not be confirmed in a type declaration statement or by an additional SAVE statement.

9.1.13 TARGET Attribute

The TARGET attribute is used to specify that a data entity may be associated with a pointer, which then points at this target.

```
REAL, TARGET :: k_array, t, right_part    ! type and attribute
```

or

```
REAL    k_array, t, right_part            ! type
TARGET k_array, t, right_part             ! attribute
```

When the TARGET attribute is specified, the following rules apply:

An entity that has the TARGET attribute must not have an additional PARAMETER, POINTER, EXTERNAL, or INTRINSIC attribute.

The TARGET attribute may be specified only for variables.

If a data object has the TARGET attribute, any subobject of this parent object (implicitly) also has the TARGET attribute. This rule applies to character substrings, for structure components, for array elements, and for array sections without vector subscripts.

Note that an allocatable array may have the TARGET attribute.

9.2 Type Declaration Statements

A type declaration statement is used to explicitly declare one or more data entities. In addition to the data type of a data entity, attributes may be specified such as its value, its accessibility, its rank and its shape, its properties as an actual argument or as a dummy argument, its status as a (local) variable in a subprogram, whether it is a pointer or possibly a target, etc.

type [, attribute]... [::] data_entity [, data_entity]...

where **type** is one of the following *type specifications*:

INTEGER	INTEGER (kind_type_parameter)	
REAL	REAL (kind_type_parameter)	DOUBLE PRECISION
COMPLEX	COMPLEX (kind_type_parameter)	
LOGICAL	LOGICAL (kind_type_parameter)	
CHARACTER	CHARACTER type_parameter	
TYPE (type_name)		

The **type_name** is the name of a derived type. And the **kind_type_parameter** is a *kind type parameter specification* of the form [**KIND =**] **kind**, where **kind** is a scalar integer *initialization expression* that evaluates to a nonnegative kind type parameter value. The allowed values of the kind type parameter are processor-dependent.

And each **attribute** is one of the following specifications:

ALLOCATABLE	for allocatable arrays,
DIMENSION (...)	for a *leading* array specification,
EXTERNAL	for names of external functions,
INTENT (IN)	for input arguments,
INTENT (OUT)	for output arguments,
INTENT (INOUT)	for input/output arguments,
INTRINSIC	for names of intrinsic functions,

OPTIONAL	for *optional* dummy arguments,
PARAMETER	for named constants,
POINTER	for pointers,
PRIVATE	for *private* entities in modules,
PUBLIC	for *public* entities in modules,
SAVE	for *saved* local entities in subprograms, and
TARGET	for targets at which pointers may point.

And each **data_entity** is one of the following:

– The name of a data object, optionally followed by its *own* array specification, optionally followed by its *own* length specification, and optionally followed by an initialization specification. If the data object being declared is not a pointer, the optional initialization specification is written as an equals with a subsequent *initialization expression*; if it is a pointer, the initialization specification is written as a pointer assignment symbol "=>" with a subsequent function reference NULL( ):

name [(**dim** [, **dim**]...)] [* **own_length**] [= **initialization_expression**]
name [(**dim** [, **dim**]...)] [* **own_length**] [=> **NULL()**]

– The name of a function, optionally followed by an *own* length specification:

fname [* **own_length**]

The length specification * **own_length** is allowed only in CHARACTER statements. The double colon "::" is required only if an **attribute** or an initialization is specified.

```
INTEGER number, years
INTEGER (4) max, min
REAL, PARAMETER :: zero=0., eins=1.
CHARACTER (LEN=8), DIMENSION (18:48), SAVE :: sweden, king (18)
TYPE (date), DIMENSION (25, 13) :: student, parents
```

Type declaration

The type of a named data entity may be declared *explicitly* by a type declaration statement or *implicitly* by an IMPLICIT statement.

Type parameters: If no kind type parameter is specified in a type declaration statement for an intrinsic type, this is a declaration for an entity of type *default* integer, *default* real, double precision real, *default* complex, *default* logical, or *default* character. The leading length specifications and the own length specifications are allowed only in CHARACTER statements.

Default implicit type declaration: If a data entity is neither affected by a type declaration statement nor by an IMPLICIT statement, the data type of

9.2 Type Declaration Statements

this entitity depends on the first letter of its name; that is, if its name begins with one of the letters I, J, K, L, M, or N (or i, j, k, l, m, or n), the entity is of type default integer; otherwise it is of type default real.

This (default) implicit typing may be confirmed or changed for some or for all letters. In each scoping unit, there is a mapping between each of the letters A, B, ..., Z, (and a, b, ..., z) and a type including its kind type parameter and (if applicable) its character length. This mapping may be null. An IMPLICIT statement is used to specify the mapping.

If an entity is affected by an implicit declaration and by a type declaration statement, the type declaration statement has precedence.

A program unit or a subprogram may include more than one IMPLICIT statement and more than one type declaration statement (even for the same type).

Explicit Initialization

Nonpointer: If = **initialization_expression** is specified for a **data_entity** in a type declaration statement, either a named constant is defined or a nonpointer variable is explicitly initialized. In the last case, the specified variable on the left-hand side of the equals becomes explicitly initialized with the result of the initialization expression; that is, an initial value is assigned to the variable when the execution of the program begins. The assignment of the initial value to the variable is done according to the rules for intrinsic assignment statements. A variable or a part of it may be explicitly initialized only once in a program. The DATA staement may also be used to specify explicit initialization.

An initialized variable which is not in a common block also has the SAVE attribute even if the SAVE attribute is not specified explicitly.

An initialization expression must *not* be specified for the following entities: dummy arguments, function results, variables in a common block (except variables in a named common block in a block data program unit), allocatable arrays, pointers, names of external subprograms, names of internal subprograms, and automatic variables.

The specification of an initial value may appear in a type declaration statement within an interface block, but the specification is ineffective in this case.

Pointer: If => **NULL()** is specified for a **data_entity** in a type declaration statement, the specified pointer variable on the left-hand side of the pointer assignment symbol becomes explicitly initialized with the pointer association status "disassociated".

9.2.1 INTEGER Statement

An INTEGER statement is used to declare named constants, variables and/or function results of type *integer*.

INTEGER [**(kind_type_parameter)**] [**, attribute**]... [**::**] **data_entity** [**, data_entity**]...

where **kind_type_parameter**, **attribute**, and **data_entity** are defined as on pages 9-11 and 9-12.

```
INTEGER a, x2, nn, fe (10, 4), k
INTEGER, PARAMETER :: long = SELECTED_INT_KIND(8)
INTEGER (long), DIMENSION (0:2, 11:14, 5), SAVE :: f1, f2
```

Regardless of whether or not a kind type parameter is explicitly specified, the Fortran processor selects (if possible) a suitable processor-dependent method of internal representation of the values of the specified integer data entities.

If a kind type parameter is not explicitly specified, it is a declaration of entities of type *default integer* and the **Fortran** processor uses the kind type parameter value KIND(0). Data entities of type INTEGER (KIND(0)) also are of type default integer.

9.2.2 REAL Statement

A REAL statement is used to declare named constants, variables and/or function results of type *real*.

REAL [**(kind_type_parameter)**] [**, attribute**]... [**::**] **data_entity** [**, data_entity**]...

where **kind_type_parameter**, **attribute**, and **data_entity** are defined as on pages 9-11 and 9-12.

```
REAL measure, value, tax, direction (8, 8), height
REAL, DIMENSION (2) :: edge1 = (/1.,0./), edge2 = (/(7., i=1,2)/)
REAL, PARAMETER :: pi=3.1415, e=2.7182
REAL (10) a, b, c, d
```

Regardless of whether or not a kind type parameter is explicitly specified, the Fortran processor selects (if possible) a suitable processor-dependent approximation method for the internal representation of the values of the specified real data entities.

If a kind type parameter is not explicitly specified, it is a declaration of entities of type *default real* and the **Fortran** processor uses the kind type parameter value KIND(0.0). Data entities of type REAL (KIND(0.0)) also are of type default real.

9.2.3 DOUBLE PRECISION Statement

A DOUBLE PRECISION statement is used to declare named constants, variables and/or function results of type *double precision real*.

DOUBLE PRECISION [**, attribute**]... [**::**] **data_entity** [**, data_entity**]...

where **attribute** and **data_entity** are defined as on pages 9-11 and 9-12.

```
DOUBLE PRECISION distance, defect, diff
DOUBLE PRECISION :: mkl (10, 100), thickness
DOUBLE PRECISION, OPTIONAL, DIMENSION (n, n) :: mata, rest, probe
```

The Fortran processor selects a processor-dependent approximation method for the internal representation of the values of the specified double precision real entities. The kind type parameter is the same as that of real entities with the kind type parameter value KIND(0.0D0).

9.2.4 COMPLEX Statement

A COMPLEX statement is used to declare named constants, variables and/or function results of type *complex*.

COMPLEX [**(kind_type_parameter)**] [**, attribute**]... [**::**] **data_entity** [**, data_entity**]...

where **kind_type_parameter**, **attribute**, and **data_entity** are defined as on pages 9-11 and 9-12.

```
COMPLEX point, vector, two, plane (100), cc
COMPLEX, PRIVATE :: c1, c_denominator, circle
COMPLEX (KIND=KIND(0.0D0)) c1, c2
```

Regardless of whether or not a kind type parameter is explicitly specified, the Fortran processor selects (if possible) a suitable processor-dependent approximation method for the internal representation of the values of the real and imaginary parts of the specified complex data entities.

If a kind type parameter is not explicitly specified, it is a declaration of entities of type *default complex*, and the Fortran processor uses the same kind type parameter value as for entities of type default real, namely KIND(0.0). Data entities of type COMPLEX (KIND(0.0)) also are of type default complex.

Double precision complex entities: If the value for **kind** in the kind type parameter specification is KIND(0.0D0), the approximation method for the internal representation of the real and imaginary parts of the specified entities is the same as that of double precision real entities.

9.2.5 LOGICAL Statement

A LOGICAL statement is used to declare named constants, variables and/or function results of type *logical*.

LOGICAL [(**kind_type_parameter**)] [, **attribute**]... [::] **data_entity** [, **data_entity**]...

where **kind_type_parameter**, **attribute**, and **data_entity** are defined as on pages 9-11 and 9-12.

```
LOGICAL my, lambda, omega, switch (10), wf
LOGICAL, DIMENSION (-n:n, 1987:1988) :: s1, jn, hit (10)
LOGICAL (1), ALLOCATABLE ::  pattern1 (100)
LOGICAL (KIND=bit), DIMENSION (10) :: pattern1, pattern2
```

Regardless of whether or not a kind type parameter is explicitly specified, the Fortran processor selects (if possible) a suitable processor-dependent method for the internal representation of the values of the specified logical data entities.

If a kind type parameter is not explicitly specified, it is a declaration of entities of type *default logical* and the Fortran processor uses the kind type parameter value KIND(.FALSE.). Data entities of type LOGICAL (.FALSE.) also are of type default logical.

9.2.6 CHARACTER Statement

A CHARACTER statement is used to declare named constants, variables and/or function results of type *character*.

CHARACTER [**type_parameter**] [, **attribute**]... [::] **data_entity** [, **data_entity**]...

where **attribute** and **data_entity** are defined as on page 9-11. An **own_length** specification of a character entity may be:

- A specification expression in parentheses;
- An asterisk in parentheses, namely (*); or
- A scalar integer literal constant without an explicitly written kind type parameter value.

And the **type_parameter** specification may be:

* **length** [,]
([**LEN =**] **leading_length**)
(**KIND = kind**)
(**leading_length** , [**KIND =**] **kind**)
(**LEN = leading_length** , **KIND = kind**)
(**KIND = kind, LEN = leading_length**)

where **length** is a *leading* length specification which has the same form as

9.2 Type Declaration Statements

own_length (see above). **leading_length** is either a specification expression or an asterisk *. And **kind** is a scalar integer *initialization expression* that evaluates to a processor-dependent nonnegative kind type parameter value.

The additional comma after a leading length specification is only allowed in a CHARACTER statement *without* the "::".

```
CHARACTER (LEN=3), PARAMETER :: one='one', two='two', five='five'
CHARACTER *8 abcfld (100), atom, text *20
CHARACTER *(n) t1 (10), t2 *(2*n), t3 *80, t4
CHARACTER (KIND=2, LEN=40) hz1, hz2
```

Regardless of whether or not a kind type parameter is explicitly specified, the Fortran processor selects (if possible) a suitable processor-dependent method for the internal representation of the values of the specified character data entities.

If a kind type parameter is not explicitly specified, it is a declaration of entities of type *default character* and the Fortran processor uses the kind type parameter value KIND('A'). Data entities of type CHARACTER ('A') also are of type default character.

9.2.6.1 Character Length

The length specifications **own_length**, **length**, and **leading_length** indicate the number of characters in the specified character entities. A length specification that appears immediately after a **data_entity** is an **own length specification** which applies only to this single data entity. A length specification which is part of the **type_parameter** after the keyword CHARACTER is a **leading length specification**. A leading length specification applies to all entities without an own length specification in this CHARACTER statement. If neither a leading length nor an own length is specified for a character data entity, the length one is assumed as its default length.

The character length may be zero. If the specification expressions **own_length**, **length**, or **leading_length** are negative, the character length zero is used. The length specification for an array object is interpreted as the length of its single array elements.

If a character entity is affected both by a CHARACTER statement and by an IMPLICIT statement, the explicit type declaration has precedence over the implicit type declaration and the length specification in the CHARACTER statement has precedence over the length specification in the IMPLICIT statement.

The length specification for a character entity may be a specification expression; in a simple case, it is an integer literal constant without a sign. If the length specification appears in a main program, or if it is a length specification for a statement function or for a dummy argument of a statement function, the specification expression must be a scalar integer *constant expression*.

If the specification expression in a length specification is *not* a constant expression, the length is determined immediately before the execution of the subprogram. And a later redefinition of any operand in the specification expression has no effect on the current length, as far as this invocation and execution of the subprogram is concerned. The length will be reevaluated only when the subprogram is invoked the next time. Note that if this nonconstant length is specified for a variable which is no dummy argument, it is an *automatic* character variable.

Length specifications *, *(*), and LEN = *

If an asterisk * appears in a length specification instead of a specification expression, this is a length specification for a named constant, the *assumed length* of a dummy argument (see chapter 13), or the assumed length of the main entry point or of another entry point of a function (see chapter 13).

Named constant: If the length * is specified for a named constant in a CHARACTER statement, the named constant automatically has the length of the result of the initialization expression defining the named constant.

```
CHARACTER (LEN=*) PARAMETER :: j='January', m='March', a='April'
CHARACTER name *(*)
PARAMETER (name='Smith')
```

The named constant **name** has the length 5. The named constants j, m, and a have the lengths 7, 5, and 5, respectively.

9.2.7 TYPE Declaration Statement

A TYPE declaration statement is used to declare named constants, variables and/or function results of a *derived type*.

TYPE (type_name) [, **attribute**]... [::] **data_entity** [, **data_entity**]...

where **type_name** is the name of an available derived type. And **attribute** and **data_entity** are defined as on pages 9-11 and 9-12.

A declaration of derived type entities is only possible if the type definition for this derived type appears lexically earlier in the same scoping unit or if the derived type definition is accessible in the scoping unit with the type declaration by USE association or host association.

```
TYPE page1
  CHARACTER (LEN=60), DIMENSION (125) :: p
END TYPE page1
```

```
TYPE page2
  CHARACTER (LEN=80), DIMENSION (25) :: p
END TYPE page2
```

In a scoping unit where these derived type definitions are available (i.e., specified or accessible), the following declarations may be written:

```
TYPE (page1) printer1, printer2
TYPE (page2) screen
```

9.3 Attribute Specification Statements

Attributes may be either specified in a type declaration statement or by an attribute specification statement. Specifying an attribute by an attribute specification statement has the same effect as the specification in a type declaration statement.

9.3.1 ALLOCATABLE Statement

The ALLOCATABLE statement may be used to specify the ALLOCATABLE attribute for arrays.

ALLOCATABLE [::] **allocatable_array** [, **allocatable_array**]...

Each **allocatable_array** is either the name of an allocatable array or the name of an allocatable array followed by its array specification.

```
REAL a, b (:, :, :), c
DIMENSION c (:, :)
ALLOCATABLE a (:), b, c
```

The array specification for the allocatable array **a** is given in the ALLOCATABLE statement, the array specification for **b** in the REAL statement and the array specification for **c** in the DIMENSION statement.

9.3.2 DATA Statement

The DATA statement may be used to specify explicit initialization for variables.

DATA variable_list /constant_list/ [[,] **variable_list /constant_list/**]...

Each **variable_list** specifies variables which are to be initialized. A variable list may include named and unnamed variables and implied-DOs (see below). Nonpointer variables become initialized with an initial value. Pointer variables become initialized as nullified pointers, i.e. with the association status "disassociated".

And each **constant_list** specifies the initializations belonging to the elements of the preceding variable list. The constant list may include initial values, namely scalar constants (even BOZ constants), numeric signed literal constants, scalar constant subobjects, and constant structure constructors; and it may include the function reference NULL( ).

In addition, such an initialization item in a constant list may appear with a leading *repetition factor* (and an asterisk): **k∗item**. The **repetition factor k** must be a scalar unsigned integer constant or a subobject of such a constant; the form **k∗item** is interpreted as **k** repetitions of the **item**. The repetition factor must be nonnegative. Initialization items with repetition factor zero are noneffective, i.e. they are ignored.

A DATA statement may appear in the specification part of a scoping unit or (in contrary to all other attribute specification statements) after the specification part in the execution part of a scoping unit. A DATA statement must not appear in an interface block.

A variable which is neither affected by a type declaration statement nor by an IMPLICIT statement is implicitly typed according to the rules of default implicit typing. Therefore, the real or integer type of such a variable may be confirmed but not changed after the DATA statement.

If a whole array, an array section, or an array element is initialized by a DATA statement, the array specification must appear before the DATA statement.

```
INTEGER a (3);   REAL b
LOGICAL c;       CHARACTER *8 d
DATA a, b /100, 200, 300, 3./, c, d /.TRUE., 'game'/
```

If a named constant or a structure constructor appears in a constant list, this named constant and this derived type, respectively, must have been defined in this scoping unit before the DATA statement, or they must have been made accessible by USE association or host association. This rule also applies to a repetition factor appearing in a constant list.

And if a structure constructor appears in a constant list, each component of the structure constructor must be an initialization expression.

A variable appearing in a variable list or in an implied-DO within a DATA statement must not be a dummy argument, must not be accessible by USE association or by host association, must not be a variable in a named common block (except in a block data program unit), a variable in the blank common, a name of a function or of a function result variable, an automatic data object, or an allocatable array.

Only for the initialization of the variables in the variable list with the items from the corresponding constant list, both the variable list and the constant list are

9.3 Attribute Specification Statements

interpreted as a sequence of pointers and *scalar* data objects (see below). And the lengths of the "expanded" variable list and the corresponding constant list must match. Zero-size arrays and implied-DOs with iteration count zero may be specified but they do not count; that is, the variables are not initialized with constants of the constant list. But a zero-length character variable is interpreted (i.e. counts) as a "normal" variable of the variable list. A structure object is *not* interpreted as a sequence of its components.

One constant must be specified in the expanded constant list for each (effective) scalar nonpointer variable in the corresponding expanded variable list. And NULL() must be specified in the constant list for each pointer variable in the corresponding variable list. The items of the expanded constant list correspond one-to-one to the variables in the expanded variable list: the first variable in the expanded variable list becomes initialized with the first item in the expanded constant list, the second variable with the second item, and so on.

Nonpointer variables

The value of a constant must be compatible with its corresponding variable according to the rules of intrinsic assignment because the variables are initialized in accordance with these rules. If a character variable is initialized, all of its character positions become defined. And if the length of the character variable is greater than the length of the corresponding character constant, the remaining character positions are filled with blanks.

Whole array: If the variable list includes an array (name) without a subscript list, this is interpreted as the specification of all array elements in array element order.

Array section: If the variable list includes an array section, this is interpreted as the specification of all array elements of this array subobject in its own array element order. The subscript expressions in the section-subscript list must be integer initialization expressions.

Array element: If a variable list includes an array element, the subscript expressions in the subscript-list must be integer initialization expressions. If the array element appears in the range of an implied-DO or of an implied-DO nest, the operands of the subscript expressions may include also the DO variable(s) of any surrounding implied-DO(s). Note that the subscript expressions may include only intrinsic operators.

Character substring: If a variable list includes a character substring, the character substring expressions (for the specification of the first and the last character position) must be integer initialization expressions.

```
TYPE date
  INTEGER :: day
  CHARACTER (LEN=3) :: month
  INTEGER :: year
END TYPE date
INTEGER i3 (3)
REAL r, rr
DOUBLE PRECISION d
COMPLEX c
LOGICAL l2 (2)
CHARACTER ch (2) *2
TYPE (date) start
PARAMETER (e=1.)
DATA i3 /3, 2, 1/, r, rr /2*e/, d, c /1.25D6, (1.5, 3.4)/
DATA ch(2), ch(1) /'kg', 'cm'/
DATA start /date(11, 'OCT', 1940)/
```

The variables are initialized as follows:

i3(1)	= 3	i3(2)	= 2	i3(3)	= 1
r	= 1.	rr	= 1.	d	= 1.25D6
c	= (1.5, 3.4)	ch(1)	= cm	ch(2)	= kg
start%day	= 11	start%month	= OCT	start%year	= 1940

Pointer variables

If a variable in a variable list of a DATA statement has the POINTER attribute, the belonging initialization item in the corresponding constant list must be NULL( ), which is a nullified pointer. Thus, according to the rules for pointer assignments, the pointer variable becomes initialized with the association status "disassociated".

9.3.2.1 Implied-DO

A variable list in a DATA statement may include an **implied-DO**. Implied-DOs may be used to initialize one or more array elements or structure components depending on one or more DO variables. Such an implied-DO may control the order and the selection of single array elements.

(variable_list, do_variable = first, last [, stride])

The **variable_list** in an implied-DO may include array elements, scalar structure components, and other implied-DOs; it is called the **range** of the implied-DO. The **do_variable** is a scalar named integer variable. And the *loop parameter expressions* **first**, **last**, and **stride** are scalar integer expressions. These

9.3 Attribute Specification Statements

expressions may include intrinsic operators, and their primary operands may be integer constants and DO variables of those implied-DOs which contain this implied-DO. If the **stride** is absent, the value one is assumed as the default stride.

```
DIMENSION f (3), ff (4, 5)
DATA (f(i), i=1,3) /3 * 1.5/
DATA ((ff(i, j), i=1,4), j=1,5) /20 * 8.5/
```

Implied-DOs in DATA statements are very similar to "normal" DO constructs; the value of the *iteration count* and the value(s) of the DO variable are determined in the same way as for a DO construct. If the iteration count is zero before the first iteration is executed, *no* variable in the range of this implied-DO is initialized.

A DO variable appearing in a variable list in a DATA statement has no effect on the definition status of any other variable having the same name.

A scalar structure component **part**$_1$ [**%part**$_i$]... **%part**$_n$ within the range of an implied-DO must contain at least on **part**$_k$ that contains a subscript list.

If an implied-DO appears in a variable list within a DATA statement, the elements of the variable list within the range of the implied-DO are initialized for each iteration of the implied-DO in such a way that the actual value of the DO-variable is used where the name of the DO-variable appears in subscript expressions of array elements or in structure components in the range of the implied-DO.

```
DIMENSION f (1000)
DATA (f(i), i=1,200) /200 * 1.0/, (f(i), i=201,500) /300 * 2.0/
```

The first 200 array elements of array f are initialized with the value 1.0 and the next 300 array elements are defined with the initial value 2.0. The other elements of the array are not initialized and remain undefined.

```
DIMENSION a (3, 4)
DATA ((a(i,j), i=1,3), j=1,4) /3 * 2.5, 3 * 5., 3 * 7.5, 3 * 10./
```

This DATA statement initializes the complete array a. The statement is equivalent to the following one:

```
DATA a /3 * 2.5, 3 * 5., 3 * 7.5, 3 * 10./
```

9.3.3 DIMENSION Statement

The DIMENSION statement may be used to specify arrays.

DIMENSION [::] **array (dim** [**, dim**]... **)** [**, array (dim** [**, dim**]... **)**]...

Each **array (dim** [**, dim**]... **)** is the name of an array followed by its array specification.

```
PARAMETER (n=10)
INTEGER b
DIMENSION a (3, 5, 4), b (10)
DIMENSION e (n**3), f (n, n*n), g (n, n, n)
```

9.3.4 EXTERNAL Statement

The EXTERNAL statement may be used to specify the EXTERNAL attribute.

EXTERNAL [::] **name** [, **name**]...

Each **name** identifies an external subprogram, an entry point of an external subprogram, a dummy argument, or a block data program unit.

A name may appear only once in all EXTERNAL statements in a scoping unit. Such a name must not appear as a specific name in an interface block in the scoping unit.

```
      EXTERNAL sin
10    y = sin(x)
      ⋮
      END
      FUNCTION sin (x)
      ⋮
      END FUNCTION sin
```

The name sin is specified in the main program as the name of an external subprogram. Therefore, the user-defined function and not the intrinsic function of the same name is referenced in statement 10.

9.3.5 INTENT Statement

The INTENT statement may be used to specify an INTENT attribute.

INTENT (IN) [::] **dummy_argument** [, **dummy_argument**]...

INTENT (OUT) [::] **dummy_argument** [, **dummy_argument**]...

INTENT (INOUT) [::] **dummy_argument** [, **dummy_argument**]...

Each **dummy_argument** is the name of a dummy argument.

```
SUBROUTINE abc (in_arg1, in_arg2, out_arg, inout_arg)
   INTENT (IN)     in_arg1, in_arg2
   INTENT (OUT)    out_arg
   INTENT (INOUT)  inout_arg
   ⋮
END SUBROUTINE abc
```

9.3.6 INTRINSIC Statement

The INTRINSIC statement may be used to specify the INTRINSIC attribute.

INTRINSIC [::] **name** [, **name**]...

Each **name** is the generic name or a specific name of an intrinsic function.

```
INTRINSIC SIN, SQRT, EXP
```

The names SIN, SQRT, and EXP are treated in this scoping unit as the names of intrinsic functions.

9.3.7 OPTIONAL Statement

The OPTIONAL statement may be used to specify the OPTIONAL attribute.

OPTIONAL [::] **dummy_argument** [, **dummy_argument**]...

Each **dummy_argument** is the name of a dummy argument.

```
SUBROUTINE op (da1, da2, da3)
  REAL      da3
  OPTIONAL  da3
  ⋮
END SUBROUTINE op
```

9.3.8 PARAMETER Statement

The PARAMETER statement may be used to specify the PARAMETER attribute.

PARAMETER (name = expression [, **name = expression**]... **)**

Each **name** identifies a named constant. And each **expression** is an *initialization expression*; in a simple case, it is a constant.

The named constant is defined with the result of the initialization expression according to the rules of intrinsic assignment.

Normally, the data type, the kind type parameter, the length (if applicable), and the shape of a named constant must be specified in the same scoping unit lexically before the PARAMETER statement or they must be accessible by USE association or host association.

But if, as an exception, the named constant is not explicitly typed before the PARAMETER statement, the type and the kind type parameter of the named constant are declared according to the rules of default implicit typing. In this case, the type (and the kind type parameter) of the named constant may be confirmed but not changed by a type declaration statement following the PARAMETER statement.

Character length: If the named constant is a character constant that has a character length not equal to one, then not only its type (and its kind type parameter) but its length must also be explicitly or implicitly specified lexically before the PARAMETER statement. If a named constant is implicitly typed, subsequent length specifications for the name of the named constant may confirm but not change the implicit length specification.

A simple form for the length specification is *(*), which may appear in an explicit or implicit type declaration. In this case, the named constant automatically receives the correct length, which results from the evaluation of the character initialization expression in the PARAMETER statement.

```
CHARACTER fmt *(*)
PARAMETER (n = 5, nn = n*n, nnn = nn*n, fmt = '(1X,10(F10.2,3X))')
DIMENSION f (n), ff (n, n), fff (n, n, n)
DATA f /n*0./, ff /nn*1./, fff /nnn*2./
DO, k=1,n
  :
END DO
WRITE (*, fmt) fff, ff, f
```

The length of fmt is given by the definition of the named constant. The other named constants are used to "parameterize" the program. Such programs may be rearranged for the use of smaller or greater arrays by changing only one line of the program.

9.3.9 POINTER Statement

The POINTER statement may be used to specify the POINTER attribute.

POINTER [::] pointer [, pointer]...

Each **pointer** is either the name of a scalar pointer, the name of an array pointer, or the name of an array pointer followed by its array specification.

If the array specification of an array pointer does not appear in the POINTER statement, the DIMENSION attribute and the array specification (: [,:]...) must be specified elsewhere in the scoping unit. The type declaration statement or the DIMENSION statement containing the array specification may appear before or after the POINTER statement.

```
REAL     p1, p2 (:, :, :), p3, p4
DIMENSION p3 (:, :)
POINTER  p1, p2, p3, p4 (:)
```

9.3.10 PRIVATE Statement

The PRIVATE statement may be used to specify the PRIVATE attribute for certain entities in a module.

The PRIVATE statement is a nonexecutable statement which may appear only in the scoping unit (that is, in the specification part) of a module or in derived type definitions within a module.

PRIVATE

PRIVATE [::] **module_entity** [, **module_entity**]...

Each **module_entity** is the name of a variable, the name of a named constant, the name of a derived type, the name of a module subprogram, the name of a namelist group, the generic name of a user-defined subprogram, the designator **OPERATOR(operator)** specifying a (generic) operator interface block for the operator **operator**, or the designator **ASSIGNMENT(=)** specifying a (generic) assignment interface block.

A module may contain more than one PRIVATE statement. The scoping unit of a module may contain one PRIVATE statement *without* a list of module entities; in this case, *all* PUBLIC statements in this scoping unit must include lists of module entities.

The PRIVATE statement *without* a list of module entities specifies the following:

1. In the scoping unit of a module, all those variables, named constants, derived types, module subprograms, namelist groups, and generic interface blocks of the module are private (that is, cannot be made accessible outside the module) which do not have the PUBLIC attribute (explicitly specified in a type declaration statement or by a PUBLIC statement).

2. In the scoping unit of a derived type definition, no component of the derived type can be made accessible outside the module.

```
MODULE mol
PRIVATE
PUBLIC a, b, c
```

In this example, the default accessibility is PRIVATE. Only the entities a, b, and c, which have the explicitly specified PUBLIC attribute, may be made accessible outside the module mol.

9.3.11 PUBLIC Statement

The PUBLIC statement may be used to specify the PUBLIC attribute for certain entities in a module.

The PUBLIC statement is a nonexecutable statement which may appear only in the specification part (that is, in the scoping unit) of a module.

PUBLIC

PUBLIC [::] **module_entity** [, **module_entity**]...

where each **module_entity** is defined as for the PRIVATE statement.

A module may contain more than one PUBLIC statement. The scoping unit of a module may contain one PUBLIC statement *without* a list of module entities; in this case, *all* PRIVATE statements in this scoping unit must include lists of module entities.

The PUBLIC statement *without* a list of module entities specifies in the scoping unit of a module that all those variables, named constants, derived types, module subprograms, namelist groups, and generic interface blocks of the module are visible (that is, can be made accessible outside the module) which do not have the PRIVATE attribute (explicitly specified in a type declaration statement or by a PRIVATE statement).

Note that the default for the accessibility is "public"; that is, the PUBLIC statement without a list of module entities confirms this default.

The PUBLIC attribute must not be specified for a dummy argument or for a function result if they are of a private derived type.

```
MODULE m
PUBLIC
PRIVATE x, y, z
```

All entities in the module which may have an accessibility attribute are public, except x, y, and z, which are unaccessible outside module m.

9.3.12 SAVE Statement

The SAVE statement may be used to specify the SAVE attribute for certain variables.

SAVE

SAVE [::] **saved_entity** [, **saved_entity**]...

Each **saved_entity** is either the name of a local variable or the name of a named common block enclosed in slashes.

```
SAVE nbr, a, /bl/, fld
```

The following names must *not* be specified in a SAVE statement: names of dummy arguments, the main program name, names of functions or function result variables, subroutine names, entry point names, names of single variables in a common block, namelist group-names, names of automatic variables, and named constants.

The SAVE statement *without* a list of saved entities has the same effect as though the names of *all* those variables and common blocks in this scoping unit were specified in the SAVE statement that might appear in a SAVE statement. Therefore, in this scoping unit, no other SAVE statement must appear and the SAVE attribute must not be specified in a type declaration statement.

Named common blocks

If a SAVE statement includes the name of a common block enclosed in slashes, *all* variables in the common block are affected by this SAVE statement in this scoping unit, because they implicitly receive the SAVE attribute. If this happens in a scoping unit of a program outside the main program, the SAVE attribute must be specified for this named common block in all scoping units where the common block appears in a COMMON statement (except in the main program). Note that if a common block contains a variable with default initialization specified in its type definition, the name of this common block is not allowed to appear in a SAVE statement.

Main program: If a particular common block is specified in the scoping unit of a main program, the values of the objects in the common block storage sequence are accessible in the program within all scoping units that specify this common block in a COMMON statement. In this case, the specification of SAVE statements in the main program and in other scoping units is insignificant and therefore superfluous.

Subprogram: Suppose a named common block is specified both in a COMMON statement and in a SAVE statement in the scoping unit of a subprogram within a program. In this case, when a RETURN statement or the END statement of this subprogram is executed, the current values of the objects in the common block storage sequence are made available to the next scoping unit (in the execution sequence of the program) that specifies the common block or has access to the common block.

```
CALL sub1
CALL sub3
END PROGRAM
SUBROUTINE sub1
  COMMON /z/ a, b
  SAVE /z/
  CALL sub2
  CALL sub2
  ⋮
END SUBROUTINE sub1
```

```
SUBROUTINE sub2
  COMMON s, t, u
  COMMON /z/ f, g
  SAVE x
  DATA iend /1000/
  DO, i=1,iend
    ⋮
    x = x + 1
    ⋮
  END DO
  ⋮
END SUBROUTINE sub2
SUBROUTINE sub3
  COMMON /z/ c, d
  SAVE /z/
  ⋮
END SUBROUTINE sub3
```

The SAVE attribute of variable x in subroutine sub2 causes the last value of the local variable x to be saved for the next call of sub2. The values of f and g in the named common block z are saved because the named common block z also is specified in the subprogram sub1 in a COMMON statement and in a SAVE statement. The value of iend also is saved because the DATA attribute causes automatically the SAVE attribute. The values stored in the common block /z/ in sub1 are saved because of the SAVE statement in sub3.

9.3.13   TARGET Statement

The TARGET statement may be used to specify the TARGET attribute.

TARGET [::] **target** [, **target**]...

Each **target** may be the name of a scalar variable, the name of an array, or the name of an array followed by its array specification.

```
REAL       z1, z2 (5, 10, 3), z3, z4
DIMENSION  z3 (10, 4:9)
TARGET     z1, z2, z3, z4 (1939:1945)
```

9.4   Additional Specification Statements

The COMMON and EQUIVALENCE statements allow the programmer to specify physical storage and to determine the order of the data objects that are in

9.4 Additional Specification Statements

a common block. The COMMON statement is used to storage associate data objects which are declared in *different* scoping units. The EQUIVALENCE statement is used to storage associate data objects which are declared in *one and the same* scoping unit.

The IMPLICIT statement is used to declare the type of data entities depending on the first letter of their names.

And the NAMELIST statement is used to specify groups of variables for namelist input/output.

9.4.1 COMMON Statement

The COMMON statement is used to specify blocks of physical memory. These blocks are called *common blocks* because they can be made accessible to each scoping unit of a program. In each scoping unit which specifies a particular common block, named variables may be specified that are *in* this common block. The storage units of these variables are then parts of the physical memory of the common block. Common blocks are global. Therefore, variables of different scoping units may become storage associated if they are in the same common block.

A common block may have a name. A common block having an explicitly specified name is called a **named common block**. There is one common block having no name, the so-called **blank common**.

COMMON variable_list [[,] /[block-name]/ variable_list]...

COMMON /[block-name]/ variable_list [[,] /[block-name]/ variable_list]...

Each **block-name** is the name of a *named* common block. Each **variable_list** is a list of those local variables in this scoping unit that are in the blank common or in a named common block. An element in this list may be the name of a scalar variable, an array name, and an array name immediately followed by its (explicit-shape) array specification.

```
COMMON a, b, c, /z/ u, v, w, // d, e, f
```

The variables a, b, c, d, e, and f are in the blank common. And the variables u, v, and w are in the named common block z.

Common block name

If a **block-name** is specified, all variables of the scoping unit which are specified in the subsequent **variable_list** are in the *named* common block **block-name**. If the **blockname** is absent or if two slashes // occur, all variables specified in the subsequent **variable_list** are in the *blank* common.

Common blocks (even the same common block) may be specified in a particular scoping unit more than once either in the same COMMON statement or in different COMMON statements. For a particular common block, the variable list in a subsequent specification is treated as the continuation of the last variable list for this common block.

Local names: The name of a common block may be the same as a local name in a scoping unit if this local entity is no named constant, no intrinsic function, and no local variable identifying the main entry point or another entry point of an external function.

If the common block name is the same as the name of a local entity in a scoping unit, this name identifies the local entity except in COMMON statements and in SAVE statements. The common block name may be the same as the name of an intrinsic function if the intrinsic function is not referenced in the scoping unit.

Common block variable list

A particular variable name may be specified only once in all COMMON statements in a scoping unit.

A particular common block variable list is valid only in the scoping unit which includes the COMMON statement with this list. In another scoping unit, another variable list may be specified for the common block.

Zero-size arrays and zero-length character strings may be specified in a COMMON statement. But the names of the following entities must *not* be specified in a COMMON statement: dummy arguments, allocatable arrays, automatic variables, functions, function entry points, and function result variables.

Data type: Variables of different intrinsic and derived types may be specified in a common block variable list in a scoping unit. But there are certain restrictions on derived type variables (see below: "Derived type entities"). And there are additional restrictions within a program if variables in different scoping units become storage associated (see below: "Storage association").

Derived type entities: The name of a derived type variable may be specified in a COMMON statement if the derived type has the SEQUENCE attribute. A variable of such a sequence type has a storage sequence. The specification of a variable of a numeric sequence type or of a character sequence type in a COMMON statement is interpreted as the specification of a sequence of single structure components in the order of the definition of the corresponding type components.

A variable in the blank common must not have a component for which default initialization is specified in the type definition. If a variable with default initialization specified in its type definition is in a common block, the name of this common block must not appear in a SAVE statement.

Arrays: If a common block variable list includes an array name immediately followed by an array specification, the array bounds must be integer initialization expressions. Thus an explicit-shape array can be specified. The array specification of an array pointer must not appear in a COMMON statement.

Modules: A public variable which is accessible by USE association must *not* appear in COMMON or EQUIVALENCE statements in a scoping unit which references the module. This rule also applies to those scoping units that are included in the scoping unit which references the module.

Empty common block storage sequence: The storage sequence of a common block is *empty* if it contains no storage unit, that is, if the variable list includes only variables with empty storage sequences such as zero-size arrays or zero-length character strings.

Storage association

The common block name identifies the first storage unit of a common block. The name is global. Therefore, the nonempty storage sequences of this common block begin in all scoping units of a program (which specify this common block) at the same storage unit. And the nonempty storage units of the blank common begin in all scoping units (which specify the blank common) at the same storage unit, which is different from any other storage unit occupied by another common block.

The *empty* storage sequences of a particular named common block which is specified in different scoping units of a program are storage associated and are associated with the first storage unit of the nonempty storage sequences of the named common block. And the empty storage sequences of the blank common in different scoping units of a program are storage associated and are associated with the first storage unit of the nonempty storage sequences of the blank common.

If a particular common block occurs in different scoping units within an executable program, the variables in a common block variable list within one scoping unit are (storage) associated in positional order with the variables in the common block variable list within another scoping unit. The association of the variables in the different scoping units occurs by means of the physical storage of the common block and not by means of the local names specified in the variable lists of the COMMON statements. Note that USE association or host association may cause these associated objects to be accessible in the same scoping unit.

If a scoping unit does not need all storage units of a common block, "place holding variables" must be specified in the COMMON statement at the places of the unused storage units. Thus it is guaranteed that the subsequent variables in this common block are correctly storage associated.

Nonpointers: A nonpointer variable may be storage associated only with a nonpointer variable. A nonpointer variable which is of type default integer, default real, double precision real, default complex, default logical, or of a numeric sequence type may be storage associated only with a variable of the same type or with a variable of any of the other mentioned types.

A nonpointer variable of type default character or of a character sequence type may be storage associated only with a variable of the same type or with a variable of the other type.

A nonpointer variable of a derived type that is neither a numeric sequence type nor a character sequence type may be storage associated only with a variable of the same type.

A nonpointer variable of nondefault intrinsic type may be storage associated only with a variable of the same type which must have the same kind type parameter and the same length (if applicable).

A variable with the TARGET attribute may become storage associated only with another object that has the TARGET attribute and the same data type, kind type parameter, and, if applicable, character length.

Pointers: A pointer variable may be storage associated only with a pointer variable. And the data type, kind type parameter, length (if applicable), and rank of the two variables must agree.

Scalar character variables: If a scalar variable is of type default character or of a character sequence type, and if the other storage associated scalar object is of any of the mentioned types, then their character lengths may be different and their storage sequences need not be totally associated.

Scalar double precision real und complex variables: If a variable is of type double precision real, of type default complex, or of a numeric sequence type, and if the other storage associated scalar object is of type default integer, default real, default logical, default complex, or of a numeric sequence type, then their storage sequences need not be totally associated.

```
PROGRAM cent
  REAL    a (4)
  COMMON a
  ⋮
END PROGRAM
```

9.4 Additional Specification Statements

```
SUBROUTINE sub
  COMPLEX c (2)
  COMMON   c
  ⋮
END SUBROUTINE
```

The real array elements a(1), a(2), a(3), and a(4) are storage associated with the complex array elements c(1) and c(2):

$$a(1) \iff \Re(c(1))$$
$$a(2) \iff \Im(c(1))$$
$$a(3) \iff \Re(c(2))$$
$$a(4) \iff \Im(c(2))$$

COMMON and EQUIVALENCE

Variables which are equivalenced to a variable in a common block are forced also into the common block; that is, all such equivalenced variables are in the common block. Equivalencing may cause a common block storage sequence to be extended at its end. Common block storage sequences must not be extended by adding storage units preceding the first storage unit of the first variable of the common block.

The **size of a common block** is given by the size of the storage sequence of the common block and by all extensions caused by EQUIVALENCE statements.

Differences

There are the following differences between the blank common and named common blocks:

A program may contain at most one blank common, but it may contain any number of named common blocks.

When the END statement of a subprogram or a RETURN statement is executed, the variables of a named common block may become undefined if the name of the common block is not specified in a SAVE statement. But the variables of the blank common preserve their definition status in such a case.

A particular named common block must have the same size in all scoping units of a program. But the blank common may have different sizes in different scoping units of a program. In this case, the size of the blank common is given by the maximum size of all occurrences of the blank common within the program.

Variables in a named common block may be explicitly initialized only within a block data program unit. But variables in the blank common must not be initialized at all.

A variable of a derived type with components for which default initialization is specified in its type definition must not appear in blank common.

9.4.2 EQUIVALENCE Statement

The EQUIVALENCE statement is used to specify that two or more objects in the same scoping unit share storage units. Such equivalenced variables are storage associated.

An EQUIVALENCE statement may be used if physical storage is to be saved or if a variable is to be identified by two or more names.

EQUIVALENCE (var, var [, var]...) [, (var, var [, var]...)]...

Each **var** is the name of a variable, an array element designator, or a character substring designator.

Zero-size arrays and *named zero-length character variables* may be specified in an EQUIVALENCE statement.

Such a list of equivalenced variables (enclosed in parentheses) is called an **equivalence set**. An equivalence set must include at least two variables.

The following entities must *not* be specified in an equivalence set: dummy arguments, pointers, zero-length character substrings, variables with TARGET attribute, allocatable arrays, nonsequence derived type variables, derived type variables having a pointer component at any component level, automatic variables, function names, entry point names of functions, function result variables, named constants, structure components, and subobjects of the above objects.

Data type, kind type parameter: If an equivalence set includes a variable of type default integer, default real, double precision real, default complex, default logical, or a variable of a numeric sequence type, the other variables in this equivalence set must be of the same type or of any of the other mentioned types.

If an equivalence set includes a variable of type default character or a variable of a character sequence type, the other variables in this equivalence set must be of the same type or of the other type.

If an equivalence set includes a variable of a derived type that is neither a numeric sequence type nor a character sequence type, all variables in this equivalence set must be of the same derived type.

If an equivalence set includes a variable of an intrinsic type which is not of type default integer, default real, double precision real, default complex, default logical, or default character, all variables in this equivalence set must be of the same type with the same kind type parameter.

9.4 Additional Specification Statements

A derived type variable may be specified in an equivalence set only if the derived type has the SEQUENCE attribute. If the equivalence set contains only variables of this sequence type, it is not be necessary that these variables have a numeric storage sequence or a character storage sequence.

If a derived type variable in an equivalence set is of a numeric sequence type or of a character sequence type, the specification of this variable in the equivalence set is equivalent to the (imaginary) specification of the single structure components in a common block in the order of their definition.

Array element, character substring: Subscript expressions included in an array element designator and substring expressions for the specification of the first and last character positions of a character substring must be integer initialization expressions. If an array element is specified in an equivalence set, the number of subscript expressions must correspond to the rank of the array.

Array: The specification of a nonzero-size array is interpreted as the specification of the first array element.

If a scalar and an array are equivalenced, the scalar does not receive array properties and the array does not receive the properties of a scalar.

Association

The variables in an equivalence set are storage associated in the sense that their storage sequences are associated. All *nonempty* storage sequences of the variables in an equivalence set occupy the same first storage unit. The *empty* storage sequences of the variables in an equivalence set are storage associated and are associated with the first storage unit of the nonempty storage sequences in the equivalence set. In this context, an **empty** storage sequence is a storage sequence of size zero.

```
REAL a (4), b (3, 2)
EQUIVALENCE (a(2), b(2, 2)), (c, b(3, 2))
```

Storage unit	Associated data object
1	$b(1, 1)$
2	$b(2, 1)$
3	$b(3, 1)$
4	$a(1) \iff b(1, 2)$
5	$a(2) \iff b(2, 2)$
6	$a(3) \iff b(3, 2) \iff c$
7	$a(4)$

A total number of 7 numeric storage units is occupied instead of 11. The data objects a(3), b(3, 2), and c are storage associated with the sixth numeric storage unit.

If variables are equivalenced which are themselves parts of a chain of storage sequences (such as array elements or character substrings), not only the explicitly specified variables in the equivalence set are associated but possibly (implicitly) also other data objects (see last example).

Character variables: The character variables in an equivalence set may have different character lengths but they must have the same kind type parameter. All equivalenced character variables with a nonempty storage sequence occupy the same first character storage unit.

Within an equivalence set, the character variables with *empty* storage sequences are storage associated and are associated with the first character storage unit of the nonempty storage sequences in the equivalence set.

```
CHARACTER a *5, b *4, c (3) *3
EQUIVALENCE (a, c(1)), (b, c(2))
```

The storage space is shared such that only 9 character storage units are occupied instead of 18. Equivalenced character variables may overlap. The first character storage unit of all variables in an equivalence set are associated. The association of the other character storage units of the variables in the equivalence set results automatically from the storage sequences of the variables.

Storage unit	Associated data objects
1	a (1:1) $\iff$ c (1) (1:1)
2	a (2:2) $\iff$ c (1) (2:2)
3	a (3:3) $\iff$ c (1) (3:3)
4	a (4:4) $\iff$ c (2) (1:1) $\iff$ b (1:1)
5	a (5:5) $\iff$ c (2) (2:2) $\iff$ b (2:2)
6	c (2) (3:3) $\iff$ b (3:3)
7	c (3) (1:1) $\iff$ b (4:4)
8	c (3) (2:2)
9	c (3) (3:3)

```
REAL a (2), re, im
COMPLEX b
EQUIVALENCE (a, b), (re, a(1)), (im, a(2))
```

Storage unit	Associated data objects
1	$\Re(b)$ $\iff$ a (1) $\iff$ re
2	$\Im(b)$ $\iff$ a (2) $\iff$ im

9.4 Additional Specification Statements   9-39

This way of associating the real part and the imaginary part of a complex variable with real variables facilitates the use of the single parts of the complex data object.

Restrictions

An EQUIVALENCE statement must not include specifications with an effect that is in contrary to the following rules for the use of storage:

- A variable can occupy storage units not more than once in a storage sequence.
- An array of default intrinsic type, a scalar variable of type double precision real, a scalar variable of type default complex, and a character string of type default character have storage sequences that occupy consecutive storage units.

```
REAL a (3)
EQUIVALENCE (a(1), b), (a(3), b)        ! <-- invalid
REAL a (4), b (3, 2)
EQUIVALENCE (a, b), (a(2), (b(1, 2))    ! <-- invalid
```

Note that an equivalence set must not include variables from different scoping units.

9.4.2.1 EQUIVALENCE and COMMON

Normally, an EQUIVALENCE statement is used to storage associate variables from the same scoping unit. But if a variable that is specified in an equivalence set is additionally specified in a COMMON statement, variables from other scoping units in the program may also be storage associated.

EQUIVALENCE statements may affect and may be affected by COMMON statements:

An EQUIVALENCE statement must not equivalence two objects in different common blocks.

```
COMMON /one/ a, /two/ b
EQUIVALENCE (a, b)         ! <-- invalid
```

An EQUIVALENCE statement must not cause a common block to be extended by adding storage units preceding the first storage unit of the common block.

```
COMMON /z/ a (2)
REAL b (5)
EQUIVALENCE (a, b(3))      ! <-- invalid
```

An EQUIVALENCE statement must not cause a derived type object with default initialization specified in its type definition to be associated with an object in a common block.

9.4.3 IMPLICIT Statement

The IMPLICIT statement supports **implicit typing** of data entities depending on the first letter of their names.

IMPLICIT NONE

IMPLICIT type (letter [**, letter**]... **)** [**, type (letter** [**, letter**]... **)]**...

Each **type** is a *type specification* such as in a type declaration statement. This type specification may include a kind type parameter specification and a length specification (if applicable) (see 9.2). And each **letter** either is a single letter or specifies a sequence of letters (see below).

IMPLICIT NONE

The IMPLICIT NONE statement is used to specify that *implicit type declarations* are not allowed in the scoping unit. In this case, even the rules of *default implicit typing* are ineffective and every named data entity must be declared by an explicit type declaration. If IMPLICIT NONE is specified, no PARAMETER statement must appear before the IMPLICIT NONE statement in the scoping unit, and the scoping unit must not contain any additional IMPLICIT statement.

Default implicit typing

If a named data entity in a program unit or in an interface block is neither affected by a type declaration statement nor affected by an IMPLICIT statement, the data type of this entitity depends on the first letter of its name. If its name begins with one of the letters I, J, K, L, M, or N (or i, j, k, l, m, or n), the data entity is of type default integer; otherwise it is of type default real.

This (default) implicit typing may be confirmed or changed for some or for all letters. There is a mapping between each of the letters A, B, ... , Z (and a, b, ... , z) and a type (and its kind type parameter and possibly its length) in each scoping unit. This mapping may be null.

Note that the default implicit typing being effective in an internal subprogram or module subprogram is the mapping in the host scoping unit.

9.4 Additional Specification Statements

Implicit type declaration

A type, a kind type parameter, and a character length may be specified in an IMPLICIT statement (without NONE) for all those data entities whose names begin with a given letter. Such letters may be specified as a single letter, as a list of single letters, or as a sequence of letters (see below). A particular letter must be specified only once in all IMPLICIT statements within a scoping unit (as a single letter or belonging to a list or sequence of letters).

An IMPLICIT statement (without NONE) must appear before (nearly) all other specification statements in a scoping unit. Only PARAMETER and USE statements may appear before an IMPLICIT statement.

```
IMPLICIT LOGICAL (b, l), DOUBLE PRECISION (d), COMPLEX (c)
```

If the affected data entities are *not* explicitly declared, this example means: all data entities whose names begin with b, B, l, or L are of type default logical; all data entities whose names begin with d or D are of type double precision real; all data entities whose names begin with c or C are of type default complex. Data entities whose names begin with another letter are implicitly declared according to the rules of default implicit typing.

Letter sequence: A sequence of letters is specified by two letters separated by a minus. This is interpreted as the sequence of letters beginning with the first specified letter and ending with the last specified letter in alphabetic order. The specification of such a form of a letter sequence has the same effect as the specification of a list of single letters.

```
IMPLICIT INTEGER (i-k)   is interpreted as   IMPLICIT INTEGER (i, j, k).
```

```
IMPLICIT INTEGER (i-n), REAL (a-h), (o-z)
```

This implicit type declaration confirms the rules of default implicit typing.

Explicit typing has precedence over implicit typing in a scoping unit (see below: "Scope").

```
IMPLICIT LOGICAL (l, o)
REAL mass, long, last, omega, my
```

All entities whose names begin with l, L, o, or O are implicitly of type default logical. The entities long, last, and omega are explicitly of type real; they are not of type logical.

Character entities: A character type specification in an IMPLICIT statement may also include a length specification. If the length specification is absent, the length one (that is, one character) is assumed as the default length.

The length specification in an IMPLICIT statement has the same form as in a CHARACTER statement, except that the optional comma at the end of a

leading length specification must be absent and that the length specifications
*, *(*), and LEN=* are not allowed.

```
PARAMETER (l = 4)
IMPLICIT CHARACTER *2 (h), CHARACTER (l) (c), CHARACTER (LEN=8) (d)
```

As long as no explicit type declaration is effective, this means: all data entities whose names begin with h or H are of type default character with character length 2; all entities whose names begin with c or C are of type default character with character length 4; and all data entities whose names begin with d or D are of type default character with character length 8. For the letters C, D, and H (and c, d, and h), these implicit type declarations have precedence over the rules of default implicit typing.

Scope

IMPLICIT statements do not affect the type of the intrinsic functions. The specifications of an IMPLICIT statement are valid only in the scoping unit containing the IMPLICIT statement and possibly in embedded scoping units (see below).

Outside the scope of an IMPLICIT NONE statement, any data entity which is not declared by a type declaration statement, which is not an intrinsic function, and which is currently not accessible by USE association or host association, has a type, a kind type parameter, and (if applicable) a character length that depend on the first letter of its own name or of the name of its parent object. This type and optionally its kind type parameter and its length are given by an IMPLICIT statement or by the rules of default implicit typing.

Data entities which are accessible by USE association or host association are subject to the rules of explicit or implicit typing being effective in the module or in the host, respectively. The local data entities in a module subprogram or in an internal subprogram are at first subject to the IMPLICIT statements in their respective host scoping unit; then, the IMPLICIT statements in the module subprogram and internal subprogram, respectively, may override, confirm, or extend this default. Note that an embedded interface block is not affected by the IMPLICIT statements in its host.

```
MODULE mod
  IMPLICIT NONE
  ⋮
  INTERFACE
    FUNCTION fun (i)      ⟵ Not all data entities need be
      INTEGER fun            declared explicitly.
    END FUNCTION fun
  END INTERFACE
```

9.4 Additional Specification Statements

```
CONTAINS
  FUNCTION jfun (j)        ←— All data entities must be declared explicitly.
    INTEGER jfun, j
    ⋮
  END FUNCTION jfun
END MODULE mod
SUBROUTINE sub
  IMPLICIT COMPLEX (c)
  c = (3.0, 2.0)           ←— c is implicitly of type default complex.
  ⋮
CONTAINS
  SUBROUTINE sub1
    IMPLICIT INTEGER (a, c)
    c = (0.0, 0.0)         ←— c is default complex by host association.
    z = 1.0                ←— z is implicitly of type default real.
    a = 2                  ←— a is implicitly of type default integer.
    cc = 1                 ←— cc is implicitly of type default integer.
    ⋮
  END SUBROUTINE sub1

  SUBROUTINE sub2
    z = 2.0                ←— z is implicitly of type default real and is
    ⋮                          different from variable z in sub1.
  END SUBROUTINE sub2

  SUBROUTINE sub3
    USE mod                ←— access to function fun by USE association.
    q = fun(k)             ←— q is implicitly of type default real.
    ⋮                          k is implicitly of type default integer.
  END SUBROUTINE sub3
END SUBROUTINE sub
```

9.4.4 NAMELIST Statement

The NAMELIST statement is used to specify *group-names* for *namelist groups*, that is, lists of names. This group-name plays the part of the input/output list in a *namelist input/output* statement.

NAMELIST /group-name/ var [**, var**]... [[**,**] **/group-name/ var** [**, var**]...]...

Each **group-name** identifies a **namelist group var** [**, var**]..., where each **var** is the name of a variable.

```
CHARACTER *10 jan, feb, mar, apr, may, jun, &
         & jul, aug, sep, oct, nov, dec
NAMELIST /summer/ jun, jul, aug, sep
  :
READ (*, summer)
```

Group-name: The group-name is a local name. If a group-name (in a module) has the PUBLIC attribute, no variable in the namelist group must have the PRIVATE attribute or have private components.

Within a particular scoping unit, the same group-name may occur in two or more NAMELIST statements. In this case, the subsequent specifications are interpreted as a continuation of the specified namelist group. Thus the complete namelist group includes all variables which are specified for the same group-name in the scoping unit.

Namelist group: Variables in a particular namelist group may also appear in other namelist groups in the same scoping unit.

The names of the following entities must *not* be included in a namelist group: a dummy argument array having a *nonconstant* array bound, an assumed-size array, an assumed-shape array, an allocatable array, a character variable having a *nonconstant* character length, an automatic variable, a pointer, and a derived type variable having an ultimate pointer component.

A variable in a namelist group must be either accessible by USE association or by host association, or its type, kind type parameter, character length (if applicable), and shape must be declared in the same scoping unit by a type declaration statement or by an IMPLICIT statement and optionally by other specification statements *before* the NAMELIST statement; otherwise the variable is declared according to the rules of default implicit typing. If for a particular variable neither a type declaration statement nor an IMPLICIT statement is effective *before* the NAMELIST statement, an explicit or implicit declaration *after* the NAMELIST statement is allowed but may only confirm the default implicit typing of the variable.

On namelist input, the order of the variable names in a namelist group is not significant. But on namelist output, this is the order in which the values are transferred to the file.

10 EXECUTION CONTROL

Normally, the statements in a main program or subprogram are executed line by line in the order in which they appear in the source program. Execution control statements are used to alter, to interrupt, or to stop this sequential execution sequence. The execution sequence is called the *control flow*. Any change of the sequential execution sequence is a **branching** or **transfer of control**. There are *simple* execution control statements, and there are execution control *constructs* consisting of several statements.

The following execution control statements are **simple** executable statements: CALL, CONTINUE, END, unconditional GO TO, computed GO TO, arithmetic IF, logical IF, RETURN, and STOP, and finally EXIT and CYCLE, which may appear only in DO constructs.

The following execution control statements are used to form *constructs*:

- DO and END DO form a DO construct (that is, a DO loop);
- IF THEN, ELSE IF, ELSE, and END IF form an IF construct; and
- SELECT CASE, CASE, and END SELECT form a CASE construct.

Such a construct may be named. If a statement contains a construct name, it belongs to the named construct; otherwise it belongs to the innermost construct in which it appears. Note that such a construct name is no statement label; therefore, it must appear elsewhere in position 7 to 72 when written in fixed source form.

GO TO statements and the arithmetic IF statement are branching statements, which specify or determine the *branch target statement* which is to be (or may be) executed next. DO constructs, IF constructs, and CASE constructs are compound statements consisting of one or more statement sequences. Such a sequence of statements is called a **block**. A construct has an internal control flow, which automatically branches within the construct.

A **branch target statement** may be one of the following executable statements: a *simple* executable statement, an IF THEN statement, an END IF statement, a SELECT CASE statement, an END SELECT statement, a DO statement, a do-termination statement, a WHERE construct statement, and a FORALL construct statement. A branch target statement must be an executable statement in the same scoping unit as the branching statement. The END IF statement, the END SELECT statement and the do-termination statement may be branch targets only for a branch appearing in the interior of the IF construct, CASE construct and DO construct, respectively. When WHERE constructs and FORALL constructs are nested, only the outmost WHERE construct statement and FORALL construct statement, respectively, may be used as a branch target.

If the control flow depends on certain values, these values must be scalar.

The (nonexecutable) CONTAINS statement in a main program or in a subprogram and any subprogram definitions between CONTAINS and the END statement of the main program or subprogram, respectively, are ignored by the sequential control flow.

Note for all branching statements, control must not be transferred from outside a construct to the interior of the construct. And control must not be transferred from outside a case-block, if-block, else-block, elseif-block, where-block, or elsewhere-block to the interior of the block.

10.1 GO TO Statements

There are two kinds of GO TO statements which may be used to branch to an executable statement in the same scoping unit: the *unconditional GO TO* statement and the *computed GO TO* statement.

10.1.1 Unconditional GO TO Statement

The form of an "unconditional GO TO" is:

GO TO label

The **label** is the statement label of the branch target statement. The execution of an unconditional GO TO statement causes program execution to continue with the statement labeled **label**.

```
10  a = b + c
    GOTO 50         ! jump forward
20  b = c + d
    GOTO 50
30  c = d + e
    GOTO 10         ! jump backward
50  CONTINUE
```

10.1.2 Computed GO TO Statement

The form of a "computed GO TO" is:

GO TO (label$_1$ [, label$_2$]...) [,] integer_expression

Each **label** is the statement label of a possible branch target statement. A particular **label** may appear more than once in the list of statement labels. The **integer_expression** must be scalar.

10.2 IF Statements

The execution of a computed GO TO causes the scalar integer expression to be evaluated. If the result is i such that $1 \leq i \leq n$, where n is the number of labels in the list of statement labels, then program execution continues with the statement labeled **label**$_i$. If $i < 1$ or $i > n$, the computed GO TO is ignored and program execution continues with the (executable) statement immediately following the computed GO TO.

```
      k = 9
      l = 4
    5 GOTO (10, 20, 30, 20) k/l
    7 k = k + 1
      ⋮
   10 CONTINUE           ! k/l = 1
      ⋮
   20 CONTINUE           ! k/l = 2 or 4
      ⋮
   30 CONTINUE           ! k/l = 3
```

Control is transferred to statement 20. Statement 7 is executed only when $k/l < 1$ or when $k/l > 4$.

10.2 IF Statements

There are the following three kinds of IF statements which may be used to branch to an executable statement in the same scoping unit, to conditionally execute a single statement, or to select for execution (at most) one block of its constituent blocks: the *arithmetic IF* statement, the *logical IF* statement, and the *IF THEN* statement. In any case, a scalar expression is evaluated first, and then, depending on its result, another statement or block is executed. An IF THEN statement is the first statement of an *IF construct*. An IF construct contains an additional END IF statement and may contain ELSE statements and ELSE IF statements.

10.2.1 Arithmetic IF Statement

The form of an "arithmetic IF" is:

IF (numeric_expression) label$_1$, label$_2$, label$_3$

The **numeric_expression** must be a scalar numeric but no complex expression. Each **label** is the statement label of a possible branch target statement. A particular **label** may appear more than once in the list of statement labels.

The execution of an arithmetic IF causes the scalar numeric expression to be evaluated. Depending on its result, control is transferred as follows:

> if **numeric_expression** < 0 :   GO TO **label**$_1$,
> if **numeric_expression** = 0 :   GO TO **label**$_2$, and
> if **numeric_expression** > 0 :   GO TO **label**$_3$.

```
   IF (n/5 - 2) 20,30,40
20 WRITE (*,*) 'N/5-2 is negative'
   ⋮
30 WRITE (*,*) 'N/5-2 equals zero'
   ⋮
40 WRITE (*,*) 'N/5-2 is positive'
```

10.2.2   Logical IF Statement

The "logical IF" controls the conditional execution of one simple executable statement.

IF (logical_expression) executable_statement

The **logical_expression** must be scalar. The **executable_statement** must be a *simple* executable statement except another logical IF, an END, an END FUNCTION, an END PROGRAM, and an END SUBROUTINE statement.

The execution of a logical IF causes the scalar logical expression to be evaluated. Depending on its result, control is transferred as follows:

If the scalar logical expression is *true*, the **executable_statement** is executed. If the scalar logical expression is *false*, the **executable_statement** is *not* executed and control is transferred to the next executable statement following the logical IF.

```
   IF (x .LT. -4.5 .OR. x .GT. 4.5) CALL error
10 CONTINUE
```

If $|x| > 4.5$, the `error` subroutine is invoked; otherwise program execution continues with statement 10.

Note that if the **logical_expression** includes a function reference, the operands of the **executable_statement** may be affected by the invocation of this function. In this case, side effects are allowed.

10.3 IF Construct

An IF construct may contain one or more blocks. It controls the execution of at most one of its constituent blocks. The conditional execution of such a block may depend on one or more conditions.

The first statement of an IF construct is an IF THEN statement. And the last statement of an IF construct is an END IF statement. Additional ELSE statements and ELSE IF statements may be used to control the control flow within the IF construct:

[if-name :] IF (logical_expression) THEN

ELSE [if-name]

ELSE IF (logical_expression) THEN [if-name]

END IF [if-name]

The **if-name** is a name identifying the IF construct. The **logical_expression** must be scalar.

If the IF THEN statement includes an **if-name**, the corresponding END IF statement *must* include the same name, and the additional ELSE IF and ELSE statements of this construct, if any, *may* include this name. If the IF THEN statement does not include an **if-name**, neither the END IF statement nor the ELSE IF and ELSE statements must include an **if-name**.

```
IF (colour .EQ. 'red') THEN         ! IFTHEN statement
:                                   ! if-block
ELSE IF (colour .EQ. 'green') THEN  ! corresponding ELSEIF stmt
:                                   ! elseif-block
ELSE                                ! corresponding ELSE statement
:                                   ! else-block
END IF                              ! corresponding ENDIF stmt
```

An IF THEN statement is the first statement of an IF construct.

An ELSE statement is the first statement of an alternative block, which may be conditionally executed after the execution of the corresponding IF THEN statement or after the execution of a corresponding ELSE IF statement. An IF construct may include not more than one corresponding ELSE statement. An ELSE statement may be labeled, but it must not be used as a branch target statement.

An ELSE IF statement combines the effect of an IF THEN statement and an ELSE statement. An ELSE IF statement is the first statement of an alternative block, which may be conditionally executed after the execution of the corresponding IF THEN statement or after the execution of a preceding ELSE IF statement. An ELSE IF statement may be labeled, but it must not be used as a branch target statement.

An END IF statement is the last statement of an IF construct. A corresponding END IF statement must appear for each IF THEN statement. An END IF statement may be labeled. It may be a branch target statement for a branch appearing in the interior of the IF construct but not outside the IF construct. The execution of an ENDIF statement has no effect.

10.3.1 Simple IF Constructs

The basic form of an IF construct containing only one block is:

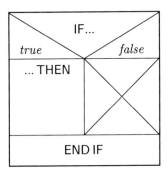

This IF construct controls the conditional execution of one block.

[if-name :] IF (logical_expression) THEN
⋮   ⟵ if-block
END IF [if-name]

If the **logical_expression** is *true*, program execution continues with the first executable statement in the if-block; otherwise program execution continues with the corresponding END IF statement.

```
stop: IF (colour .EQ. 'red') THEN
   time = 30;  CALL red(time)
END IF stop
```

If the value of the scalar character variable `colour` is `'red'`, the two statements in the if-block are executed; otherwise they are ignored.

The basic form of an IF construct containing two alternative blocks is:

10.3 IF Construct

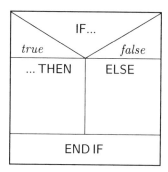

Exactly one of the two blocks of this IF construct is conditionally executed.

[**if-name :**] **IF (logical_expression) THEN**
⋮   ← if-block
ELSE [**if-name**]
⋮   ← else-block
END IF [**if-name**]

If the **logical_expression** is *true*, program execution continues with the first executable statement in the if-block; otherwise program execution continues with the first executable statement in the else-block. Suppose, there is no branch to outside the IF construct, when the execution of the block is completed, program execution continues with the corresponding END IF statement.

```
IF (colour .EQ. 'red') THEN
   time = 30;   CALL red(time)        ! if-block
ELSE
   time = 20;   CALL green(time)      ! else-block
END IF
```

If the value of the scalar character variable `colour` is 'red', the two statements in the if-block are executed, the two statements in the else-block are ignored, and program execution continues with the corresponding END IF statement. But if the character variable `colour` does *not* have value 'red', the two statements in the if-block are ignored, the ELSE statement, the two statements in the else-block, and finally the corresponding END IF statement are executed.

The basic form of an IF construct containing two alternative blocks, where the first block depends on one condition and the second block depends on two conditions, is:

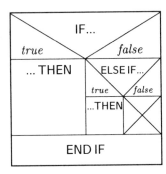

At most one of the two blocks of this IF construct is conditionally executed.

An IF construct may include more than one ELSE IF statement if the execution of blocks depends on more than one condition. Note that such a form of an IF construct is not a nested IF construct:

[**if-name** :] **IF (logical_expression$_1$) THEN**
⋮   ⟵ if-block
ELSE IF (logical_expression$_2$) THEN [**if-name**]
⋮   ⟵ first elseif-block
ELSE IF (logical_expression$_3$) THEN [**if-name**]
⋮   ⟵ second elseif-block
END IF [**if-name**]

If the **logical_expression$_1$** is *true*, program execution continues with the first executable statement in the if-block; otherwise program execution continues with the first ELSE IF statement. If the **logical_expression$_2$** in the first ELSE IF statement is *true*, program execution continues with the first executable statement in this elseif-block; otherwise the second ELSE IF statement is executed. If the **logical_expression$_3$** in the second ELSE IF statement is *true*, program execution continues with the first executable statement in this elseif-block; otherwise program execution continues with the corresponding END IF statement. Suppose, there is no branch to outside the IF construct, when the execution of the selected block is completed, program execution continues with the corresponding END IF statement.

An IF construct containing more than two corresponding ELSE IF statements is analogously executed.

An IF construct with ELSE IF statements may also include an ELSE statement. This form of an IF construct is different from that of the last one as follows: when the logical expressions in the IF THEN statement and in all ELSE IF statement(s) are *false*, program execution continues with the first executable statement in the else-block.

10.3 IF Construct

```
IF (colour .EQ. 'red') THEN
  time = 30;   CALL red(time)
ELSE IF (colour .EQ. 'green') THEN
  time = 20;   CALL green(time)
ELSE IF (colour .EQ. 'yellow') THEN
  time = 5;    CALL yellow(time)
ELSE
  CALL error
END IF
```

The ELSE statement causes one block of the IF construct to be executed in any case.

Branching : An if-block, an else-block, and an elseif-block may be left before the completion of the block either by branching to the corresponding END IF statement or by branching to outside the IF construct.

10.3.2 Nested IF Constructs

IF constructs may be nested; that is, an IF construct may appear within any block of another IF construct.

If the inner IF construct is in the else-block or in one of the elseif-blocks of the outer IF constructs, this is a piece of program which may also be written without nesting IF constructs by the use of an additional ELSE IF statement.

```
a: IF (c .EQ. 'red') THEN        ! a: IF (c .EQ. 'red') THEN
     time = 30                   !      time = 30
     CALL red(time)              !      CALL red(time)
   ELSE a                        !    ELSE IF (f .EQ. 'green') THEN a
     b: IF (f .EQ. 'green') THEN !      time = 20
        time = 20                !      CALL green(time)
        CALL green(time)         !    ELSE IF (c .EQ. 'yell') THEN a
     ELSE b                      !      time = 5
        c: IF (c .EQ. 'yell') THEN !    CALL yellow(time)
           time = 5              !
           CALL yellow(time)     !
        END IF c                 !
     END IF b                    !
   END IF a                      ! END IF a
```

In the above example, both IF constructs are equivalent.

If the inner IF construct is in the if-block of the outer IF construct, the execution of the if-block of the inner IF construct depends on both the logical expression in the outer IF THEN statement and the logical expression in the inner IF THEN statement.

An inner IF construct must be entirely contained within the if-block, within the else-block, or within one of the elseif-blocks of the outer IF construct.

Branching: During the execution of a nested IF construct, control may be transferred from the inner IF construct either to the surrounding block of the outer IF construct or to outside the nested IF construct. A branch to the END IF statement of an IF construct has the same effect as a branch to outside the (possibly nested) IF construct containing this END IF statement.

10.4 CASE Construct

A CASE construct selects for execution at most one of its constituent blocks. The first statement of a CASE construct is a SELECT CASE statement. The last statement of a CASE construct is an END SELECT statement. And the CASE and CASE DEFAULT statements are used to control the control flow within the CASE construct. The CASE DEFAULT statement is a special case of a CASE statement.

[**case-name :**] **SELECT CASE (case_expression)**

CASE selector [**case-name**]

END SELECT [**case-name**]

The **case-name** is a name identifying the CASE construct. The **case_expression** is a scalar integer, logical, or character expression. In a CASE statement, the **selector** designates the **case values** for the selection of the *case-block* following the CASE statement.

If the SELECT CASE statement includes a **case-name**, the corresponding END SELECT statement *must* include the same name, and the corresponding CASE statements *may* include this name. If the SELECT CASE statement does not include a **case-name**, neither the END SELECT statement nor the CASE statements must include a **case-name**.

Normally, a CASE construct consists of several alternative blocks. Such an alternative block is called a **case-block**.

The execution of the SELECT CASE statement causes the **case_expression** to be evaluated; its result is called the **case index**. Then, this case index is used to determine whether or not any case-block in the CASE construct is to be executed or which case-block is to be executed. If a CASE construct includes a CASE DEFAULT statement, *exactly* one case-block is selected for execution.

10.4 CASE Construct

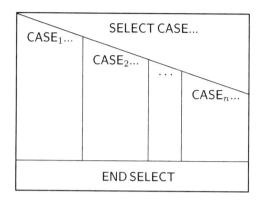

```
SELECT CASE (colour)     ! SELECTCASE statement
CASE ('red')             ! corresp. CASE statement with selector
   :                     ! first case-block, case 'red'
CASE ('green')           ! corresp. CASE statement with selector
   :                     ! second case-block, case 'green'
CASE DEFAULT             ! corresp. CASE stmt. with selector DEFAULT
   :                     ! third case-block, case 'otherwise'
END SELECT               ! corresponding ENDSELECT statement
```

The SELECT CASE statement containing the case expression is the first statement of a CASE construct.

```
SELECT CASE (n*3 - J)              ! integer
SELECT CASE (SIN(beta) - pi/2 /= 0.)  ! logical
SELECT CASE ('light ' // colour)   ! character
```

A CASE statement is the first statement of a case-block. The form of the **selector** is either **DEFAULT** or **(cv [, cv]...)**, where each **cv** is a single *case value* or a *case value sequence*. The form of a **case value sequence** is as follows:

case_value :
: case_value
case_value : case_value

Each case value included in a selector must be a scalar *initialization expression* of type integer, logical, or character. Note that the colon is not allowed in logical case value sequences.

The colon notation is interpreted as a sequence of single case values:

Selector	Corresponding *case values*
case_value$_1$: case_value$_2$	**case_value$_1$** $\leq$ *case value* $\leq$ **case_value$_2$**
case_value$_1$:	**case_value$_1$** $\leq$ *case value*
: case_value$_2$	*case value* $\leq$ **case_value$_2$**

DEFAULT is a special selector. At most one corresponding CASE DEFAULT statement may appear within a CASE construct. The selector DEFAULT matches all values of the case expression for which there is no matching case value in any other CASE statement in the CASE construct.

A CASE statement may be labeled, but it must not be used as a branch target statement.

The type and the kind type parameter of each case value in a particular CASE construct must be those of the case expression. If the case expression is of type character, the lengths of the case values may be different from that of the case expression.

```
CASE (11)
CASE (n/2 +1)
CASE (1,8,10:12,19)              ! CASE (1, 8, 10, 11, 12, 19)
CASE (:1933, 1939:1942, 1945:)
CASE (switch .AND. night)
CASE ('one', 'two', 'three')
CASE DEFAULT
```

The case values of the selector in a single CASE statement must *not* overlap. And the case values of the selectors within a particular CASE construct must *not* overlap; that is, for a particular result of the case expression, there must be at most one matching case value in all CASE statements within a CASE construct.

The END SELECT statement is the last statement of a CASE construct. An END SELECT statement may be labeled. Note that it may be a branch target statement only for a branch appearing in the interior of the CASE construct.

Each "case" of a CASE construct consists of a sequence of executable statements, a case-block. The default-block is a special case-block.

A CASE construct may include one, none, or more than one corresponding CASE statements. A case-block or the default-block of a CASE construct may be empty.

The CASE DEFAULT statement, if any, need not be the last CASE statement in a CASE construct; that is, the CASE statements may appear in any order within the CASE construct.

[**case-name :**] **SELECT CASE (case_expression)**
CASE selector [case-name]
⋮   ⟵ first case-block
CASE selector [case-name]
⋮   ⟵ second case-block, etc.
CASE DEFAULT [case-name]
⋮   ⟵ default-block
END SELECT [case-name]

Execution

The execution of a CASE construct causes the scalar case expression to be evaluated. When the case index matches a case value belonging to a selector in any CASE statement within this CASE construct, the case-block following this CASE statement is selected for execution. The execution of this selected case-block completes the execution of the CASE construct. If the case index does not match a case value, program execution continues with the next executable statement following the END SELECT statement.

10.4.1 Simple CASE Constructs

A simple CASE construct includes exactly one CASE statement and one case-block. The conditional execution of the block depends on one condition.

```
cylinder: SELECT CASE (n)            ! IF (n .EQ. 1) THEN
CASE (1)                             !
   o = 2*pi*r*h + pi*r**2            !    o = 2*pi*r*h + pi*r**2
   v = pi/2*r**2*h                   !    v = pi/2*r**2*h
END SELECT cylinder                  ! END IF
```

In this case, the CASE construct and the IF construct are equivalent.

The following example implements actions for every meaningful and meaningless value of the case expression. The second case-block is empty. And the third case-block is the default-block, which will be executed when **time** is out of a meaningful range:

```
tower_clock: SELECT CASE (time)
CASE (8:18)
   CALL bell(time)
CASE (1:7, 19:24)
CASE DEFAULT
   CALL error
END SELECT tower_clock
```

Branching: A case-block or the default-block may be left before the completion of the block either by branching to the corresponding END SELECT statement or by branching to outside the CASE construct.

10.5 DO Construct

A DO construct is a sequence of statements which may be executed more than once. The first statement of a DO construct is a DO statement. The subsequent block is the *range* of the DO construct. The last statement of a DO construct is the *do-termination statement*.

The range may contain CYCLE statements and EXIT statements, which are used to control the control flow within the range of the DO construct. Normally, the do-termination statement is an END DO statement, but some other statements also are allowed.

There are different kinds of DO constructs; they may be classified as *count loops*, *WHILE loops*, and *endless loops*. The general form is:

[**do-name :**] **DO** ...
⋮   ⟵ range of the DO construct
do-termination [**do-name**]

If the DO statement includes a **do-name**, the corresponding do-termination statement *must* include the same name, and additional CYCLE and EXIT statements of this construct, if any, *may* include this name. If the DO statement does not include a **do-name**, neither the do-termination statement nor the CYCLE and EXIT statements must include a **do-name**.

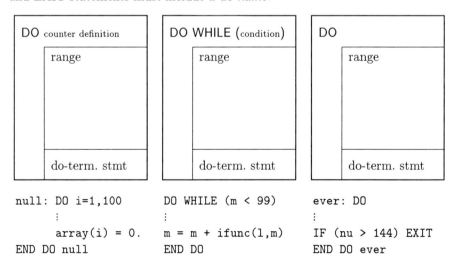

10.5 DO Construct

The **range** of a DO construct includes all executable statements from the first statement after the DO statement to the do-termination statement, inclusively; it is also called the **body** of the DO construct.

10.5.1 DO Statement

The execution of a DO statement causes the internal control flow of the DO construct to be initiated.

[**do-name :**] **DO** [**label**]
[**do-name :**] **DO** [**label**] [**,**] **WHILE (execution_condition)**
[**do-name :**] **DO** [**label**] [**,**] **do_variable = first, last** [**, stride**]

The **do-name** identifies the DO construct. **label** is the statement label of the corresponding do-termination statement. The **execution_condition** is a scalar logical expression. **do_variable** is the name of a scalar variable of type integer. And **first**, **last**, and **stride** are scalar integer expressions.

Loop parameters: The scalar results of **first**, **last**, and **stride** are converted, if necessary, to the kind type parameter of the DO variable. These values are called the **initial parameter**, **terminal parameter**, and **increment parameter**, respectively, of the DO construct. The value of the optional **stride** must be nonzero. If the **stride** is absent, the value one is assumed as the default increment parameter.

```
DO
year: DO 30
DO WHILE (n < 100)
row: DO 55 WHILE (m == 2)
DO, l=1,1+n
DO 10 i=1,100,2
```

10.5.2 Do-Termination Statement

Normally, the do-termination statement of a DO construct is an END DO statement.

END DO [do-name]

The **do-name** identifies the DO construct. If this name is present, the same name must be included in the DO statement.

If both the **do-name** is absent in the DO statement and the DO statement includes the **label** of the corresponding do-termination statement, then the do-termination statement may be any simple executable statement except an unconditional GO TO, an assigned GO TO, a computed GO TO, or a RETURN, STOP, EXIT, CYCLE, END, END FUNCTION, END PROGRAM, END SUBROUTINE statement, and an arithmetic IF statement.

If the DO statement does not include the **label** of the corresponding do-termination statement, the corresponding do-termination statement must be an END DO statement.

If the DO statement includes the **label** of the do-termination statement, the do-termination statement must be labeled. And conversely, if the do-termination statement is labeled, this statement label must be included in the DO statement as the **label** of the corresponding do-termination statement.

10.5.3 Forms of DO Constructs

If the first statement of a DO construct is a DO statement with a DO variable, it is a "count loop". A count loop is controlled by an *iteration count* (see below), which automatically terminates the execution of the DO construct when the maximum limit for the loop iterations is reached.

If the first statement of a DO construct is a DO statement without a DO variable but with the keyword WHILE and with an execution condition, it is a "WHILE loop". The execution of a WHILE loop automatically terminates before the execution of the next iteration when the execution condition of the DO statement is not satisfied.

If the first statement of a DO construct is a DO statement that includes neither a DO variable nor an execution condition, it is an "endless loop". The execution of such a DO construct can be terminated only by a branching or terminating statement within its range.

```
      DO 10 i=1,100              ! count loop, no do-name, label
        odd(i) = 2*i + 1
   10 CONTINUE                   ! no do-name, label


      DO                         ! endless loop, no do-name, no label
        IF(x .GT. y) THEN
          xmax = x
          EXIT                   ! exit this loop
        END IF
        READ *, x
      END DO                     ! no do-name
```

```
x_temp = 0.
m: DO WHILE (x_temp < 75)  ! WHILE loop, do-name, no label
   READ 15, day_temp
   x_temp = x_temp + day_temp
END DO m                   ! do-name
x_temp = x_temp/365
```

10.5.4 Execution of a DO Construct

When a DO construct is executed, the following steps are performed: loop control is initiated, then the range is executed (generally several times), and finally, loop execution is terminated.

Loop control initiation

Count loop: First, the loop parameters are determined (10.5.1). Suppose, $m1$ is the value of the initial parameter, $m2$ is the value of the terminal parameter, and $m3$ is the value of the increment parameter. Then the DO variable is defined with $m1$. Finally, the *iteration count* is evaluated:

$$\text{Iteration count} = \text{MAX}\left((m2 - m1 + m3)/m3,\ 0\right)$$

Execution cycle

1. **Count loop:** The iteration count is tested. If the iteration count is greater than zero, program execution continues with the first (executable) statement in the range. If the iteration count is zero, the count loop is "satisfied" and the execution of the DO construct terminates (see below).

2. **WHILE loop:** The execution condition is evaluated. If the result of the logical expression is *true*, program execution continues with the first (executable) statement in the range. If the result of the logical expression is *false*, the WHILE loop is "satisfied" and the execution of the DO construct terminates (see below).

3. Beginning with the first executable statement in the range, (executable) statements are executed until the do-termination statement is executed, until a CYCLE statement is executed, or until the loop is terminated during this execution cycle.

4. **Count loop:** The iteration count is decremented by one, and the DO variable is incremented or decremented, depending on the sign of the increment parameter m_3, by the absolute value of the increment parameter.

5. The next iteration of the loop is executed (step 1).

These steps are repeated until the execution of the loop terminates (see below).

Loop termination

The execution of a DO construct terminates:

- When the iteration count is zero.
- When the execution condition of a WHILE loop is *false*.
- When a RETURN statement is executed in the range.
- When an EXIT statement is executed in the range.
- When an EXIT or CYCLE statement of a surrounding DO construct is executed in the range.
- When control is transferred from the range to outside the range.
- When a STOP statement is executed or when the execution of the program is terminated for any other reason.

When a DO construct (but not the program) terminates, control is transferred to the next surrounding DO construct having the same do-termination statement. If there is no such DO construct, the execution of the program continues with the next (executable) statement following the do-termination statement of the terminated DO construct.

10.5.4.1   Additional Details about Count Loops

Iteration count: The iteration count may already be zero before the first iteration, for example, if both $m1 > m2$ and $m3 > 0$ are true, or if both $m1 < m2$ and $m3 < 0$ are true. In such a case, the range is not executed and the DO construct is skipped.

The maximum number of iterations is determined when the DO construct is initiated. This maximum number of iterations cannot be modified within the range. It is equal to the actual number of iterations if the DO construct is not terminated until it is satisfied.

DO variable: When a DO construct terminates before the first iteration is executed, the DO variable retains the value of the initial parameter. When a DO construct terminates before the iteration count is zero, the DO variable retains its last defined value. When the DO construct terminates because the iteration count is zero, the DO variable also retains its last defined value, that is, its value during the last iteration plus the value of the increment parameter $m3$.

10.5 DO Construct

Except for the automatic incrementation or decrementation, the DO variable must neither be redefined nor become undefined during the execution of the DO construct.

```
   never = 100
   DO 10 i=never,85
     ⋮
10 END DO
```

During the execution of the DO 10 statement, the initial parameter is greater than the termination parameter; that is, the iteration count already is zero before the first iteration. Therefore, the range is not executed.

```
back: DO, k=10,1,-1
   a(k) = 11 - k
END DO back
```

This DO construct is interpreted as:

```
a(10) = 1
a(9)  = 2
a(8)  = 3
  ⋮
a(3)  = 8
a(2)  = 9
a(1)  = 10
k     = 0
```

If the increment parameter is negative, the DO variable is not incremented but decremented.

```
   DO 30 m=1,11,3
      IF (f(m) .GE. 0.0) GOTO 30
      f(m) = - f(m)
30 PRINT *, m, f(m)
70 CALL sub
```

The range of this DO construct is executed four times because the DO variable becomes defined with 1, 4, 7, and 10. After termination of the DO construct, the DO variable retains the value 13. And the execution of the program continues with statement 70.

10.5.4.2 Additional Details about WHILE Loops

Execution condition: The execution condition may already be *false* before the first iteration. In this case, the range is *not* executed; that is, the DO construct is skipped.

If there is no other termination of the DO construct (except for the execution condition), the execution of the DO construct can be terminated only if such variables are redefined or if such functions return different results that affect the evaluation of the execution condition.

10.5.4.3 Additional Details about Endless Loops

An endless loop has no upper limit for the number of iterations. A termination condition may appear only in the range.

```
ever: DO
  IF (age < 100) EXIT      ! termination condition
  ⋮
END DO ever
```

10.5.4.4 CYCLE Statement and EXIT Statement

The CYCLE statement may be used to terminate the current iteration of the corresponding DO construct (but not the complete loop) immediately.

CYCLE [do-name]

The execution of a CYCLE statement has the effect such as a branch to the END DO statement of a DO construct. Precisely: after the execution of a CYCLE statement, no other statement in the corresponding DO construct is executed during this iteration and the execution of the DO construct continues with the next iteration.

The EXIT statement may be used to terminate the execution of the corresponding DO construct immediately.

EXIT [do-name]

Mostly (as in the following example), the execution of an EXIT statement has the effect such as a branch to the next executable statement after the corresponding do-termination statement of the DO construct. In any case, after the execution of an EXIT statement, no other statement in the corresponding DO construct is executed.

```
   year = 0;   READ *, salary
   DO 10
     year = year + 1
     salary = 1.04 * salary
     IF (year > 15) EXIT          ! IF (year > 15) GO TO 20
     IF (salary > 6500.) GOTO 10  ! IF (salary > 6500.) CYCLE
     salary = salary + base
10 END DO
20 CONTINUE
```

10.5.5 Nested DO Constructs

If the range of a DO construct includes one or more other DO constructs, the construct is a **nested DO construct**.

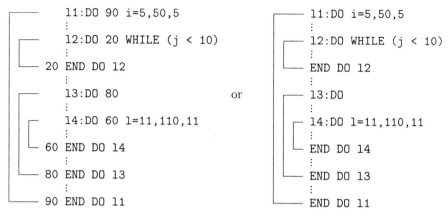

DO constructs within the range of an outer DO construct may be nested or may appear one after the other at the same nesting level. An *inner* DO construct must be contained entirely within the range of an *outer* DO construct.

Shared do-termination statement: Nested DO constructs may share their do-termination statements. Then all contained DO statements must include the same statement label of the do-termination statement and the shared do-termination statement must be labeled. Note that this shared do-termination statement must not be an END DO statement.

```
        DO 99 k=1,100
        ⋮
        DO 99 j=1,100
        ⋮
        DO 99 i=1,100
        ⋮
        IF (i .EQ. j .AND. j .EQ. k) GOTO 99
        e(i, j, k) = 0.
     99 CONTINUE
```

A shared do-termination statement may be used as a branch target statement only for a branch appearing in the range of the innerst DO construct.

Branching: Control may be transferred from the range of an inner DO construct to a branch target statement appearing either outside the complete DO nest or in the range of a surrounding DO construct. The do-termination statement may be used as a branch target statement only for a branching statement appearing in the range of the corresponding DO construct. And control must not be transferred from outside a DO construct into its range.

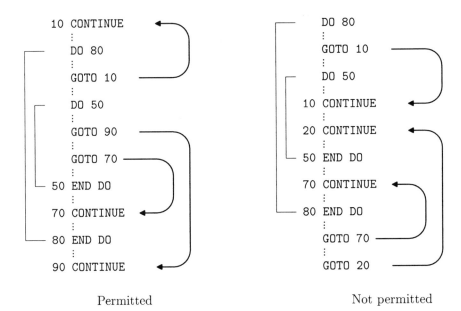

Permitted   Not permitted

10.6 Nested Constructs

IF constructs, CASE constructs, and DO constructs may be nested in each other, but they must not overlap; that is, one block of such a construct may contain another complete construct.

```
CHARACTER line (80)
INTEGER level
level = 0
DO, i=1,80
  SELECT CASE (line(i:i))
  CASE ('(')
    level = level + 1
  CASE (')')
    level = level - 1
    IF (level < 0) THEN
      PRINT *, - level, ' right parentheses excess'
      EXIT
    END IF
  CASE DEFAULT    ! only parentheses are checked
  END SELECT
END DO
IF (level > 0) PRINT *, level, ' right parentheses lack'
```

10.7 CONTINUE Statement

The execution of a CONTINUE statement has no effect.

label CONTINUE

A CONTINUE statement must be labeled. It may be used to place a statement label in a particular line of the execution part of a main program or subprogram such that other statements in the same scoping unit can refer to this line by means of this statement label. For example, if a particular statement cannot be used as a do-termination statement, a CONTINUE statement may be written.

```
      DO 10 i=1,10
      ⋮
    5 CONTINUE
      ⋮
      IF (expr) 5, 20, 10
   10 CONTINUE
   20 CONTINUE
```

The arithmetic IF statement cannot be used as a do-termination statement. Therefore, the CONTINUE statement 10 is inserted.

10.8 STOP Statement

The execution of a STOP statement causes the termination of execution of the program.

STOP [n]

The **n** is either a sequence of at most 5 digits or a scalar constant of type default character. If **n** is a sequence of digits, leading zeros are not significant.

The execution of a STOP statement causes the program to terminate immediately without any possibility to resume execution. If a sequence of digits or a character constant appears in the statement, this information is made available in a processor-dependent manner before the program stops.

A main program or subprogram may contain several STOP statements. STOP statements containing different sequences of digits or different character constants may be used to determine the actual position where program execution has stopped.

```
STOP 'Normal program termination'
```

10.9 CALL, END, and RETURN Statements

CALL, END, and RETURN statements are presented in chapter 13.

Note that input/output statements may contain certain specifiers that may cause a transfer of control when certain input/output conditions occur (chapter 11).

11 INPUT/OUTPUT

Input statements are used to transfer data from an external or internal file to internal storage; this is called *reading*. **Output statements** are used to transfer data from internal storage to an external or internal file; this is called *writing*. In addition to the data transfer statements, there are input/output statements that inquire about, specify, or modify the properties of a file or of a *unit*. And finally, there are input/output statements which manipulate the external medium such as for the positioning of external files.

Each of the following terms characterizes a special kind of input/output:

Sequential io	= input/output with sequential access
Direct io	= input/output with random access
Formatted io	= input/output with format specifications
Unformatted io	= input/output without format specifications
List-directed io	= io-statements with * as FMT= specifier
Namelist io	= io-statements without input/output lists
Nonadvancing io	= input/output without automatic positioning
Internal io	= input/output from storage to storage

11.1 Records

The basic constituent of the Fortran file system is the *record*. A **record** is a sequence of values or a sequence of characters. For example, normally, a line on the screen is a record. A record need not be a physical record. There are three kinds of records: *formatted* records, *unformatted* records, and *endfile* records.

A **formatted record** consists of a sequence of characters representable in the Fortran processor. A Fortran processor may prohibit certain control characters from appearing in a formatted record. Formatted records may contain characters of a nondefault character type.

The length of a formatted record is measured in characters. The length is processor-dependent and depends primarily on the number of characters written into the record. The length may be zero.

Formatted records may be read or written only by formatted input/output statements. They may also be produced by means other than Fortran, such as the keyboard of a terminal.

An **unformatted record** consists of a sequence of values in a processor-dependent internal representation. An unformatted record may contain data of any intrinsic or derived type. The length of an unformatted record is measured in processor-dependent units; its length is processor-dependent and depends on the output list when the record is written. The length may be zero.

Unformatted records may be read or written only by unformatted input/output statements.

An **endfile record** may appear only as the last record of a file. An endfile record has no length (from the point of view of the programmer). An endfile record may be written to a file opened for sequential input/output, and it may be recognized during reading. The (internal) representation of an endfile record is processor-dependent.

An endfile record may be (explicitly) written by the execution of an ENDFILE statement, but it may also be written implicitly. If the file is opened for sequential access and if the most recent input/output statement for that file is a data transfer output statement and no positioning statement, an endfile record is written implicitly (that is, automatically) when a REWIND or BACKSPACE statement is executed for the connected unit, when the file is explicitly closed by the execution of a CLOSE statement, when the file is implicitly closed at program termination, or when the file is implicitly closed by the execution of another OPEN statement for the same unit.

11.2 Files

A **file** is a sequence of records. There are *external files* and *internal files*.

External files are files that exist in a medium external to the program such as a magnetic disk or a tape storage. But an external file also may be a device such as the printer, the screen, or the keyboard of a terminal. Files are normally organized by the operating system environment in which the *program* is embedded.

Internal files are character variables (that is, internal storage of the program) playing the role of a file for internal input/output.

11.3 File Attributes of External Files

An external file may have the following attributes which are partly processor-dependent: the set of allowed names (processor-dependent), the existence, the set of allowed access methods (processor-dependent), the set of allowed forms (processor-dependent), the set of allowed record lengths (processor-dependent), the set of allowed actions, and the position. Certain file attributes are only established if the file is connected to a unit, that is, if it is opened.

11.3 File Attributes of External Files

11.3.1 File Names

An external file may have a name. This name can appear only in OPEN and INQUIRE statements.

If a unit is not (explicitly) connected to a file by the execution of an OPEN statement *with* FILE= specifier, the Fortran processor possibly generates a processor-dependent file name.

11.3.2 Access Methods

Records may be read and written in sequential order or in random order. There are two access methods: *sequential access* and *direct access*. Input/output data transfer with sequential access is described as *sequential input/output*, and input/output data transfer with direct access is described as *direct input/output*.

External files may be processed with sequential or direct access. Internal files may be processed only with sequential access.

The access method for an external file which is not preconnected is determined when the file is connected to a unit. For the time of this connection, the access method must not be changed.

11.3.2.1 Sequential Access

If a file is processed only with sequential access, the order in which the records are written determines the order in which they may be read; that is, the nth record can be read only after the first record, the second record, and so on, and the $(n-1)$th record have been read.

If the Fortran processor also allows direct access to the file, the order of the records for the sequential access is given by the record numbers which are associated with the records for the direct access.

External sequential files may be connected to a unit by the execution of an OPEN statement. Such explicit connection is not necessary if the operating system performs the connection, such as in the case of *preconnected units*.

11.3.2.2 Direct Access

If an external file is opened for direct access, the order of the records depends on the order of the *record numbers*. In contrary to sequential input/output, the records need not be read or written in any order. For example, the record with record number 3 may be written though the records with the record numbers 1 and 2 have not (yet) been written.

Input/output actions such as the rewinding of a file at its initial point or as the positioning of a file at its terminal point are meaningless for a file which is opened for direct access. An endfile record, if any, is not considered as a part of the file as long as the file is opened for direct access.

There is formatted and unformatted direct input/output. List-directed input/output with direct access, namelist input/output with direct access, and nonadvancing input/output with direct access are not allowed. Internal files cannot be opened for direct access.

If an external file is to be opened for direct access, an OPEN statement must be executed specifying both ACCESS='DIRECT' and the RECL= specifier with an appropriate record length.

OPEN (11, FILE=..., ACCESS='DIRECT', FORM='FORMATTED', RECL=80)

A file is connected to unit 11 for direct access. The file name is processor-dependent (and is not written here). All records are formatted and have the same length of 80 (characters).

Record number

Each record is unambiguously identified by a natural number (= positive integer value), the **record number**. The record number of a particular record is specified by the programmer and established when the record is written. This record number cannot be changed. A record with a particular record number cannot be deleted, but it can be rewritten.

As long as a file is opened for direct access, READ and WRITE statements for the connected unit must include the REC= specifier (with a record number). A record of a file which is opened for direct access may be read only if it has been written since the file was created.

Record length

The record length of a direct access file must be specified in the same OPEN statement which specifies the direct access attribute. All records of a direct access file have the same length. This length is independent of the length of the output list when the records are written. If the length of the output list in a formatted direct WRITE statement is less than the record length, the rest of the record is filled with blanks. But if the length of the output list in an unformatted direct WRITE statement is less than the record length, the rest of the record is undefined. Note that the length of the output list in a direct WRITE statement must never exceed the record length specified in the OPEN statement for the connected unit.

11.3 File Attributes of External Files

```
CHARACTER *8 text (9)
OPEN (7, FILE=... , ACCESS='DIRECT', FORM='FORMATTED', RECL=72)
  :
DO i=10,5,50
WRITE (7, FMT='(9A)', REC=i) text
  :
END DO
```

The records 10, 15, 20, ..., 45, and 50 are written. These records may be read immediately after they have been written. There are no records written for any other record numbers; therefore, other records cannot be read.

11.3.3 Form of a File

The records of a file must be either *all* formatted or *all* unformatted, except that the last record may be an endfile record. Therefore, files may be classified as *formatted files* and *unformatted files*. The set of allowed forms for a file is processor-dependent.

11.3.4 File Position

A file which is connected to a unit has a **position**. The position may become *indeterminate*.

The **initial point** of a file is the position just before the first record. The **terminal point** is the position just after the last record.

If the file is positioned *within* a record, this record is the **current record**; otherwise there is no current record.

The record before the current record is the **preceding record**. If the file is positioned between two records, the first of these records is the preceding record.

The record after the current record is the **next record**. If the file is positioned between two records, the second of these records is the next record. If the file is positioned at its initial point, the first record is the next record.

File position prior to data transfer

Sequential input: If the file is positioned *within* a record, the file position is not changed. Otherwise, the file is positioned at the beginning of the next record. And this record becomes the current record.

Sequential output: If the file is positioned *within* a record, the file position is not changed. Otherwise, a new record is created. This record becomes the last and current record of the file, and the file is positioned at the beginning of this new record.

Direct input/output: The file is positioned at the beginning of the record with the specified record number. This record becomes the current record.

File position after data transfer

If an error condition occurs, the file position is indeterminate.

Nonadvancing input: If no error condition, no endfile condition, but an end-of-record condition occurred, the file is positioned after the record just read.

Nonadvancing input/output: If no error condition, no endfile condition, and no end-of-record condition occurred, the file position is not changed.

Advancing input/output: If no error condition and no endfile condition occurred, the file is positioned after the record just read or written.

Sequential input: If no error condition occurred while the endfile record is read, but if the endfile condition occurred, the file is positioned after the endfile record. At this position no READ, WRITE, and PRINT statements must be executed for the connected unit. However, the file may be repositioned (before any data transfer) by the execution of a BACKSPACE or REWIND statement.

11.4 Units

If an input/output statement refers to an external file, a *unit* must be specified which is *connected* to the file. These units are the Fortran specific processor-independent means of referring to a file.

A unit may be **connected** to an external file; this operation is described as the *opening* of the file.

Data transfer statements and positioning statements may be executed only for such external files which are connected to a unit. OPEN, CLOSE, and INQUIRE statements may also be executed for external files which are *not* connected to a unit.

A unit may be connected to an external file either by the execution of an OPEN statement, this is described as *explicit opening*, or they may be *preconnected*. If they are preconnected, the connection is automatically established when the program is executed; this is described as *implicit opening*.

A unit must not be connected to more than one file at the same time. And a file must not be connected to more than one unit at the same time. A unit and a file may be disconnected by the execution of a CLOSE statement. After a unit has been disconnected, it may be connected to the same file or to another external file, and the file may be connected to the same unit or to another unit.

The set of allowed unit numbers is processor-dependent. For additional informations, see the description of the UNIT= specifier in 11.6.1.1.

11.5 Preconnected Units and Predefined Files

A unit and an external file are **preconnected** if they are automatically connected at the beginning of the execution of the program. This means that a preconnected unit may be used in all input/output statements without prior execution of an OPEN statement for this unit and this file.

In the operating system environment which controls the execution of the program, there may be particular files with processor-dependent function such as *device files* and *standard files*.

Usually, device files and standard files are connected (processor-dependent) to *standard units* for sequential formatted input/output; for example, many **Fortran** processors use unit number 5 for the standard input (unit) and unit number 6 for the standard output (unit).

The **standard input unit** (abbreviated: standard input) is the unit which may be specified in a READ statement (with a control information list) as the unit asterisk * or which is implicitly used when a READ statement without a control information list is executed.

The **standard output unit** (abbreviated: standard output) is the unit which may be specified in a WRITE statement (with a control information list) as the unit asterisk * or which is implicitly used when a PRINT statement is executed.

It is processor-dependent which devices are associated with standard input and standard output. In an interactive environment, the standard input device is usually the keyboard and the standard output device is the screen. And in a batch environment, the standard input is usually connected to a predefined file, and the standard output is associated with the system printer.

Often, external files are preconnected for sequential input/output. Some **Fortran** processors allow a unit to be connected to a file outside the **Fortran** program in the operating system environment before the execution of the program. In these cases, possibly no explicit opening of the file is necessary. If the file has no name, the **Fortran** processor may generate a processor-dependent file name.

11.6 Input/Output Statements

The input/output statements for data transfer are PRINT, READ, and WRITE. The input/output statements for inquiring about, specifying, and changing of properties of files and units are CLOSE, INQUIRE, and OPEN. The input/output statements for the positioning of external files are BACKSPACE, ENDFILE, and REWIND.

```
PRINT 2222, ' test variable XYZ', xyz
READ (UNIT=5, FMT=*, END=100) a, b, c
WRITE (9) n, x(1:100,5), x/n
CLOSE (UNIT=7, IOSTAT=ivar, STATUS='KEEP')
INQUIRE (14, PAD=pad)
OPEN (UNIT=77, ACCESS='DIRECT', RECL=150)
BACKSPACE 10
ENDFILE (77, ERR=180)
REWIND (UNIT=13, ERR=200)
```

Side effects are not allowed during the execution of an input/output statement. That is, a function reference must not appear anywhere in an input/output statement if the invocation of the function causes another input/output statement to be executed.

11.6.1 Input/Output Specifiers

Each input/output statement includes one or more **input/output specifiers**. These specifiers are used to specify the unit, the file name, the form of the input/output, the access method, the error handling, and so on (see the last example). In the following, those input/output specifiers are described that may appear in the *control information lists* of READ and WRITE statements and, partly, in other input/output statements.

Source or destination of the data to be transferred:	[**UNIT =**] u
Kind of editing of the external data:	[**FMT =**] format
Identification of a record for namelist io:	[**NML =**] group-name
Identification of a record for direct io:	**REC** = record_number
Advancing or nonadvancing input/output:	**ADVANCE** = yes_or_no
End-of-record condition handling:	**EOR** = label
Output of input/output status informations:	**IOSTAT** = status_variable
Error condition handling:	**ERR** = label
End-of-file condition handling:	**END** = label
Output of number of read characters:	**SIZE** = number_of_chars

11.6.1.1 UNIT= Specifier

The UNIT= specifier is used to specify a unit (referring to an external file) or to specify an internal file.

[**UNIT =**] **u**

The **u** may be:

- A nonnegative scalar integer expression; its result is the **unit number**;
- An asterisk * specifying a standard unit which is preconnected for formatted sequential input or for sequential formatted output; or
- A variable of type default character identifying an internal file.

Unit number: The unit number is global to the program; that is, a particular unit number identifies in different program units of an program the same unit for the reference to an external file. The set of allowed unit numbers is processor-dependent.

11.6.1.2   FMT= Specifier

The FMT= specifier is used to specify a format or list-directed formatting for a formatted input/output statement.

[**FMT =**] **format**

The **format** may be:

- The label of a FORMAT statement in the same scoping unit;
- A variable of type default character (but no assumed size array) defined with a format specification;
- Another character expression of type default character which evaluates to a valid format specification; or
- An asterisk * specifying list-directed formatting.

11.6.1.3   NML= Specifier

The NML= specifier is used in a namelist input/output statement to specify a namelist group-name.

[**NML =**] **group-name**

```
NAMELIST /address/name, first_name, street, no, zip, town
  :
READ (*, NML=address)
```

11.6.1.4 REC= Specifier

The REC= specifier is used in a direct input/ouput statement to specify the record number of the record to be transferred.

REC = record_number

The **record_number** must be a scalar positive integer expression.

```
WRITE (16, REC=i+25) sort, location, year
```

11.6.1.5 ADVANCE= Specifier

The ADVANCE= specifier is used to specify whether it is an advancing or nonadvancing READ or WRITE statement.

ADVANCE = yes_or_no

yes_or_no is a scalar default character expression which must evaluate to 'YES' if it is an advancing (that is, a "normal" record oriented sequential formatted) input/output statement or which must evaluate to 'NO' if it is a nonadvancing input/output statement. If the ADVANCE= specifier is absent, the default is ADVANCE='YES'.

The result of the evaluation of the character expression may include trailing blanks, but they are not significant; lower-case characters are allowed, but they are processor-dependent.

11.6.1.6 End-of-Record Condition and EOR= Specifier

The **end-of-record condition** occurs when a nonadvancing READ statement attempts to transfer data from a position beyond the end of the record. The end-of-record condition is no error condition. When an end-of-record condition occurs during the execution of a nonadvancing READ statement:

- The record is padded with blanks if the file is opened with PAD='YES' and if the record contains fewer characters than are needed according to the input list and the associated format specification;
- The execution of the READ statement terminates; and
- The file is positioned after the read record.

If the end-of-record condition occurs during the execution of a READ statement without an IOSTAT= specifier and without an EOR= specifier, the execution of the program is terminated.

11.6 Input/Output Statements

EOR = label

The EOR= specifier is used in a nonadvancing READ statement to specify the label of an executable statement in the same scoping unit. This is the statement label of the branch target statement which will be executed when during the execution of the READ statement the end-of-record condition but no error condition occurs.

```
READ (10, '(A1)', ADVANCE='NO', EOR=400) chars
```

The EOR= specifier is allowed only in a nonadvancing READ statement.

When the end-of-record condition but no error condition occurs during the execution of a nonadvancing READ statement that includes the EOR= specifier:

1. The record is padded with blanks if the file is opened with PAD='YES' and if the record contains fewer characters than are needed according to the input list and the associated format specification.
2. The execution of the READ statement terminates.
3. All DO variables of all implied-DOs in the input list become undefined.
4. The file is positioned after the record just read.
5. The IOSTAT= specifier (if present) becomes defined with a processor-dependent negative integer value.
6. The SIZE= specifier (if present) becomes defined with the number of transferred characters.
7. And program execution continues with the branch target statement labeled **label**.

Note that the input list items which have been transferred already do *not* become undefined when the end-of-record condition occurs. And the remaining input list items which have not been transferred retain their definition status.

11.6.1.7   IOSTAT= Specifier

The IOSTAT= specifier is used to specify a scalar variable of type default integer for input/output status informations. These status informations are partly processor-dependent.

IOSTAT = status_variable

After execution of an input/output statement that includes the IOSTAT= specifier, the specified status variable becomes defined as follows:

    0 (zero) : when no error condition, no end-of-file condition, and no end-of-record condition occurs.

> 0 : when an error condition occurs.

< 0 : when a sequential READ causes the occurrence of the end-of-file condition but no error condition.

< 0 : (with a negative value different from the end-of-file value) when a nonadvancing READ causes the occurrence of the end-of-record condition but no error condition.

These possible values of the status variable are processor-dependent except the value zero.

When during the execution of an input/output statement that includes the IOSTAT= specifier the end-of-file condition, the end-of-record-condition, or an error condition occurs, the execution of the program is *not* terminated. If an additional END=, EOR=, or ERR= specifier is present, program execution continues with the specified branch target statement. If both the IOSTAT= specifier is present and the END=, EOR=, and ERR= specifiers are absent, program execution continues with the next executable statement immediately after the input/output statement.

```
INTEGER io_status
READ (UNIT=12, FMT='(F10.2)', IOSTAT=io_status) measurement
IF (io_status < 0) THEN
    CALL eof                    ! end-of-file condition on unit 12
ELSE IF (io_status > 0) THEN
    CALL error(io_status)       ! error condition on unit 12
ENDIF
```

11.6.1.8 Error Conditions and ERR= Specifier

The set of error conditions that may occur during the execution of an input/output statement is processor-dependent. When an error condition occurs during the execution of an input/output statement, the execution of the statement terminates, all DO variables of all implied-DOs in the input/output list of the statement become undefined, all items in the input list become undefined, and the position of the file becomes indeterminate.

When an error condition occurs during the execution of an input/output statement that includes neither the IOSTAT= nor the ERR= specifier, the execution of the program is terminated.

ERR = label

The ERR= specifier is used to specify the label of an executable statement in the same scoping unit. This is the label of the branch target statement that will be executed when an error condition occurs during the execution of the input/output statement.

11.6 Input/Output Statements

```
      READ (25, ERR=400) x, y, z
        ⋮
400   CALL io_error
```

When an error condition occurs during the execution of an input/output statement that includes the ERR= specifier:

1. The execution of the input/output statement terminates.
2. The file position becomes indeterminate.
3. The status variable of the IOSTAT= specifier (if present) becomes defined with a processor-dependent positive integer value.
4. The SIZE= specifier (if present in a READ statement) becomes defined with the number of transferred characters.
5. All DO variables of all implied-DOs in the input/output list become undefined.
6. The items in the input list in a READ statement become undefined.
7. And program execution continues with the branch target statement labeled **label**.

11.6.1.9 End-of-File Condition and END= Specifier

The **end-of-file condition** occurs when a sequential READ statement attempts to transfer data from a position beyond the end of the file. This may happen for an external file and for an internal file. In the case of a sequential external file, the end-of-file condition occurs when the READ statement encounters an endfile record.

When the end-of-file condition occurs, the execution of the READ statement terminates, all DO variables of all implied-DOs in the input list become undefined, and all items in the input list become undefined.

When the end-of-file condition occurs during the execution of a READ statement that includes neither the IOSTAT= nor the END= specifier, the execution of the program is terminated.

END = label

The END= specifier is used in a READ statement to specify the label of an executable statement in the same scoping unit. This is the label of the branch target statement that will be executed when the end-of-file condition but no error condition occurs during the execution of the READ statement.

```
READ (15, END=600) x, y, z
  ⋮
600 STOP 'no data'
```

When the end-of-file condition but no error condition occurs during the execution of a READ statement that includes the END= specifier:

1. The execution of the READ statement terminates.

2. The file is positioned after the endfile record if it is an external file.

3. The IOSTAT= specifier (if present) becomes defined with a processor-dependent negative integer value.

4. And program execution continues with the branch target statement labeled **label**.

11.6.1.10 SIZE= Specifier

The SIZE= specifier is used in a nonadvancing READ statement to specify a scalar integer variable which becomes defined with the count of the characters transferred by the execution of this READ statement.

```
INTEGER chars
  ⋮
READ (37, FMT=10, ADVANCE='NO', SIZE=chars) syllable, letter
IF (chars > 4) CALL word_division
```

When the nonadvancing READ statement terminates, the SIZE= specifier becomes defined with the count of the characters which have been transferred under format control, that is, under control of data edit descriptors. *Padding characters* are not counted.

11.6.2 Input/Output Lists

An **input/output list** specifies the entities whose values are transferred by READ, WRITE, or PRINT statements. The list items must be specified in the same order from the left to the right in which they are to be processed. An input/output list may specify any number of list items; it may be empty.

[**io_list_item** [, **io_list_item**]...]

An *input list* (see below) must specify only input list items and an *output list* (see below) must specify only output list items.

Noneffective list items: A zero-size array, an array section of size zero, or an implied-DO with iteration count zero may be specified as an input/output list item but they are **not effective**; that is, they are ignored as list items during the data transfer.

Arrays: If an input/output list specifies an array, it is treated as though all elements (if any) were specified in array element order. If the input/output list item is an allocatable array, it must be currently allocated. The name of an assumed-size array must not be specified as a list item.

Derived type entities: If a derived type entity appears as an input/output list item in a formatted input/output statement, it is treated as though all of its components were specified in the same order as in the derived type definition.

If a derived type entity appears as an input/output list item in an unformatted input/output statement, it is treated as a single value in processor-dependent form.

A derived type entity may appear as an input/output list item only if all components ultimately contained within the entity are accessible within the scoping unit containing the input/output statement.

A derived type entity which ultimately contains a pointer component must not appear as an input/output list item.

Pointers: If an input/output list item is a pointer, data are transferred between the associated target and the file.

Processing of list items: The input/output list is processed from the left to the right; that is, the processing of the next input/output list item does not begin until the value(s) to or from the current input/output list item are transferred. The significance of this rule is demonstrated by the following example:

```
READ (n) n, a(n)
```

The old value of n identifies the unit. When the data transfer for this unit begins, the new value for n is transferred. Then the position of a(n) is determined using this new value of n before the value for a(n) is transferred.

Input list

The input/output list in a READ statement must be an *input list*. An **input list item** must be either a variable or an implied-DO (see below). The range of such an implied-DO must contain only input list items.

```
READ (8, 1000) a, b, c, d
READ (UNIT=11, FMT=*) a, e(i), f(5, 10), g
READ 2000, x, y(5*j+3, 10*k-7), l
READ (13) ff, ge(k, 1, m+3)
```

Empty input list: If the input list is empty or if all list items are noneffective, one or more records are read without transferring data. An unformatted or list-directed READ statement skips exactly one record. For a formatted (but not list-directed) READ statement, the number of skipped records is one plus the number of slash edit descriptors in the format specification.

There are the following restrictions on input lists:

- An expression enclosed in parentheses must not appear as an input list item.
- A constant, an expression with operators, or a function reference must not appear as an input list item.
- An input list item (or an associated variable) in a formatted READ statement must not contain any part of the format specification.
- If the input list of an internal READ statement includes an item of type character, this item must be of type *default* character.
- A variable in the range of an implied-DO must neither be the DO variable of the implied-DO nor be associated with the DO variable.
- If an input list item is a pointer, this pointer must be currently associated with a definable target when the READ statement is executed.
- If an input list item is an array section, no array element must affect the value of any expression in the section-subscript list of the input list item and no array element must appear more than once within this array section.

```
INTEGER, DIMENSION (100) :: a
READ (5, *) a(a)                  ! not allowed
READ (5, *) a(a(1) : a(10))       ! not allowed
```

Output list

The input/output list in a WRITE or PRINT statement must be an *output list*. An **output list item** must be either an expression or an *implied-DO* (see below). The range of such an implied-DO must contain only output list items.

```
WRITE (UNIT=6, FMT=*) 4, g(i, k, k+2), a, p
PRINT 3000, t, v, r, r*SIN(v), (pi), pi
```

Empty record: If the output list in an unformatted WRITE statement or in a list-directed WRITE or PRINT statement is empty or if all list items are noneffective, one (empty) record of length zero is written. In the case of a formatted (but not list-directed) WRITE or PRINT statement, the number of written records depends on the edit descriptors in the format specification. The

11.6 Input/Output Statements

record is empty if the output list is empty or if all list items are noneffective, and if, in addition, the form of the format specification is ( ). An empty record does not contain any data, but it is counted when a READ statement or a BACKSPACE statement is executed.

There are the following restrictions on output lists:

- An output list item must be defined before the data transfer begins.
- If the output list of an internal WRITE statement includes a character entity, this entity must be of type *default* character.
- If an output list item is a pointer, this pointer must be currently associated with a target when the WRITE or PRINT statement is executed.

11.6.2.1   Implied-DO

An input/output list item in a READ, WRITE, or PRINT statement may be an *implied-DO*. Such an **implied-DO** is a shorthand form for the specification of several single list items.

(io_list, do_variable = first, last [, stride])

The **io_list** is an input/output list which is called the **range** of the implied-DO. The **do_variable** is a named scalar variable of type integer. The loop parameter expressions **first**, **last**, and **stride** are scalar expressions of type integer. They are evaluated and possibly converted to the kind type parameter of the DO variable.

The DO variable and the loop parameters have the same meaning as the corresponding entities in a DO statement. And loop control is executed as for a DO construct.

If an input/output list item is an implied-DO, each appearance of the DO variable within the range of the implied-DO is replaced by the current value of the DO variable. The list items in the range of an implied-DO are transferred once for each execution cycle of the loop.

```
      READ (*, 100, END=99) e, f, (i, y(i), z(i), i=1,10,2)
100   FORMAT (1X,2(F5.2,2X),5(I3,2X,2(F5.2,2X)))
```

This READ statement reads two values for e and f and 15 values for the implied-DO from the standard input file.

If the iteration count of an implied-DO is zero when the implied-DO is initiated, the implied-DO is a noneffective input/output list item which is ignored.

When an effective implied-DO is terminated in the normal way because its iteration count is finally zero, its DO variable retains its last value, that is, its value for the last execution cycle plus the value of its increment parameter. When an implied-DO is terminated before its iteration count is zero, the DO variable becomes undefined (in contrary to a DO construct).

Often, implied-DOs are used when only parts of arrays are to be transferred or when array elements are to be transferred in another order than array element order. Note that the transfer of whole arrays is most efficient when only the name of the array is specified as an input/output list item.

```
READ (5, 2000) a(1), (a(i), i=2,10,2), a(11)
```
is equivalent to
```
READ (5, 2000) a(1), a(2), a(4), a(6), a(8), a(10), a(11)
```

```
DIMENSION b (10, 2)
WRITE (*, *) (b(i, 2), i=1,10)
```

Only the second column of array b is transferred.

The operands of the loop parameter expressions of an implied-DO may appear in the range of the implied-DO. If they appear as list items in the range of an implied-DO in a READ statement, this does not affect the execution of the implied-DO.

```
l = 5
READ (77, *, END=100) (l, y(i), i=1,50+l)
```

The values read successively into variable l have no effect on the number of execution cycles of the implied-DO.

Nested implied-DOs

Implied-DOs may be nested; that is, a list item in the range of an implied-DO may be another implied-DO. Often, a nested implied-DO is used to transfer a multi-dimensional array.

```
DIMENSION b (3, 4, 5)
READ (5, 1000) (((b(i, j, k), i=1,3), j=1,4), k=3,5)
```

Values are read only for the planes 3, 4, and 5 of array b. These values are transferred column-wise.

If implied-DOs are nested, each of these implied-DOs must have its own DO variable. These DO variables must not be associated with each other.

During the execution of an implied-DO, the first (innerst) DO variable iterates most rapidly and the last (outest) one most slowly; that is, the innerst implied-DO is executed first.

```
DIMENSION f (5, 4, 7)
WRITE (11, 7000) (((f(i, j, k), k=1,7), j=1,4), i=1,5)
```

The entire array f is transferred. The array elements are not transferred in array element order, but they are transferred row-wise. This way of transferring complete arrays is correct but not efficient.

11.6.3 Data Transfer Statements

The READ statement is the data transfer *input* statement. The WRITE statement and the PRINT statement are the data transfer *output* statements. These three statements are used to perform all kinds of data transfer, as there are sequential, direct, formatted, unformatted, list-directed, namelist, nonadvancing, and internal input/output.

READ (control_information_list) [input_list]

READ format [, input_list]

WRITE (control_information_list) [output_list]

PRINT format [, output_list]

The READ and PRINT statements *without* a control information list are *sequential formatted* data transfer statements. If **format** is an asterisk *, they are *list-directed* data transfer statements. In any case, the standard units are used. Because such a statement cannot include an IOSTAT=, ERR=, EOF=, or EOR= specifier, the occurrence of an error condition, end-of-file condition, or end-of-record condition causes the program to be terminated.

The data transfer statements *with* a control information list may be used to perform input/output from and to standard units (and standard files) and to perform input/output from and to external files and internal files.

The control information list in a READ or WRITE statement *must* include at least the UNIT= specifier. If the UNIT= specifier is the first specifier in the control information list, the keyword UNIT= may be absent. The other specifiers *may* be written (not more than once), but they need not be specified. The appearance or nonappearance of the other input/output specifiers determines the kind of data transfer, the access method, the kind of error handling, of end-of-record handling, and of end-of-file handling to be performed by the READ or WRITE statement.

If the control information list includes the FMT= or the NML= specifier, the statement is a *formatted data transfer statement*; otherwise it is an *unformatted data transfer statement*. If the FMT= specifier **format** is an asterisk *, the statement is a *list-directed data transfer statement* (because formatting is list-directed).

If the FMT= or the NML= specifier is the second specifier in the control information list, if, in addition, the UNIT= specifier is the first specifier, and if this first specifier is written without the keyword UNIT=, then the FMT= specifier and the NML= specifier may be written without the keyword FMT= or NML=, respectively.

If the control information list includes a REC= specifier, the statement is a *direct* data transfer statement; otherwise it is a *sequential* data transfer statement. If the REC= specifier is present, this control information list must not include an additional END= specifier and the FMT= specifier **format** must not be an asterisk *.

If ADVANCE='NO' is specified, it is a *nonadvancing* data transfer statement. This kind of data transfer reads or writes a record (character-wise) continuously such that one record may be processed by several READ or WRITE statements.

If each data transfer statement includes either no ADVANCE= specifier or ADVANCE='YES', then the READ, WRITE, and PRINT statements, possibly except the first one, start transferring data at the beginning of each record. This also is true if the last READ statement did not transfer all data elements of the record. In this case, the remaining data elements are skipped if the file is opened for sequential access. Such a kind of sequential formatted input/output is called *advancing input/output*.

Normally, a data transfer statement reads, writes, or skips at most one record. Only a formatted input/output statement may process more than one record.

Data are transferred between the records in a file and the items in an input/output list or the members of a namelist group. If an input/output list item or a member of a namelist group is a subobject, its current position is determined immediately before the transfer of this subobject begins. And the loop parameters and the DO variable of an implied-DO are evaluated and defined, respectively, immediately before the transfer of the list items within the range of the implied-DO begins.

The execution of a data transfer statement terminates:

- When format control encounters a data edit descriptor for which there is no effective input/output list item left;
- When a namelist WRITE statement has written the values of all members of the specified namelist group;
- When an unformatted or list-directed data transfer has processed all items in the input/output list;
- When an error condition occurs;
- When the end-of-record condition or the end-of-file condition occurs; or
- When a slash "/" is encountered as a value separator in the record being read by a list-directed or namelist READ statement.

11.6.3.1 Formatted Input/Output

During formatted data transfer, data are transferred with editing, because, normally, their internal and external representations are different. There are special kinds of formatted input/output such as list-directed, namelist, nonadvancing, and internal input/output.

A formatted WRITE or PRINT statement writes data in character-oriented form to the file, and the formatted READ statement reads data in character-oriented form from the file. This character-oriented data might be read also by a human reader. The conversion (or editing) between the internal form and the external character-oriented form may be very expensive. This kind of input/output is usefull only if the data in the file are to be read by a human reader or are to be interchanged with another computer model.

READ (control_information_list) [input_list]
READ format [, input_list]
WRITE (control_information_list) [output_list]
PRINT format [, output_list]

The following input/output specifiers must be or may be present for formatted input/output: [UNIT =] **u**, [FMT =] **format**, [NML =] **group-name**, REC = **record_number**, ADVANCE=**yes_or_no**, EOR=**label**, IOSTAT=**status_variable**, ERR = **label**, END = **label**, and/or SIZE = **number_of_chars**. The UNIT= specifier **u** and either the FMT= specifier or the NML= specifier *must* be present; all other specifiers *may* be present.

Statements without a control information list

The READ and PRINT statements *without* a control information list are *sequential formatted* data transfer statements. They use the standard units. If **format** is an asterisk ∗, they are *list-directed* data transfer statements.

Statements with a control information list

The kind of formatted input/output is given by the presence of suitable input/output specifiers. The following table shows which specifier(s) *must* be present, which *may* be present, and which must *not* be present:

Formatted io:	UNIT	FMT	NML	REC	ADVANCE	EOR	IOSTAT	ERR	END	SIZE
list-directed	must	asterisk	not	not	not	not	may	may	5)	not
namelist	1)	not	must	not	not	not	may	may	5)	not
nonadvancing	2)	must	not	not	'NO'	5)	may	may	5)	5)
internal	3)	must	not	not	not	not	may	may	5)	not
direct	1)	4)	not	must	not	not	may	may	not	not
advancing	must	must	not	not	6)	not	may	may	may	not

Notes:
1) must; but no character variable
2) must; but no asterisk * and no character variable
3) character variable
4) must; but no asterisk *
5) may; but only in READ statement
6) may; but only with 'YES'

```
CHARACTER text *80, word *8;   DIMENSION s (10), y (100), word (10)
NAMELIST /measurement/ x, n
READ (10, FMT=1000) sort, area, quantity   ! sequ. form. input
PRINT *, a, b, c                            ! list-directed output
READ (17, NML=measurement, END=20)          ! namelist input
WRITE (23, 2222, ADVANCE='NO') l, b, z      ! nonadvancing output
READ (text, '(10A)') word                   ! internal input
WRITE (22, '(10I5,100F7.2)', REC=5) s, y    ! direct form. output
```

If the ADVANCE= specifier is absent, a formatted data transfer statement processes one or more records and transfers data (according to a format specification) between the items of the input/output list or the members of the namelist group and the specified unit, which is connected to a file. If ADVANCE='NO' is present, a single record may be processed character-wise by the execution of more than one data transfer statement.

If a formatted direct access data transfer statement processes more than one record, the record number is increased by one as each succeeding record is read or written.

The input/output list items must match the corresponding data edit descriptors. If a file is opened for unformatted input/output, formatted input/output is not allowed.

Formatted input
(*without* list-directed and namelist formatting)

An input list item (or an associated variable) must not contain any part of the format specification.

If the input list and the corresponding format specification attempt to read more data from the record than the record contains, padding characters (i. e., blanks) are supplied for the missing characters if this happens during formatted input from an internal file or during formatted input from an external file which is opened with PAD='YES' or which is opened without an explicitly specified PAD= specifier. If an external file is opened with PAD='NO', the end-of-record condition occurs when a nonadvancing READ statement attempts to read beyond the end of the record; but in the case of an advancing READ statement, the input list must match the format specification such that the READ statement does not attempt to read more characters than the record contains.

If the record contains characters of a nondefault character type, these characters may be transferred as character values only for an input list item of this nondefault character type.

When a sequential READ statement encounters the endfile record in an external file or when it attempts to read beyond the end of an internal file, the execution of the program is terminated if neither an END= specifier nor an IOSTAT= specifier are present in the READ statement.

Formatted output
(*without* list-directed and namelist formatting)

If the characters specified by the output list and the format of a direct or internal WRITE statement do not fill the record, the rest of the record is filled with blanks.

If the record should contain characters of a nondefault character type, these characters may be written as the values only of output list items of this nondefault character type.

In the case of an external file, the output list must match the format specification such that the output statement does not attempt to write more characters in the record than have been specified by the RECL= specifier in the OPEN statement.

And an internal WRITE statement must not attempt to write more characters in the record than the record length of the internal file.

```
DIMENSION tree (10), y (100)
WRITE (22, '(10I5,100F7.2)', REC=152) tree, y
```

The WRITE statement writes (with direct access and controlled by the specified format) the 10 values of array `tree` and the 100 values of array `y` into the record with record number 152.

11.6.3.2 Unformatted Input/Output

An unformatted data transfer statement does not need a format specification, because the data are transferred in their internal binary form from and to external files. That is, there is no difference between the internal and the external representations of the data.

Because unformatted input/output is simpler and cheaper than formatted input/output, this kind of data transfer is usefull for intermediate or temporary files. Usually, unformatted files created by one computer cannot be processed by another computer model.

READ (control_information_list) [input_list]

WRITE (control_information_list) [output_list]

The following input/output specifiers must be or may be present for unformatted input/output: [UNIT =] **u**, REC = **record_number**, IOSTAT = **status_variable**, ERR = **label**, and/or END = **label**. The UNIT= specifier **u** *must* be present; all other specifiers *may* be present.

The kind of unformatted input/output is given by the presence of suitable input/output specifiers. The following table shows which specifier(s) *must* be present, which *may* be present, and which must *not* be present:

Unformatted io:	UNIT	FMT	NML	REC	ADVANCE	EOR	IOSTAT	ERR	END	SIZE
sequential	1)	not	not	not		not	not	may	may	2)	not
direct	1)	not	not	must		not	not	may	may	not	not
Notes:	1) must; but no asterisk * and no character variable									
	2) may; but only in READ statement									

```
DIMENSION x (100), z (20, 10)
READ (12, END=120) n, (x(i), i=1,n)     ! sequ. unform. input
WRITE (UNIT=47, REC=12, ERR=140) fe, z  ! direct unform. output
```

If a file is opened for formatted input/output, unformatted data transfer statements are not allowed for that file.

Unformatted input

An unformatted READ statement may read only such records which have been written before by unformatted WRITE statements.

Each value in the input record must be of the same type as the corresponding input list item, except two real values may correspond to one complex input list item, and one complex value may correspond to two real input list items. The kind type parameters of the value and of the corresponding list item must

11.6 Input/Output Statements

agree. In the case of character data, the character lengths of the value and of the corresponding list item also must agree.

The number of values required by the input list must be less than or equal to the number of values in the record. If the input list requires fewer values than the record contains, the rest of the values is ignored.

```
DIMENSION x (100)
READ (12, END=120, IOSTAT=ista) n, (x(i), i=1,n)
```

The unformatted READ statement reads with sequential access the file which is connected to unit 12. A value is read in the scalar integer variable n. Then n values are transferred into the first n elements of array x. If the endfile record is encountered, program execution continues with statement 120. After the execution of the READ statement, the variable ista becomes defined with a status value.

Unformatted output

A direct access WRITE statement must not transfer more values into the record than have been specified by the RECL= specifier in the OPEN statement.

If the file is opened for sequential access and if the maximum record length is specified by the RECL= specifier in the OPEN statement, then a WRITE statement must not attempt to transfer more values than have been specified by the RECL= specifier.

If an unformatted direct access WRITE statement writes fewer values than the record length, the rest of the values in the record is undefined. If the output list is absent, an empty record is written.

```
DIMENSION z (20, 10)
WRITE (UNIT=47, REC=12, ERR=140) fe, z
```

The unformatted WRITE statement writes (with direct access to the file which is connected to unit 47) the value of the variable fe and the 200 values of array z into the record with the record number 12. When an error condition occurs, the execution of the program continues with statement 140.

11.6.3.3 List-Directed Input/Output

List-directed input/output is a special kind of sequential formatted input/output from and to external files or internal files. The correct term is *formatted input/ output with list-directed formatting*. Though formatted records are processed, the Fortran programmer need not and cannot specify a format, because, depending on the type of the input/output list items, certain internal predefined edit descriptors are used.

READ (control_information_list) [input_list]

READ * [**, input_list**]

WRITE (control_information_list) [output_list]

PRINT * [**, output_list**]

The following input/output specifiers must be or may be present for list-directed input/output: [UNIT =] **u**, [FMT =] *, IOSTAT = **status_variable**, ERR = **label** and/or END = **label**. The UNIT= specifier **u** and the FMT= specifier *must* be present; all other specifiers *may* be present.

Statements without a control information list

The READ statement and PRINT statement *without* a control information list use the standard units.

Statements with a control information list

The kind of list-directed input/output is given by the presence of suitable input/output specifiers. The following table shows which specifier(s) *must* be present, which *may* be present, and which must *not* be present:

List-directed io:	UNIT	FMT	NML	REC	ADVANCE	EOR	IOSTAT	ERR	END	SIZE
external	1)	asterisk	not	not	not	not	may	may	3)	not
internal	2)	asterisk	not	not	not	not	may	may	3)	not
Notes:	\multicolumn{10}{l}{1) must; but no character variable}

Notes:
1) must; but no character variable
2) character variable
3) may; but only in READ statement

```
CHARACTER address *40
READ (*, FMT=*, END=100) current, field, voltage
READ *, current, field, voltage
WRITE (UNIT=*, FMT=*, ERR=120, IOSTAT=ivar) alpha, beta, gamma
PRINT *, alpha, beta, gamma
```

List-directed input

The order of the data in the input record may be almost arbitrary and is not subject to any specifications (like edit descriptors in a FORMAT statement). An input record contains a sequence of values, which are separated by *value separators*.

A **value separator** is one of the following:

- One or more contiguous blanks, except one of the two values is a *null value* (see below) without a repeat factor. A sequence of two or more blanks is treated as a single blank;
- A comma optionally enclosed in one or more blanks; or
- A slash optionally enclosed in one or more blanks.

A list-directed READ statement may read more than one record. The end of a record has the same effect as a blank character, unless a character value is continued to the next record. That is, if more values are to be transferred than fit in the record, the rest of the values may be continued to the next record. Normally, a record must contain complete values, but in the case of a complex value or a character value, the end of the record may appear within the value.

Blanks are not interpreted as zeros. Because blanks are used as value separators, they must not appear within values, except significant blanks in character values and insignificant blanks in complex values.

There are some rules governing the form of the external representation of the input values. Generally (with few exceptions), input values which can be read with explicit formatting can also be read with list-directed formatting. The form of an input value must be acceptable by the next effective input list item.

A value in the input record is either an *integer value*, a *real value*, a *double precision real value*, a *complex value*, a *logical value*, a *character value*, a *null value*, or such a value preceded by a repeat factor (see below):

Integer value: An integer value must have the same form as an integer literal constant without an explicitly specified kind type parameter. The corresponding effective input list item must be of type integer.

Real value: A real value may have the same form as a real literal constant without an explicitly specified kind type parameter. It also may have the same form as an integer value; then the input value is interpreted as the significand of a real value with no fractional digits. The corresponding effective input list item must be of type real or double precision real.

Double precision real value: A double precision real value must have the same form as a double precision real literal constant. The corresponding effective input list item must be of type real or double precision real.

Complex value: A complex value must have the same form as a complex literal constant. Precisely, it consists of two real values, which are separated by a comma and which are enclosed in parentheses. The first value is the real part and the second value is the imaginary part of the complex value. The parentheses are not interpreted as separators. Each of the two values may be enclosed in blanks. The end of the record may occur between the real part and the comma or between the comma and the imaginary part. The corresponding effective input list item must be of type complex.

Logical value: A logical value may have the same form as a logical literal constant without an explicitly specified kind type parameter. Allowed are: an optional point, followed by the letter T or F, followed by any characters except slashes, blanks, equals, and commas. The corresponding effective input list item must be of type logical.

Character value: A character value may have the same form as a character literal constant without an explicitly specified kind type parameter. If the apostrophe-delimited or quote-delimited character value includes a doubled apostrophe or doubled quotation mark, respectively, these may not be separated by blanks or by the end of the record. The character value has implicitly the same kind type parameter value as the corresponding effective input list item.

An apostrophe-delimited or quote-delimited character value may contain blanks, commas, and slashes.

Character values may be continued from the end of one record to the beginning of the next record. If the character value does not have the same length as the input list item, the effect of the data transfer is as though the character value were assigned to the list item in a character intrinsic assignment statement.

If the corresponding effective input list item is of type default character and:

1. The character value does not contain a blank, a comma, or a slash; and
2. The character value is not continued to the next record(s); and
3. The first nonblank character is neither an apostrophe nor a quotation mark; and
4. The leading characters are no digits followed by an asterisk *; and
5. The character value contains at least one character,

then the delimiting apostrophes or quotation marks are not required. In this case, the character value is terminated by the first blank, comma, slash, or end of record; and apostrophes and quotation marks within the character value are *not* to be doubled.

Null value: A null value is represented either by two consecutive value separators without intervening blanks, or by no characters appearing before the first value separator in the first record read by a list-directed READ statement, or by the form **n**∗ (see below).

The rest of a record after the last comma or slash, with or without blanks, is not interpreted as a null value. A null value has no effect on the definition status or the value of the corresponding effective input list item. A null value must not be

11.6 Input/Output Statements

used as the imaginary part or as the real part of a complex value, but a single null value may be used as an entire complex value. The effective input list item corresponding to a null value may be of any type.

Other possibilities of forming an input record:

Repeat factor: The input values may specify a *repeat factor* **n**, which must be an unsigned integer literal constant without an explicitly specified kind type parameter. The following forms are allowed:

n∗   is the repetition of the null value. This form is the same as **n** successive appearances of the null value.

n∗c   is the same as **n** appearances of the integer, real, double precision real, complex, logical, or character value **c**. Embedded blanks are not allowed except where permitted in **c**. **c** has implicitly the same kind type parameter value as the corresponding effective input list item.

Slash: When during the execution of a list-directed READ statement a *slash* as a value separator is encountered, the execution of this READ statement terminates after the transfer of the previous value. Any remaining characters in the input record are ignored. The effect is as though null values were transferred to those input list items (if any) which have not been processed. Therefore, a DO variable of an implied-DO in the input list is defined as though enough null values were transferred for the remaining input list items.

```
  50 READ (*, *, END=100) l, m
     WRITE (9, 44) l, m
  44 FORMAT (I5,I6)
     GO TO 50
 100 CONTINUE
```

Input records: 5 70
                    ,   1
                    5*2
                    ,,
                    /   7
                    12345678901   ← record position

Output records: 5   70
                    5   1
                    2   2
                    2   2
                    2   2
                    12345678901   ← record position

List-directed output

A list-directed WRITE or PRINT statement writes the values of the output list items in principle in a form to the output record that corresponds to the form of the input values. The **Fortran** processor uses its own (sensible) internal edit descriptors.

Character values written to an internal file or to an external file which is opened without a DELIM= specifier or with DELIM='NONE' are not delimited by apostrophes or quotation marks.

The programmer has nearly no influence on the form of the external representation of the output data and on the form of the output record. Therefore, this kind of output is usefull mainly during the test phase of the program development cycle.

11.6.3.4 Internal Input/Output

Internal input/output is a special kind of sequential formatted input/output either with explicit formatting or with list-directed formatting. Internal input/output supports the transfer of data from one place in internal storage to another place in internal storage without involvement of an input/output device or an external file.

During internal data transfer, the representation of the data may be changed. Depending on the kind of the internal input/output, the kind of data conversion and the "external" form of the data is explicitly specified by a format specification or is implicitly specified by the **Fortran** processor. An *internal file* is used instead of an external file.

READ (control_information_list) [input_list]

WRITE (control_information_list) [output_list]

The following input/output specifiers must be or may be present for internal input/output : [UNIT =] **u**, [FMT =] **format**, IOSTAT = **status_variable**, ERR = **label**, and/or END = **label**. The UNIT= specifier **u** and the FMT= specifier *must* be present; all other specifiers *may* be present.

The kind of internal input/output is given by the presence of suitable input/output specifiers. The following table shows which specifier(s) *must* be present, which *may* be present, and which must *not* be present:

11.6 Input/Output Statements

Internal io:	UNIT	FMT	NML	REC	ADVANCE	EOR	IOSTAT	ERR	END	SIZE
formatted	1)	must	not	not	not	not	may	may	2)	not
list-directed	1)	asterisk	not	not	not	not	may	may	2)	not
Notes:	1) character variable of type default character									
	2) may; but only in READ statement									

```
CHARACTER 140 *40, ad80 *80
READ (140, '(I8,6X,E8.2,E10.4,I8)', END=100) max, r_min, sel, nz
WRITE (ad80, FMT=*, IOSTAT=k_stat) value, 'no instruction', year
```

Internal file

An **internal file** is a variable of type *default* character, which is specified by the UNIT= specifier **u**. If the internal file is a whole array (that is, if **u** is an array name), it must not be an assumed-size array. If it is an array section, it must not have a vector subscript.

Internal files must not and cannot be opened explicitly or implicitly. If an internal file is an allocatable array which is currently not allocated, or if the internal file is a subobject of such an array, or if the internal file is a pointer which is currently not associated with a target, then the internal file is inaccessible and must not be involved in data transfers.

Record: A record of an internal file is a *scalar* character object. If the internal file is a scalar variable, the internal file consists of a single record with a record length which is given by the length of the character variable.

If the internal file is an *array* (that is, a whole array or an array section), the internal file is treated as a sequence of array elements in array element order, and each array element is a record of the internal file. In this case, all records of the internal file have the same record length, which is the character length of the single array elements.

A record of an internal file must be read only if it is defined.

Prior to data transfer, an internal file is always positioned before the first character position of its first record.

Input/output statements: File positioning statements (BACKSPACE, REWIND, ENDFILE) and file status statements (OPEN, CLOSE, INQUIRE) are not applicable to internal files.

Input/output list: If the input/output list includes character items, they must be of type *default* character. And an input/output list item must neither be a part of the internal file nor be associated with the internal file.

Internal input

Beginning with the first character position of the internal file, data are read and transferred to the input list items under control of the supplied format specification or under control of list-directed formatting. During internal input, the data may be converted from the external representation to an internal representation.

Blanks: During list-directed formatting, blanks are normally treated as value separators. Under control of an explicit format specification, blanks are treated in the same way as in the case of an external file that is opened with BLANK='NULL'.

Padding characters: If an internal READ attempts to read more characters than the record contains, missing characters are treated in the same way as in the case of an external file that is opened with PAD='YES'.

```
CHARACTER line *40
REAL price, kg
INTEGER weight
DATA line /'15.00 $  for  750 g,   20.00 $ per kg'/
 :
READ (line, FMT='(F5.2,8X,I4,6X,F5.2)') price, weight, kg
```

The internal file `line` contains one record with record length 40. After execution of the READ statement, the variables `price`, `weight`, and `kg` are defined such:

    price = 15.   weight = 750   kg = 20.

Internal output

The values of the output list items are written to the internal file under control of the specified format specification or under control of list-directed formatting. During internal data transfer, the data may be converted from the internal representation to an external representation. The data transfer begins always at the first character position of the file, even if the file has more than one record.

If the number of output characters is less than the record length, the rest of the record is filled with blanks. The number of output characters must not exceed the record length.

During list-directed formatting, character values are written without delimiters in the same way as in the case of an external file that is opened with DELIM='NONE'.

The format specification for the internal WRITE statement must not be contained in the internal file or in an associated data object.

11.6 Input/Output Statements   11-33

```
      CHARACTER (LEN=3) digits (8), a, b, c
      DATA digits /8*'***'/
      DATA a, b, c /'uno', 'due', 'tre'/
      WRITE (digits, FMT=11) a, b, c
   11 FORMAT (A/)
```

The internal file digits contains 8 records with record length 4. After execution of the WRITE statement, these 8 records are defined as follows:

bafw(1) = 'uno'   bafw(2) = ␣␣␣   bafw(3) = 'due'   bafw(4) = ␣␣␣
bafw(5) = 'tre'   bafw(6) = ␣␣␣   bafw(7) = ***   bafw(8) = ***

When format control for an internal (or direct) WRITE statement encounters the characters "/" in the format specification, a new record is defined (created) which is filled with blanks.

11.6.3.5 Namelist Input/Output

Namelist input/output is a special kind of sequential formatted input/output from and to external files. The correct term is *formatted input/output with namelist formatting*.

No input/output lists but *group-names* of namelists of variables are specified in the data transfer statements for namelist input/output. The group-name and the variables that are members of such a namelist group must be specified in a NAMELIST specification statement. This specification must be available in the scoping unit with the namelist data transfer statement. The specification of such a namelist group must not include the names of variables whose shape, size, or length become not determined until the program is executed.

Though formatted records are processed, the Fortran programmer need not and cannot specify a format, because, depending on the type of each group object, particular internal predefined edit descriptors are used.

READ (control_information_list)

WRITE (control_information_list)

The following input/output specifiers must be or may be present for namelist input/output: [UNIT =] **u**, [NML =] **group-name**, IOSTAT = **status_variable**, ERR = **label**, and/or END = **label**. The UNIT= specifier **u** and the NML= specifier *must* be present; all other specifiers *may* be present.

Namelist input/output is given by the presence of suitable input/output specifiers. The following table shows which specifier(s) *must* be present, which *may* be present, and which must *not* be present:

Namelist io:	UNIT	FMT	NML	REC	ADVANCE	EOR	IOSTAT	ERR	END	SIZE
external	1)	not	must	not	not	not	may	may	2)	not
Notes:	1) must; but no character variable									
	2) may; but only in READ statement									

```
DIMENSION no (5)
INTEGER one, two, three, four
NAMELIST /store/ one, two, three, four
NAMELIST /date/ year, month, day
NAMELIST /start/ no
 ⋮
DATA no /7, 2, 1, 5, 4/
READ (17, NML=store)
READ (17, date)
WRITE (*, date)
WRITE (UNIT=*, NML=start)
```

Input records:

```
&store one=1500, three=100, two=100,  ! Jan to Mar 1996
      four=1200 /
&date year=40, month=10, day=11 /
```

The first READ statement reads the next record from the file which is connected to unit 17. The first nonblank characters of this record must be &store. And the second READ statement reads the record that begins with &date. The WRITE statements writes the following lines to the standard output file (which may be the screen):

```
&date day=11,month=10,year=40 /
&start no=7,2,1,5,4 /
```

Namelist input

The order of the data in the input record may be almost arbitrary and is not subject to any specifications (like edit descriptors in a FORMAT statement). An input record contains a sequence of *name-value pairs*, each of which consists of a variable name or variable designator, followed by an equals, and followed by one or more values separated by value separators. These variables being contained in the input records play a similar role as the input list items for the other kinds of input/output.

When a namelist READ statement is executed, not only the READ statement but also the input data determine which variables are to be (re)defined.

11.6 Input/Output Statements

A namelist READ statement may read more than one record. A READ statement reads until values are transferred to each variable in the namelist group or until a slash is encountered; the remainder of the record is ignored and need not follow the rules for namelist input values. Apart from namelist comments (see below), the first nonblank record must begin with an ampersand followed by the same group-name as specified in the READ statement, and the last of these records ends with a slash. Between them are the data forming a sequence of **name-value pairs**; the form of each name-value pair is either *variable = value* or *variable = value-list*.

The form of a namelist input record is:

&group-name [variable = value [, variable = value]...] /

Each **variable** is a *group object*, that is, the name of a variable or a subobject designator of a variable in the namelist group **group-name**. And each **value** is an input value or a sequence of input values.

The name-value pairs must be separated by *value separators*. **Value separators** are defined as for list-directed formatting (11.6.3.3). For the sake of simplicity, we use a comma. Within input records, *namelist comments* (see below) are allowed to occur.

Not all variables in a namelist group need to appear in the sequence of name-value pairs. And a particular variable may appear in more than one name-value pair. The order of the name-value pairs is significant only if a variable appears on the left-hand side of more than one equals. The sequence of name-value pairs is processed in left-to-right order.

Group-name: A **group-name** appearing in the data must be specified in a NAMELIST specification statement which is available in the same scoping unit as the data transfer statement. If the Fortran processor supports upper-case and lower-case letters, the group-name in the data is without regard to case.

Group objects: If the Fortran processor supports upper-case and lower-case letters, the variable on the left-hand side of the equals is without regard to case. If the variable is a subobject of a variable in a namelist group, then subscript expressions in array element designators or array section designators, and substring range expressions must be optionally signed integer literal constants without explicitly specified kind type parameters. Vector subscripts for the designation of array sections are not allowed.

The variable on the left-hand side of the equals must be **effective**; that is, it must not be a zero-size whole array or a zero-size array section, and it must not be a zero-length named character variable or a zero-length character substring.

Input values

If the variable on the left-hand side of the equals is a scalar variable of an intrinsic type, the equals must be followed by a single value (which may be a null value). This value must match the corresponding variable.

If the variable on the left-hand side of the equals is an array object or a derived type variable, the equals may be followed by a sequence of values separated by value separators. The variable is treated as though it were expanded into a sequence of (scalar) array elements or structure components of intrinsic type. And the values on the right-hand side of the equals, which must match the variables in the expanded variable list, are transferred in left-to-right order to the items in the expanded variable list.

The elements of an array are processed in array element order. The components of a derived type variable are processed in the order of the type components in the derived type definition.

The number of values in a value sequence may be less than or equal to but not greater than the number of variables in the expanded variable list on the left-hand side of the equals. If the number of values is less than the number of variables, the remaining variables in the expanded variable list are *not* redefined.

All values following an equals are transferred before the next name-value pair is processed.

The end of a record has the same effect as a blank character, unless a character value is continued to the next record. If more values are to be transferred than fit in the record, the rest of the values may be continued to the next record. Normally, a record may contain only complete values, but in the case of a complex value or a character value, the end of the record may appear within the value. A namelist READ statement may read more than one record until a slash in encountered; the remainder of the last record after the slash is ignored. The first record for the next namelist READ statement must begin with optional blanks followed by an ampersand and the group-name.

If no value is transferred for a variable of a namelist group or for a subobject of such a variable, the definition status of this variable or subobject is not modified.

There are some rules governing the form of the external representation of the input values. Generally (with few exceptions), input values which can be read with explicit formatting can also be read with namelist formatting.

Each value of the input record is either an *integer value*, a *real value*, a *double precision real value*, a *complex value*, a *logical value*, a *character value*, a *null value*, or such a value preceded by a repeat factor (see below):

11.6 Input/Output Statements

Integer value: Defined as for list-directed input. The corresponding effective variable must be of type integer.

Real value: Defined as for list-directed input. The corresponding effective variable must be of type real or double precision real.

Double precision real: Defined as for list-directed input. The corresponding effective variable must be of type real or double precision real.

Complex value: Defined as for list-directed input. The corresponding effective variable must be of type complex.

Logical value: Defined as for list-directed input. The corresponding effective variable must be of type logical.

Character value: Defined as for list-directed input except that character values must always be delimited by apostrophes or quotation marks.

Null value: A null value is represented either by two consecutive nonblank value separators without intervening blanks, by zero or more blanks between an equals and the next value separator, by blanks between two consecutive value separators following an equals, or by the form **n∗** (see below: "Repeat Factor"). The rest of a record after the last comma or slash, with or without blanks, is not interpreted as a null value. A null value has no effect on the definition status and the value of the corresponding effective variable. A null value must not be used as the imaginary part or as the real part of a complex value, but a single null value may be used as an entire complex value. The effective variable corresponding to a null value may be of any type.

Other possibilities of forming an input record:

Repeat factor: Defined as for list-directed input.

Slash: When during the execution of a namelist READ statement a *slash* as a value separator is encountered, the execution of this READ statement terminates after the transfer of the previous value. Any remaining characters in the input record are ignored. If the previous variable is an array or a derived type variable, the remaining array elements or components, respectively, retain their definition status. The effect is as though null values were transferred to those array elements or components (if any) which have not been processed.

Namelist comments: Except within a character literal constant, a "!" character after a value separator or in the first nonblank position of an input record initiates a namelist comment. The comment extends to the end of the current input record and may contain any graphic character in the processor-dependent character set.

Blanks within namelist input records must be present, may be present, or must be absent. The following cases apply:

- Zero, one, or more blanks before the leading ampersand &;
- No blank between the ampersand & and the group-name;
- No blank within a group-name;
- One or more blanks after a group-name;
- Zero, one, or more blanks before and/or after a variable;
- No blank within a variable;
- Zero, one, or more blanks before and/or after the equals;
- One or more blanks as value separators;
- No blank within an input value except significant blanks in character values and insignificant blanks in complex values;
- No blank before and after the asterisk * if the value is written with a repeat factor; and
- Zero, one, or more blanks within namelist comments.

Namelist output

A namelist WRITE statement writes the values of the namelist group items in principle in a form in the output record that corresponds to the form of the input values. The Fortran processor uses its own (sensible) internal edit descriptors. The programmer has nearly no influence on the form of the external representation of the output data and on the form of the output record.

Character values written to an external file which is opened without a DELIM= specifier or with DELIM='NONE' are not delimited by apostrophes or quotation marks.

11.6.3.6 Nonadvancing Input/Output

Nonadvancing input/output is a special kind of sequential formatted input/output from and to external files with explicit formatting. List-directed formatting, namelist formatting, and internal files are not supported. Nonadvancing input/output transfers the characters of a record (character-wise) continuously such that one record may be processed by several data transfer statements. Nonadvancing input supports processing of records of varying (unknown) lengths.

READ (control_information_list) [input_list]

WRITE (control_information_list) [output_list]

11.6 Input/Output Statements

The following input/output specifiers must be or may be present for nonadvancing input/output: [UNIT =] **u**, [FMT =] **format**, ADVANCE = **yes_or_no**, EOR = **label**, IOSTAT = **status_variable**, ERR = **label**, END = **label**, and/or SIZE = **number_of_chars**. The UNIT= specifier **u**, the FMT= specifier, and the ADVANCE= specifier *must* be present; all other specifiers *may* be present.

Nonadvancing input/output is given by the presence of suitable input/output specifiers. The following table shows which specifier(s) *must* be present, which *may* be present, and which must *not* be present:

Nonadvancing io:	UNIT	FMT	NML	REC	ADVANCE	EOR	IOSTAT	ERR	END	SIZE
formatted	must	1)	not	not	2)	3)	may	may	3)	3)
Notes:	1) must; but no asterisk * and no character variable										
	2) must; **yes_or_no** must be 'NO'										
	3) may; but only in READ statement										

```
READ (14, 1111, EOR=140, ADVANCE='NO') byte, m    ! nonadv. input
WRITE (18, '(I1)', ADVANCE='NO') n                ! nonadv. output
```

Nonadvancing input/output differs from the "advancing" forms of input/output with regard to the way the file is positioned after the data transfer. After nonadvancing data transfer, the file position is frozen and there is *no* automatic positioning after the record just read or written. If no end-of-record condition, no error condition, and no end-of-file condition occur, the file remains positioned within the current record. Then the next data transfer statement may continue to process the current record at this position.

When a slash edit descriptor is encountered or when format control reverts, the file is positioned in the usual manner as for the advancing forms of input/output.

The X, T, TL, and TR edit descriptors are processed in the same manner as for advancing input/output, except that the file must not be positioned to the left of the *left tab limit*.

If the file is opened for formatted input/output, advancing and nonadvancing data transfer statements may be executed in any order.

Nonadvancing input

In the following, we assume that neither an error condition nor the end-of-file condition occurs.

If during the execution of a nonadvancing data transfer statement no end-of-record conditon occurs, there is a current record after the data transfer, and the record is positioned within this current record just after the last transferred character. The next READ statement begins reading at this position.

If there is no current record, the next READ statement begins reading at the first character position of the next record.

If during the execution of a nonadvancing READ statement the end-of-record condition occurs, the execution of the READ statement terminates. The already transferred input list items do *not* become undefined. But the DO variables of all implied-DOs in the input list become undefined. And depending on the absence or presence of the EOR= specifier or IOSTAT= specifier, the execution of the program is terminated or continued.

A nonadvancing READ statement may include the SIZE= specifier. After the execution of the READ statement, this SIZE= specifier is defined with the count of characters which have been transferred under control of data edit descriptors.

```
CHARACTER word *8
INTEGER w1
READ (9, '(A)', EOR=100, SIZE=w1, ADVANCE='NO') word
```

The nonadvancing READ statement reads a character value from the file which is connected to unit 9. If the end-of-record condition occurs, program execution continues with statement 100. If no end-of-record condition occurs, the variable w1 is redefined with the value 8, because exactly 8 characters have been transferred. But if the end-of-record condition occurs, the variable w1 is redefined with the number of characters which have been transferred before the READ statement has encountered the end of the record.

Nonadvancing output

In the following, we assume that no error condition occurs.

If during the execution of a nonadvancing data transfer statement no end-of-record conditon occurs, there is a current record after the data transfer, and the record is positioned within this current record just after the last transferred character. The next WRITE statement begins writing at this position.

If there is no current record, the next WRITE statement begins writing at the first character position of the next record.

```
      INTEGER, DIMENSION (10) :: nu, m
      ⋮
40    WRITE (UNIT=47, FMT='(2I10)', ADVANCE='NO') nu
60    WRITE (UNIT=47, FMT='(2I10)', ADVANCE='NO') m(1), m(3), m(5)
80    WRITE (UNIT=47, FMT='(2I10)', ADVANCE='YES') m(7), m(9)
99    WRITE (47, 1111, ADVANCE='NO') m(2), m(4), m(6), m(8)
```

11.6 Input/Output Statements

The nonadvancing WRITE statement 40 transfers 5 records, each containing 20 characters, to the file which is connected to unit 47. The 5th record is the current record after the execution of statement 40; that is, the file is positioned within this record. The statement 60 first extends the current record by 20 characters and then writes additional 10 characters in the next record. This 6th record is now the current record. The advancing WRITE statement 80 extends the current record by additional 20 characters and terminates the record. The nonadvancing WRITE statement 99 creates the first characters of the 7th record.

11.6.3.7 Printing

Sequential formatted output to particular (processor-dependent) predefined output devices is called **printing**. The first character of each output record is not printed, but it is used as a *carriage control character* for the output device. The remaining characters of the record are printed beginning at the first character position of the printed *line*.

If a formatted record will be written but never be printed, the record need not contain a carriage control character. But if it still contains a carriage control character, this character belongs to the data and is written as the first character of the data.

An empty record is printed as a new line containing only blanks.

The first character of *each* printed record must be a carriage control character. This is necessary even if a single PRINT or WRITE statement prints more than one record when format control reverts or when format control encounters a slash edit descriptor.

The **carriage control character** must be of type *default* character. It controls the vertical spacing before printing:

Character	Vertical spacing before printing
Blank	One line.
+	No advance.
1	To first line of next page.
0	Two lines; i.e., print one line with blanks, then advance one line.

The carriage control character + supports printing of more than one record in one line.

Carriage control characters may be created by different methods.

```
   11 FORMAT ('+','result = ',F10.4,I15)
   33 FORMAT ('+result = ',F10.4,I15)
      CHARACTER ccc
      ccc = '+'
      WRITE (*, FMT=44, END=100) ccc, x, i
   44 FORMAT (A,F10.4,I15)
```

11.6.4 File Status Statements

There are three file status statements: OPEN, CLOSE, and INQUIRE. They are used to specify properties (attributes) of external files and units, to modify certain properties of external files and units, and to inquire about properties of external files and units.

11.6.4.1 OPEN Statement

The OPEN statement may be used to **open** an external file, that is, to connect the file to a unit. An existing external file may be connected to a unit, a nonexisting external file may be created and then connected to a unit, an external file may be created which is preconnected to a unit, or particular properties of a connection between an external file and a unit may be specified or modified.

Most of the optional specifiers that may be supplied in an OPEN statement accept character expressions. These character expressions must be *scalar* expressions of type *default* character. Their valid results are described in the following. Trailing blanks are allowed but are ignored. Lower-case letters are allowed if the **Fortran** processor supports them. Some specifiers have (implicitly specified) default values if they are not explicitly specified.

OPEN ([UNIT =] u [, IOSTAT = iostat] [, ERR = err] [, FILE = file]
   [, STATUS = status] [, ACCESS = access] [, FORM = form]
   [, RECL = recl] [, BLANK = blank] [, POSITION = position]
   [, ACTION = action]   [, DELIM = delim]   [, PAD = pad])

u   is a scalar nonnegative integer expression specifying the number of the unit which is connected or is to be connected to the file being opened.

iostat is a scalar status variable of type default integer which will become defined with a zero value if no error condition occurs during the execution of this file status statement; otherwise it becomes defined with a processor-dependent positive error status value.

11.6 Input/Output Statements

err is the statement label of the branch target statement (in the same scoping unit) which will be executed when an error condition occurs during the execution of this file status statement.

file is a character expression specifying the name of the file being opened. This file becomes connected to unit **u**. The form of the file name is processor-dependent.

status is a character expression specifying the status of the file being connected to unit **u**.

- **OLD** The file exists already.
- **NEW** The file does not yet exist.
- **SCRATCH** The file is to be deleted at the execution of a CLOSE statement for this unit or at the termination of the program.
- **REPLACE** If the file does not exist, it is created. If it exists, it is deleted and a new file is created.
- **UNKNOWN** The status is processor-dependent.

Default: **STATUS = 'UNKNOWN'**.

access is a character expression specifying the access method.

- **SEQUENTIAL** The file is being opened for sequential access.
- **DIRECT** The file is being opened for direct access.

Default: **ACCESS = 'SEQUENTIAL'**.

form is a character expression specifying the form of the file.

- **FORMATTED** The file is being opened for formatted input/output.
- **UNFORMATTED** The file is being opened for unformatted input/output.

Default: **FORM = 'FORMATTED'**   if access is sequential and
FORM = 'UNFORMATTED' if access is direct.

recl is a scalar positive integer expression specifying either the record length in a file being connected for direct access or the maximum record length in a file being connected for sequential access.

blank is a scalar character expression specifying the interpretation of blanks in numeric fields for formatted input.

- **NULL** Blanks are ignored, but a field of all blanks is interpreted as the value zero.
- **ZERO** Blanks other than leading blanks are treated as zeros.

Default: **BLANK = 'NULL'**.

position   is a character expression specifying the position of the (existing sequential) file immediately after opening.

    REWIND   The file is positioned at its initial point.

    APPEND   The file is positioned at its terminal point (if possible, before the endfile record) such that the file can be extended.

    ASIS   The position of the file is not changed. The position is processor-dependent if the file exists but has not been connected.

    Default: **POSITION = 'ASIS'** .

action   is a character expression specifying the allowed data transfer actions for the file.

    READ   WRITE, PRINT and ENDFILE are not allowed.

    WRITE   READ is not allowed.

    READWRITE   All input/output statements are allowed.

    Default: (processor-dependent)

delim   is a character expression specifying the external representation of character values for list-directed output or namelist output.

    APOSTROPHE   A character value is delimited by apostrophes.

    QUOTE   A character value is delimited by quotation marks.

    NONE   No delimiters.

    Default: **DELIM = 'NONE'** .

pad   is a character expression specifying whether padding characters are to be transferred when the formatted input record does not contain enough characters.

    YES   Padding characters are transferred.

    NO   No padding characters are transferred; an error condition may occur.

    Default: **PAD = 'YES'**.

UNIT= specifier: The specifiers in the file status statements may be supplied in any order. If the UNIT= specifier appears as the first specifier, the keyword UNIT= may be absent.

IOSTAT= and **ERR= specifiers:**   When an error condition occurs during the execution of a file status statement that includes neither an ERR= specifier nor an IOSTAT= specifier, the execution of the program is terminated.

11.6 Input/Output Statements

FILE= specifier: The FILE= specifier must be present if the STATUS= specifier is present and has the value 'OLD', 'NEW', or 'REPLACE'. The FILE= specifier must be absent if the STATUS= specifier is present and has the value 'SCRATCH'. If both the FILE= specifier is absent and the unit **u** is currently not connected to a file, the STATUS= specifier must be present with value 'SCRATCH'; in this case, the unit becomes connected to a processor-dependent file, and the Fortran processor may generate a processor-dependent name for this file.

```
CHARACTER (LEN=3) head, tail
OPEN (7,FILE=head//tail,STATUS='OLD',BLANK='ZERO',ACTION='READ')
```

The existing file whose name results from the evaluation of the character expression `head//tail` is being connected to unit 7. Blanks in numeric input fields are interpreted as zeros. The file is being opened for sequential access. The file is formatted. WRITE and ENDFILE statements cannot be executed for this file. When an error condition occurs during the execution of this OPEN statement, the execution of the program is terminated.

STATUS= specifier: STATUS='SCRATCH' specifies that the file is a temporary file; the FILE= specifier must be absent. If STATUS='OLD', 'NEW', or 'REPLACE' is specified, the FILE= specifier must be present (exception see below: "Multiple OPENs"). The interpretation of STATUS='UNKNOWN' is processor-dependent. If during the execution of an OPEN statement with STATUS='NEW' *no* error condition occurs, the file is created and its status becomes 'OLD'.

ACCESS= specifier: For an existing file, the specified access method must be included in the set of allowed access methods for the file.

RECL= specifier: For an existing file, the specified record length or maximum record length must be included in the set of allowed record lengths for the file. The RECL= specifier must be present if ACCESS='DIRECT' is specified; otherwise the RECL= specifier may be absent. If the RECL= specifier is absent, the maximum record length is processor-dependent.

BLANK= specifier: The BLANK= specifier may be present only if the file is being opened for formatted input/output.

POSITION= specifier: Only existing files are affected by the POSITION= specifier. A sequential nonexisting file is always positioned at its initial point.

PAD= specifier: The padding character for an input list item of a nondefault character type is processor-dependent. The PAD= specifier may be present only if the file is being opened for formatted input/output. The PAD= specifier is ignored for formatted output.

Multiple OPENs: Though a unit is already connected to an existing file, an OPEN statement may be executed for this unit. If the file to be connected to the unit is the same as the file to which the unit is currently connected, then:

- The FILE= specifier may be absent in the currently executed OPEN.
- Only the BLANK=, DELIM=, PAD=, and ERR= specifiers may specify values that are different from those currently in effect.
- If the POSITION= specifier is present, the value specified must agree with the current position of the file.
- If the STATUS= specifier is present, its value must be 'OLD'.
- The execution of the OPEN statement does not cause any change in any of the absent specifiers. The ERR= and IOSTAT= specifiers from any previously executed OPEN statement have no effect on the currently executed OPEN statement.

If the file to be connected to a particular unit is not the same as in the previously executed OPEN statement for this unit, the effect is as though a CLOSE statement without a STATUS= specifier were executed for the unit immediately prior to the execution of the OPEN statement.

If an OPEN statement is executed for a nonexisting file and a unit, though the file is already preconnected to the unit, the specifications in the OPEN statement override the predefined attributes of the connection.

11.6.4.2 CLOSE Statement

The execution of a CLOSE statement **closes** an *external* file and disconnects a unit, respectively; that is, the connection between the specified unit and an external file is terminated. Depending on the value of the STATUS= specifier, the file will continue to exist or will be deleted.

CLOSE ([UNIT =] u [, IOSTAT = iostat] [, ERR = err] [, STATUS = status])

u	is a scalar nonnegative integer expression specifying the number of the unit to be disconnected.
iostat	(see OPEN statement)
err	(see OPEN statement)
status	is a scalar character expression of default character type specifying the disposition of the file to be closed.

        KEEP   The file will *not* be deleted after the execution of the CLOSE statement.

        DELETE   The file will be deleted after the execution of the CLOSE statement.

        Default: **STATUS='DELETE'** if the file has been opened with STATUS='SCRATCH'; otherwise is **STATUS='KEEP'**.

11.6 Input/Output Statements

UNIT=, IOSTAT=, and **ERR=** specifier: (see OPEN statement)

STATUS= specifier: Trailing blanks are allowed but they are ignored. Lower-case letters are allowed but they are processor-dependent. STATUS= 'KEEP' must not be specified for an external file that has been opened with STATUS='SCRATCH'.

```
CLOSE (10, ERR=200, STATUS='DELETE')
```

The file which is connected to unit 10 is closed and deleted. If an error condition occurs during the execution of the CLOSE statement, the execution of the program will continue with statement 200.

Program termination: When the execution of a program is terminated in the normal way (that is, if neither an error condition nor another condition occurs), then all files will be automatically closed which are connected in the moment of the termination. The effect is as though a CLOSE statement without a STATUS= specifier were executed for each of these files.

11.6.4.3   INQUIRE Statement

The INQUIRE statement may be used to inquire about the attributes of a particular named external file, about the connection to a particular unit, or about the length of an output list.

There are tree kinds of INQUIRE statements:

- INQUIRE with IOLENGTH= specifier (called an *inquiry by output list*),
- INQUIRE with UNIT= specifier (called an *inquiry by unit*), and
- INQUIRE with FILE= specifier (called an *inquiry by file*).

INQUIRE (IOLENGTH = iolength) output_list

INQUIRE ({ [UNIT =] u / FILE = file } [, IOSTAT = iostat] [, ERR = err] [, EXIST = exist]
[, OPENED = opened]   [, NUMBER = number]   [, NAMED = named]
[, NAME = name]   [, ACCESS = access]   [, SEQUENTIAL = sequential]
[, DIRECT = direct]   [, FORM = form]   [, FORMATTED = formatted]
[, UNFORMATTED = unformatted]   [, RECL = recl]   [, NEXTREC = nextrec]
[, BLANK = blank] [, POSITION = position] [, ACTION = action] [, READ = read]
[, WRITE = write] [, READWRITE = readwrite] [, DELIM = delim] [, PAD = pad] **)**

The specifiers of the INQUIRE statement returning values must be *scalar* variables of type *default* integer, *default* logical, or *default* character.

u	is a scalar nonnegative integer expression specifying the number of the unit being inquired about.
file	is a scalar character expression specifying the name of the file being inquired about.
iolength	is an integer variable which is assigned the record length that would result from the use of the **output_list** in an unformatted WRITE statement.
iostat	(see OPEN statement)
err	(see OPEN statement)
exist	is a logical variable which is assigned the value *true* if the specified unit exists (when inquired by unit) or if there exists a file with the specified name (when inquired by file); otherwise the variable is assigned the value *false*.
opened	is a logical variable which is assigned the value *true* if the specified unit is connected to a file (when inquired by unit) or if the specified file is connected to a unit (when inquired by file); otherwise the variable is assigned the value *false*.
number	is an integer variable which is assigned the number of the unit that is currently connected to the specified file. If the file is not connected to a unit, the variable is assigned the value -1.
named	is a logical variable which is assigned the value *true* if the file has a name; otherwise the variable is assigned the value *false*.
name	is a character variable which is assigned the name of the file if the file has a name; otherwise the variable becomes undefined.
access	is a character variable which is assigned the value:

	SEQUENTIAL	if the file is opened for sequential access.
	DIRECT	if the file is opened for direct access.
	UNDEFINED	if there is no connection.
sequential	is a character variable which is assigned the value:	
	YES	if sequential access is allowed for the file.
	NO	if sequential access is *not* allowed for the file.
	UNKNOWN	if the Fortran processor is unable to determine whether or not sequential access is allowed.
direct	is a character variable which is assigned the value:	
	YES	if direct access is allowed for the file.
	NO	if direct access is *not* allowed for the file.

11.6 Input/Output Statements

	UNKNOWN if the Fortran processor is unable to determine whether or not direct access is allowed.
form	is a character variable which is assigned the value:
	FORMATTED if the file is opened for formatted input/output.
	UNFORMATTED if the file is opened for unformatted input/output.
	UNDEFINED if there is no connection.
formatted	is a character variable which is assigned the value:
	YES if formatted is an allowed form for the file.
	NO if formatted is *not* an allowed form for the file.
	UNKNOWN if the Fortran processor is unable to determine whether or not formatted is an allowed form for the file.
unformatted	is a character variable which is assigned the value:
	YES if unformatted is an allowed form for the file.
	NO if unformatted is *not* an allowed form for the file.
	UNKNOWN if the Fortran processor is unable to determine whether or not unformatted is an allowed form for the file.
recl	is an integer variable which is assigned the record length or the maximum record length (see OPEN statement).
nextrec	is an integer variable which is assigned the value $n+1$ if n is the record number of the last record read from or written to the file connected for direct access. If the file is opened but no records have been transferred since the connection, the integer variable is assigned the value 1. If the file is not opened for direct access or if the position of the file is indeterminate because of a previous error condition, the integer variable becomes undefined.
blank	is a character variable which is assigned the value:
	NULL if blanks in numeric fields for formatted input are ignored and if an entire blank field is interpreted as the value zero.
	ZERO if blanks in numeric fields for formatted input are interpreted as zeros.
	UNDEFINED if the file is not opened for formatted input/output.

position	is a character variable which is assigned the value:	
	REWIND	if the file is connected by an OPEN statement for positioning at its initial point.
	APPEND	if the file is connected by an OPEN statement for positioning at its endfile record or at its terminal point.
	ASIS	if the file is connected without changing its position.
	UNDEFINED	if the file is not opened or if the file is opened for direct access.
action	is a character variable which is assigned the value:	
	READ	if the file is opened only for input.
	WRITE	if the file is opened only for output.
	READWRITE	if the file is opened for input and output.
	UNDEFINED	if there is no connection.
read	is a character variable which is assigned the value:	
	YES	if input is an allowed action for the file.
	NO	if input is *not* an allowed action for the file.
	UNKNOWN	if the Fortran processor is unable to determine whether or not input is an allowed action for the file.
write	is a character value which is assigned the value:	
	YES	if output is an allowed action for the file.
	NO	if output is *not* an allowed action for the file.
	UNKNOWN	if the Fortran processor is unable to determine whether or not output is an allowed action for the file.
readwrite	is a character variable which is assigned the value:	
	YES	if both input and output are allowed actions for the file.
	NO	if input and output are *not* allowed actions for the file.
	UNKNOWN	if the Fortran processor is unable to determine whether or not both input and output are allowed actions for the file.

delim   is a character variable which is assigned the value:

    APOSTROPHE   if character values for list-directed and namelist output are delimited by apostrophes.

    QUOTE   if character values for list-directed and namelist output are delimited by quotation marks.

    NONE   if character values for list-directed and namelist output are not delimited by apostrophes or quotation marks.

    UNDEFINED   if the file is not opened for formatted input/output.

pad   is a character variable which is assigned the value:

    NO   if the file is explicitly opened with PAD='NO'.

    YES   if the file is *not* opened with PAD='NO'.

```
LOGICAL ex;   INTEGER rl
REAL, DIMENSION (100) :: z
CHARACTER access *10, filename *8, name *15
INQUIRE (FILE=filename, EXIST=ex, ACCESS=access) ! inquiry by file
INQUIRE (12, NAME=name, ERR=100)                 ! inquiry by unit
INQUIRE (IOLENGTH=rl) y, (z(i), i=1,7)    ! inquiry by output list
```

When an error condition occurs during the execution of an INQUIRE statement that includes neither an ERR= specifier nor an IOSTAT= specifier, the execution of the program is terminated.

When an error condition occurs during the execution of an INQUIRE statement, the values of all output parameters, except the IOSTAT= specifier, become undefined.

The NAME= specifier and the NEXTREC= specifier become defined by the execution of the INQUIRE statement only if there is a connection; otherwise their values become undefined.

The EXIST= specifier and the OPENED= specifier become always defined by the execution of the INQUIRE statement, except when an error condition occurs.

The POSITION= specifier is defined with a processor-dependent value if the file has been repositioned since the connection.

Inquiry by output list

The value assigned to the IOLENGTH= specifier is always processor-dependent. But this value may be reused as an input value for the RECL= specifier in an OPEN statement if a file with this record length is to be opened for direct access.

Inquiry by unit

UNIT= specifier: The specifiers in the INQUIRE statement may be supplied in any order. If the UNIT= specifier appears as the first specifier, the keyword UNIT= may be absent.

The unit need not exist and need not be connected to a file. But if the unit is connected to a file, not only the properties of the unit are inquired about but also the properties of the file.

Inquiry by file

FILE= specifier: The **file** need not exist and need not be connected to a unit. The file name must be of a form acceptable to the processor. Trailing blanks are ignored.

NAME= specifier: The value assigned to the NAME= specifier may be different from the file name specified in the FILE= specifier. But its value may be reused as an input value for the FILE= specifier in an OPEN statement.

11.6.5 File Positioning Statements

There are three statements for the manipulation of external sequential files. The BACKSPACE statement may be used to position a file at most one record backwards. The REWIND statement may be used to position a file at its initial point. And the ENDFILE statement writes an endfile record.

BACKSPACE u
BACKSPACE ([UNIT =] u [, IOSTAT = status_variable] [, ERR = label])

REWIND u
REWIND ([UNIT =] u [, IOSTAT = status_variable] [, ERR = label])

ENDFILE u
ENDFILE ([UNIT =] u [, IOSTAT = status_variable] [, ERR = label])

The **u** is a scalar nonnegative integer expression specifying the number of the unit connected to the file which is to be positioned or to which an endfile record is to be written. The **status_variable** must be scalar and of type default integer. It becomes defined with the value zero if no error condition occurs during the execution of the positioning statement; otherwise it becomes defined with a processor-dependent positive error status value. And the **label** is the statement label of the branch target statement (in the same scoping unit) which will be executed if an error condition occurs during the execution of the positioning statement.

The specifiers in the file positioning statements may be supplied in any order. If the UNIT= specifier is the first specifier, the keyword UNIT= may be absent.

When an error condition occurs during the execution of a positioning statement that includes neither an ERR= specifier nor an IOSTAT= specifier, the execution of the program is terminated.

BACKSPACE statement

Execution of a BACKSPACE statement causes the file that is connected to the specified unit to be positioned before the current record if the file is currently positioned within a record. If there is no current record, the file will be positioned before the preceding record.

If the preceding record is the endfile record, the file will be positioned before the endfile record.

If the most recent input/output statement for that file is a data transfer output statement and no positioning statement, an endfile record is implicitly (that is, automatically) written before the file is backspaced before the last record.

The BACKSPACE statement must not be executed for a nonexisting file. Backspacing over such records that are written using list-directed formatting or namelist formatting is prohibited.

REWIND statement

Execution of a REWIND statement causes the file that is connected to the specified unit to be positioned at its initial point. If the file is currently positioned after the endfile record, it will also be rewound before the first record of the file.

The REWIND statement may be executed for a nonexisting file.

ENDFILE statement

Execution of an ENDFILE statement causes an endfile record to be written as the next record to the file. The file is positioned after this endfile record; that is, the endfile record becomes the last record of the file. Additional READ, WRITE, PRINT, or ENDFILE statements must not be executed for this file until a BACKSPACE or REWIND statement has been executed for the unit connected to this file.

The ENDFILE statement must not be executed for a file which is opened with ACTION='READ'.

```
      REAL, DIMENSION (12, 10) :: a, b
      INTEGER end_status
      :
      DO l=1,12
        WRITE (17, *) (a(l, m), m=1,10)
      END DO
      :
      REWIND 17                              ! <-- REWIND
      DO l=10,1,-1
        READ (17, *) (b(l, m), m=10,1,-1)
      END DO
      ENDFILE (17, IOSTAT=end_status)        ! <-- ENDFILE
      BACKSPACE (UNIT=17, ERR=100)           ! <-- BACKSPACE
100   PRINT *, 'backspacing error'
```

12 records are written to the file connected to unit 17. Then the file is rewound at its initial point. The first 10 records are read, and an endfile record is written after the 10th record. After the execution of the ENDFILE statement, the remaining two records containing the values a(11, 1:10) and a(12, 1:10) are no more available. The effect is as though they were deleted.

12 FORMATS

An explicit *format specification* (often abbreviated to "format") is used in conjunction with certain formatted input/output statements to control the input of external data and the output of internal data. During this data transfer, the data are converted from the internal representation (in the storage) to the external character representation (in the record) or vice versa.

During the execution of a formatted input/output statement that includes the FMT= specifier, normally one *edit descriptor* is needed for each effective input/output list item. Exception: two edit descriptors are needed for a complex input/output list item.

Edit descriptors are *data edit descriptors*, *control edit descriptors*, and *character string edit descriptors*.

A format specification may appear in a FORMAT statement, may be contained in a character object, or may be the result of the evaluation of a character expression.

12.1 Format Specification

A **format specification** is a list of format items enclosed in parentheses. These format items may be single edit descriptors or groups of edit descriptors. A group of edit descriptors is a list of format items enclosed in parentheses. A single repeatable edit descriptor or a group of edit descriptors may be preceded by a *repeat specification*.

A comma is used to separate format items in a format specification. This comma may be omitted between a P edit descriptor and an immediately following D, E, EN, ES, F, or G edit descriptor; it may be omitted before a slash edit descriptor if the (optional) repeat specification is absent; it may be omitted after a slash edit descriptor; and it may be omitted before and after a colon edit descriptor.

A named constant must not appear within the character string of a format specification. A format specification may be empty, such as ( ); but nevertheless, the parentheses must not be omitted.

Leading blank characters preceding the left parenthesis of a format specification are allowed. Additional blank characters may appear within a format specification, but they have no effect and are ignored except within a character string edit descriptor.

12.1.1 Format Specification in FORMAT Statement

FORMAT statements may appear after the USE statement(s), if any, and before the CONTAINS statement, if any, among the executable or nonexecutable statements in a program unit or in a subprogram. The FORMAT statement must be labeled.

label FORMAT format_specification

```
    SUBROUTINE sub
77  FORMAT (15X,F10.2,I5,E20.5)
    ⋮
    READ (9, 77, END=900) a, n, b
    ⋮
    END
```

12.1.2 Character Format Specification

If the format specification for a formatted input/output statement is contained within a character variable, or if it is the value of a character constant, or if it is the result of the evaluation of another character expression, the leading characters of the character entity must be defined with characters forming a format specification when the input/output statement ist executed.

The character positions following the right parenthesis that ends the format specification may remain undefined. They have no effect on the interpretation of the format specification and are ignored.

If the FMT= specifier **format** in a formatted input/output statement is a character array object, the format specification may be longer than the first array element. In this case, the format specification continues with the next array elements as though all array elements were chained in array element order.

```
CHARACTER :: coform *20 = '(15X,F10.2,I5,E20.5)'
READ (9, FMT=coform, END=900) a, n, b
```

has the same effect as

```
    READ (9, 77, END=900) a, n, b
77  FORMAT (15X,F10.2,I5,E20.5)
```

which has the same effect as

```
CHARACTER ff (9) *6
DATA (ff(i), i=1,4) /'   (15','X,F10.','2,I5,E','20.5)'/
READ (9, ff, END=900) a, n, b
```

Variable format specification

If **format** is a character entity but no character constant expression, it is called a "variable format specification". In this case, the format specification need not be known until the execution of the program. For example, the program may compute and generate a suitable format specification.

A variable format specification must be complete when the formatted input/output statement is executed. The format specification for an internal WRITE statement must not be contained within the internal file. An input list item or an associated data object must not contain any part of the format specification. Note that a variable format specification may be redefined during the execution of the program.

```
(F7.2,E9.1,I5,F9.3,2E12.3,I6)    ←— Input record
123456789112345678921234567 89   ←— Position
```

This input record may be read into a character variable. Then this variable may be used as the FMT= specifier **format** in a data transfer statement:

```
CHARACTER frm *30
READ (12, '(A)', END=100) frm
⋮
WRITE (15, FMT=frm) a, b, i, c, d, e, j
```

This WRITE statement has the same effect as

```
WRITE(15, FMT='(F7.2,E9.1,I5,F9.3,2E12.3,I6)') a, b, i, c, d, e, j
```

12.2 Interaction between Input/Output List and Format

The interaction between an input/output list and a format specification is a dynamic process which is called the **format control**. Format control processes the input/output list and the format specification from the left to the right, associating in order one effective input/output list item with one corresponding data edit descriptor (exception: complex editing).

When format control encounters a control edit descriptor or a character string edit descriptor, the edit descriptor is executed immediately because an input/output list item is not needed.

The format specification must include at least one data edit descriptor if the input/output list includes an effective list item. An empty format specification of the form ( ) may occur only if the input/output list is empty or if all input/output list items are noneffective. If there is no current record, an advancing READ statement with empty format specification skips one record without transferring data; and an advancing output statement with empty format specification writes an empty record. If there is a current record, the execution of a nonadvancing data transfer statement with empty format specification does not change the file position within the current record.

Even if format control has already processed all input/output list items, remaining control edit descriptors and character string edit descriptors (but no data edit descriptors) are processed. If there is no unprocessed effective input/output list item left, format control terminates when the next data edit descriptor, the right parenthesis at the end of the format specification, or a colon edit descriptor is encountered.

12.2.1 Repeat Specification, Groups of Edit Descriptors

A single repeatable edit descriptor or a group of edit descriptors (enclosed in parentheses) may be preceded by a **repeat specification**, which must be an unsigned nonzero integer literal constant without an explicitly specified kind type parameter. Such a repeat specification n has the effect as though n identical edit descriptors or n identical groups of edit descriptors were specified.

```
22 FORMAT (2I4,2(I5,2X,E7.2))      has the effect of
22 FORMAT (I4,I4,I5,2X,E7.2,I5,2X,E7.2)
```

These different forms with or without a repeat specification are equivalent during the execution of the program as long as no *reversion of format control* (see below) occurs.

A group of edit descriptors may include another group of edit descriptors (enclosed in parentheses). The maximum nesting level of such groups is processor-dependent.

Nonrepeatable edit descriptors must *not* be preceded by a repeat specification. But they may be enclosed in parentheses to form a group. Then this group may be preceded by a repeat specification.

```
44 FORMAT (3'test case'/3(I2,9X))    ! <-- not allowed
44 FORMAT (3('test case')/3(I2,9X))  ! <-- allowed
```

12.2.2 Reversion of Format Control

If format control encounters the right parenthesis at the end of the format specification and there is another unprocessed effective input/output list item, **reversion of format control** occurs as follows: the current record is terminated, the file is positioned at the beginning of the next record, and format control reverts to the beginning of the last preceding group of edit descriptors. If this group is preceded by a repeat specification, this repeat specification will also be reused. If the format specification does not include such a group, format control reverts to the beginning of the format specification. In any case, format control continues to process edit descriptors from the left to the right beginning with the found position. And if format control encounters the last right parenthesis of the format specification for a second time, reversion of format control occurs again, and so on.

Reversion of format control has no effect on scale factors, on sign control (S, SP, and SS editing) and on the interpretation of blank characters in numeric input fields (BN and BZ editing).

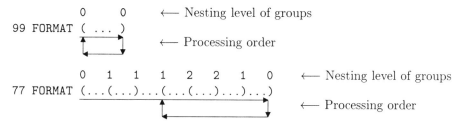

The marked group, where format control continues after reversion, is reused including its repeat specification (if any).

```
CHARACTER (LEN=20) PARAMETER :: fmt = '(I4,2(I3,F8.2),I5)'
READ (11, fmt, END=100) i, j, a, k, b, l, m, c, n, d
```

The data edit descriptors I4, I3, F8.2, I3, F8.2, and I5 are used in this order to transmit the input list items i, j, a, k, b, and l. Then the next record is read, and the data edit descriptors I3, F8.2, I3, and F8.2 are used to transmit the remaining input list items m, c, n, and d.

12.3 Edit Descriptors

There are three kinds of edit descriptors: *data edit descriptors*, *character string edit descriptors*, and *control edit descriptors*.

Data edit descriptors are used to describe the editing, that is, the kind of the conversion and the external representation of the data being transferred between input/output list items and a file.

Character string edit descriptors are used to write character strings without involvement of output list items.

Control edit descriptors are used to position in a record, to specify the interpretation of blank characters in numeric input fields, to specify sign control for numeric output fields, to terminate a record, and so on. Though no input/output list items are involved, control edit descriptors may affect subsequent data edit descriptors.

Data edit descriptors and the slash edit descriptor are **repeatable edit descriptors**. Character string edit descriptors and control edit descriptors except the slash edit descriptor are **nonrepeatable edit descriptors**.

Data edit descriptors					
Descr.	Type	Descr.	Type	Descr.	Type
A	character	ENw.d	numeric	Iw	numeric
Aw	character	ENw.dEe	numeric	Iw.m	numeric
Bw	numeric	ESw.d	numeric	Lw	logical
Bw.m	numeric	ESw.dEe	numeric	Ow	numeric
Dw.d	numeric	Fw.d	numeric	Ow.m	numeric
Ew.d	numeric	Gw.d	intrinsic	Zw	numeric
Ew.dEe	numeric	Gw.dEe	intrinsic	Zw.m	numeric

In the above table, upper-case letters specify the type of the editing. The lower-case letters **w**, **d**, **e**, and **m** are unsigned integer literal constants without an explicitly specified kind type parameter. They have the following meaning:

w specifies the *field width* of the external representation. The **field width** is the number of all characters transferred during the processing of the data edit descriptor including leading blank characters, sign, decimal point, and exponent, if any. **w** must be nonnegative.

d specifies the number of digits in the *fractional part* to the right of the decimal point. **d** may be zero.

e specifies the number of digits in the *exponent part*. **e** must be positive.

m specifies that at least **m** digits are written to the output field. **m** may be zero and must be $\leq$ **w**.

The valid values of **d**, **e**, and **m** depend on the value of the field width **w**.

Control edit descriptors				Char. string edit descr.	
Descr.	Type	Descr.	Type	Descr.	Type
BN	numeric	nX	tabulator	'	character output
BZ	input	Tn		"	
kP	scale factor	TRn			
SP	numeric	TLn			
SS	output	:	format control		
S		/	end of record		

12.3 Edit Descriptors

k is an integer literal constant (signed or unsigned) without an explicitly specified kind type parameter. It is called a **scale factor**.

n is an unsigned positive integer literal constant without an explicitly specified kind type parameter.

```
11 FORMAT (I6,2F7.3)              ! <-- Data edit descriptors
22 FORMAT (10X,'test case'/T19,'7') ! <-- Character string and
                                           control edit descriptors
```

Numeric editing

The data edit descriptors B, D, E, EN, ES, F, G, I, O, and Z may be used to specify the editing of numeric data. Note that the G edit descriptor may also be used to specify the editing of nonnumeric data.

Numeric input: Leading blank characters are not significant. Other blank characters are also ignored, except when an external file is explicitly opened with BLANK= 'ZERO' or when format control has encountered a BZ edit descriptor. A field containing only blank characters is interpreted as the value zero.

On input with a D, E, EN, ES, F, or G edit descriptor, a decimal point in the input field overrides the **d**, which otherwise would specify the position of the decimal point. The input field may contain more digits than the Fortran processor uses to approximate to the value.

Numeric output: Values of type real, double precision real, and complex may be rounded. The rounding method is processor-dependent.

The output representation is right-justified in the field. If the number of characters produced by the editing is less than the field width **w**, leading blank characters are transferred in the field. If the edit descriptors **Bw.m**, **Iw.m**, **Ow.m**, or **Zw.m** are used to edit a "small" value, at least **m** digits are transferred in the field, possibly with leading zeros. Then, the field is filled with (**w** − **m**) leading blank characters.

A zero internal value or a positive internal value may be prefixed with a plus sign. Sign control depends on the Fortran processor and on S, SP, or SS edit descriptors. A negative internal value is always prefixed with a minus. The zero internal value is never written with a leading minus.

Field width: On output, the number of transferred characters should not be greater than the field width **w**; and when the edit descriptors **Ew.dEe**, **ENw.dEe**, **ESw.dEe**, and **Gw.dEe** are used, the number of digits produced for the exponent should not be greater than **e**. Otherwise, the field or the exponent part are too small, and the Fortran processor fills the entire field with **w** asterisks.

On output with B, F, I, O, and Z edit descriptors, the specified output field width may be zero. In these cases, the Fortran processor actually selects the smallest field width such that the output value fits into the output field.

Complex editing: A complex datum is represented (internally) as a pair of real data. Therefore, the editing of a complex datum is specified by two D, E, EN, ES, F, or G edit descriptors. The first of the two edit descriptors directs the editing of the real part and the second one directs the editing of the imaginary part. The two edit descriptors may be different. Additional character string edit descriptors and control edit descriptors may appear between the two data edit descriptors.

```
COMPLEX a, b, c, d
READ (11, '(2(E9.2),F7.3,E9.3,2(F5.1,F5.2))', END=100) a, b, c, d
```

Summary:

Edit descriptors		
Data	Control	Character
A B D E EN ES F G I L O Z /	BN BZ P S SP SS T TL TR X :	" '
repeatable	nonrepeatable	

12.3.1 A Edit Descriptors: A Aw

The A edit descriptor may be used to describe the editing of character data. The corresponding input/output list item must be of type integer. The kind type parameter of all characters transferred under control of an A edit descriptor is implied by the kind type parameter of the corresponding input/output list item.

Input:

If the field width **w** is less than the length of the corresponding input list item, the input characters are transferred left-justified in the input list item, which is then filled with *padding characters*. These **padding characters** are blank characters for input list items of type default character, but they are processor-dependent for input list items of type nondefault character. If **w** equals the length *len* of the input list item, exactly *len* characters are transferred. If **w** is greater than the length *len* of the input list item, only the last *len* characters of the input field are transferred and the other (that is, the first ($\mathbf{w} - len$)) characters are ignored.

If **w** is not specified, the field width is given by the length *len* of the corresponding input list item; that is, *len* characters are transferred.

12.3 Edit Descriptors

```
CHARACTER c *9, cc *5
READ (12, '(A7)', END=100) c
READ (14, '(8X,A)', END=100) cc
```

```
circuit noises        ←— 1st input record
circuit breaker       ←— 2nd input record
123456789112345       ←— Position
```

After the execution of the READ statement, the variable c is defined with the value circuit␣␣ and the variable cc with the value break.

Output:

If the field width **w** is less than the length *len* of the corresponding output list item, only the first **w** characters are transferred. The other $(len - \mathbf{w})$ characters are ignored. If the field width **w** equals the length *len* of the output list item, exactly *len* characters are transferred. If **w** is greater than the length *len* of the output list item, the *len* characters are transferred right-justified in the output field, which is then filled with $(\mathbf{w} - len)$ leading padding characters.

If the field width **w** is not specified, the field width is given by the length of the corresponding output list item; that is, all characters of the output list item are transferred.

```
CHARACTER :: cccc *15 = 'circuit breaker'
WRITE (18, '(A7)') cccc
WRITE (18, '(A18)') cccc
```

The WRITE statements create the following output records:

```
circuit              ←— 1st output record
   circuit breaker   ←— 2nd output record
123456789112345678   ←— Position
```

12.3.2 B Edit Descriptors: Bw Bw.m

The B edit descriptor is used to describe the editing of binary data. The corresponding input/output list item must be of type integer.

Input:

The input field may contain the binary digits *0* and *1*; and it may contain blank characters. The specification of **m** has no effect on input. **w** must be positive.

```
INTEGER bi, bj, bk
READ (22, '(B3,B16,B5)', END=100) bi, bj, bk

101        0111001 1000     ←— Input record
123456789112345678921234    ←— Position
```

After the execution of the READ statement, the variables bi, bj, and bk are defined as follows: bi = 5, bj = 57, and bk = 8.

Output:

For a **Bw** edit descriptor, at least one binary digit without leading zeros is transferred to the output field. For a **Bw.m** edit descriptor, if the value to be transferred occupies fewer than **m** positions, the value is transferred right-justified with leading zeros as a binary number with **m** digits. And the remaining leading (**w** − **m**) positions of the output field are filled with blank characters.

On output of the value zero with a **Bw** edit descriptor, the digit zero is transferred right-justified with (**w** − 1) leading blank characters. On output of the value zero with a **Bw.m** edit descriptor, **m** zeros are transferred right-justified; if **m** is zero, no digits but **m** blank characters are transferred, giving a field of all blank characters. When **m** and **w** are both zero on output of the value zero, one blank character is transferred to the output field.

```
INTEGER :: bi = 6, bj = 110, bk = 46
WRITE (34, '(B7.4,B10,B8)') bi, bj, bk
```

The WRITE statement creates the following output record:

```
   0110    1101110  101110      ←— Output record
1234567891123456789212345        ←— Position
```

12.3.3 Blank Control Edit Descriptors: BN BZ

The BN and BZ edit descriptors may be used to describe the interpretation of blank characters in numeric input fields. They do not affect leading blank characters in the input field and fields containing only blank characters. And they are not in effect during the execution of an output statement.

Default: For an external file, blank control depends on the value of the BLANK= specifier in the OPEN statement. For an internal file or a preconnected file, blank characters are ignored.

These defaults remain in effect for a particular READ statement until format control encounters a BN or BZ edit descriptor. The effect of a BN or BZ edit descriptor terminates when the execution of the READ statement terminates.

BN: When format control encounters a BN edit descriptor, all nonleading blank characters in subsequent numeric input fields are ignored.

BZ: When format control encounters a BZ edit descriptor, all nonleading blank characters in subsequent numeric input fields are treated as zeros.

12.3 Edit Descriptors

```
OPEN (5, BLANK='NULL')
READ (5, '(I5,BZ,I5,BN,I5,I5)', END=100) i, j, k, l
5    10       15             ←― Input record
12345678911234567892          ←― Position
```

After the execution of the READ statement, the variables i, j, k, and l are defined as follows: i = 5, j = 10000, k = 0, and l = 15.

12.3.4 Character Constant Edit Descriptors: "..." '...'

A character constant edit descriptor may be used to describe the output of character strings without using an output list item.

The specified character string has the form of a character literal constant without an explicitly specified kind type parameter. The string may contain any representable character of the processor-dependent character set. If the delimiting apostrophe or quotation mark appears within the character string, it must be written by two consecutive apostrophes or quotation marks, respectively, without intervening blank characters. The delimiting apostrophes or quotation marks are not transferred to the output field.

The width of the output field equals the length of the character constant. Note that two included delimiters are counted as one character for the output field.

```
   WRITE (14,22)
22 FORMAT ('"That''s correct", he said.')
```

The WRITE statement creates the following record:

```
"That's correct", he said.     ←― Output record
12345678911234567892123456     ←― Position
```

12.3.5 Colon Edit Descriptor: :

The colon edit descriptor is used to terminate format control if there are no more effective input/output list items when the colon edit descriptor is encountered. The colon edit descriptor is ignored if there are more effective input/output list items.

```
  DATA i, j, k, l, me /1, 2, 3, 4, 5/
  WRITE (12, FMT="(5(' ',I3,:,' |'))") i, j, k, l, me
  WRITE (12, FMT="(4('------+'),'------')")
5 READ (*, *, END=90) i, j, k, l, me
```

```
       WRITE (12, FMT="(5(' ',I3,:,' |'))") i, j, k, l, me
       GOTO 5
90     CONTINUE
       WRITE (12, FMT="(4('------+'),'------')")
```

This piece of program produces the following table of its input data. Note that the last column is not delimited by |.

```
    1 |     2 |     3 |     4 |     5
------+------+------+------+------
   22 |   549 |     9 |    46 |   221
  142 |     5 |    66 |     8 |   519
   23 |    31 |     0 |   938 |    66
------+------+------+------+------
12345678911234567892123456789312 34        ←— Position
```

12.3.6 D Edit Descriptor: Dw.d

The D edit descriptor may be used to describe the editing of real, double precision real, and complex data. The corresponding input/output list item must be of type real, double precision real, or complex.

Input:

The input field consists of one or more subfields:

Integer part	Fractional part	Exponent part

Integer part: An optional sign followed by a sequence of digits. This sequence of digits may be omitted if the fractional part contains at least one digit.

Fractional part: An optional decimal point optionally followed by a sequence of digits. This sequence of digits may be omitted if the integer part contains at least one digit. If the decimal point is omitted, the **d** determines the number of digits in the fractional part. The integer part and the fractional part must contain together at least one digit.

The basic form of a real or double precision real input datum consists of an integer part and a fractional part. This basic form may be optionally followed by an exponent part.

Exponent part: There are the following forms for the exponent part:

- A sign, followed by one or more optional blank characters, followed by a sequence of digits (possibly with intervening blank characters).

12.3 Edit Descriptors

- The **exponent letter** D or E, followed by one or more optional blank characters, followed by an optional sign, followed by a sequence of digits (possibly with intervening blank characters).

If the input field contains a decimal point, the specification of **d** is ignored. If the decimal point is omitted, the **d** designates the position of the "imaginary" decimal point of the input value; that is, the input value is multiplied by 10^{-d}. An exponent part beginning with the exponent letter E has the same effect on input as an exponent part beginning with the exponent letter D.

On input, a D edit descriptor has the same effect as an E, EN, ES, F, or G edit descriptor; that is, it may be used instead of an E, EN, ES, F, or G edit descriptor.

Input field	Data edit descriptor	Value	Remarks
-534.9D-3	D9.1	−.5349	All parts are complete, with sign; the exponent part begins with the letter D.
-35.24E2	D8.2	−3524.	The sign of the exponent part is assumed to be +; the exponent part begins with the letter E.
-23.725+2	D9.3	−2372.5	The exponent letter is omitted, therefore, the sign of the exponent must be present.
19984.	D6.0	19984.	The fractional part contains only the decimal point; the exponent part is omitted; the omitted sign of the integer part is assumed to be +.
.456739	D7.6	.456739	The integer part and the exponent part are omitted; the omitted sign of the integer part is assumed to be +.
36988	D5.2	369.88	The decimal point is inserted between the 2nd and 3rd digit from the right; the omitted sign of the integer part is assumed to be +.

Output:

The external character representation of a real or double precision real value consists of a **significand** followed by an **exponent**. More precisely: the output field consists of leading blank characters (if necessary), followed by a sign (if necessary), followed by the zero, followed by the decimal point, followed by the **d** most significant digits of the rounded value, followed by four digits for the exponent. If another form of the significand is needed (such as one or more significant digits *before* the decimal point), an additional *scale factor* must be specified.

The following details of the output field are processor-dependent: the leading zero before the decimal point, the plus sign for a positive value, and the form of the exponent if the absolute value of the exponent is ≤ 99.

```
WRITE (12, '(D10.2,D15.6,D9.3)') a, b, c
```

12.3.7 E Edit Descriptors: Ew.d Ew.dEe

The E edit descriptor may be used to describe the editing of real, double precision real, and complex data. The corresponding input/output list item must be of type real, double precision real, or complex.

Input:

The form and the interpretation of the input field are the same as those of the D edit descriptor. The specification of **e** is ignored. An E edit descriptor may be used instead of a D, EN, ES, F, or G edit descriptor.

```
READ (12, '(E10.2,E15.6,E9.3)', END=100) a, b, c
```

Output:

The output field for an **Ew.d** edit descriptor has the same form as that for a **Dw.d** edit descriptor, except that the exponent letter is an E instead of a D. On output with an **Ew.dEe** edit descriptor, the exponent is an E, followed by a sign, and followed by **e** digits.

```
WRITE (12, '(E10.2,E15.6E4,E9.3)') a, b, c
```

12.3.8 EN Edit Descriptors: ENw.d ENw.dEe

The EN edit descriptor may be used to describe the editing of real, double precision real, and complex data. It differs from the E edit descriptor only in the form of the output field. The corresponding input/output list item must be of type real, double precision real, or complex.

Input:

The form and the interpretation of the input field are the same as those of the D edit descriptor. The specification of **e** is ignored. An EN edit descriptor may be used instead of a D, E, ES, F, or G edit descriptor.

```
READ (15, '(EN11.3,EN8.2,EN7.1)', END=100) x, y, z
```

Output:

An EN edit descriptor creates an output field in engineering notation, such that the decimal exponent is divisible by three and the absolute value of the significand is ≥ 1 and < 1000 (if the output value is nonzero). Apart from that, the output field has the same form as the output field for the E edit descriptor. The scale factor has no effect.

Value	Format specification	Output field
3.1234	SS, EN11.3	3.123E+00
1024.82	SS, EN10.4	1.0248E+03
−.72	SS, EN12.2	-720.00E-03
.006472	SS, EN14.1	6.5E-03
		12345678911234 ⟵ Position

12.3.9 ES Edit Descriptors: ESw.d ESw.dEe

The ES edit descriptor may be used to describe the editing of real, double precision real, and complex data. It differs from the E edit descriptor only in the form of the output field. The corresponding input/output list item must be of type real, double precision real, or complex.

Input:

The form and the interpretation of the input field are the same as those of the D edit descriptor. The specification of **e** is ignored. An ES edit descriptor may be used instead of a D, E, EN, F, or G edit descriptor.

```
READ (23, '(ES10.3,ES6.2,ES9.1)', END=100) x, y, z
```

Output:

An ES edit descriptor creates an output field in scientific notation, such that the absolute value of the significand is ≥ 1 and < 10 (if the output value is nonzero). Apart from that, the output field has the same form as the output field for the E edit descriptor. The scale factor has no effect.

Value	Format specification	Output field
21.234	SS, ES11.3	2.123E+01
80246.7	SS, ES10.4	8.0247E+04
−.72	SS, ES12.2	-7.20E-01
.000472	SS, ES14.1	4.7E-04
		12345678911234 ⟵ Position

12.3.10 F Edit Descriptor: Fw.d

The F edit descriptor may be used to describe the editing of real, double precision real, and complex data. The corresponding input/output list item must be of type real, double precision real, or complex.

Input:

The form and the interpretation of the input field are the same as those of the D edit descriptor. An F edit descriptor may be used instead of a D, E, EN, ES, or G edit descriptor.

```
READ (12, '(F10.2,F15.6,F9.3)', END=100) a, b, c
```

Output:

The form of the output field is as follows: leading blank characters (if necessary), followed by a sign (if necessary), followed by a sequence of digits (if necessary), followed by the decimal point, followed by **d** digits.

Value	Format specification	Output field	Remarks
45.783	(SS, F6.3)	45.783	
45.783	(SS, F11.3)	45.783	
-45.783	(SS, F6.3)	******	field width too small
.45783	(SS, F6.3)	.458	possibly with leading zero
45.783	(SS, F6.0)	45.	
		12345678901 ⟵ Position	

The following details of the output field are processor-dependent: the plus sign for a positive value, and the leading zero before the decimal point if the absolute value is less than 1 (and at least a second digit is written).

12.3.11 G Edit Descriptors: Gw.d Gw.dEe

The G edit descriptor may be used to describe the editing of data of any intrinsic type. The corresponding input/output list item must be of type integer, real, double precision real, complex, logical, or character.

Integer, logical, and character input:

The form and the interpretation of the input field are the same as those of the I, L, and A edit descriptor, respectively. The **d** and the **e** are ignored. A G edit descriptor may be used instead of an I, L, or A edit descriptor.

Real, double precision real, and complex input:

The form and the interpretation of the input field are the same as those of the D edit descriptor. The specification of **e** is ignored. A G edit descriptor may be used instead of a D, E, EN, ES, or F edit descriptor.

```
READ (12, '(G10.2,G15.6,G9.3)', END=100) a, b, c
```

Integer, logical, and character output:

The form and the interpretation of the output field are the same as those of the I, L, and A edit descriptor, respectively, with **w** being positive. **d** and **e** are ignored.

Real, double precision real, and complex output:

A G edit descriptor has the same effect as a corresponding E edit descriptor, except when **d** > 0 and the output value is zero or when the magnitude of the output value is $\geq 0.1 - .5 * 10^{-d-1}$ or $< 10^d - 0.5$. For these exceptions, a G edit descriptor has the same effect as an F edit descriptor, and either 4 or (**e** + 2) blank characters are transferred right-justified in the output field (this is the width of the exponent subfield of the corresponding E edit descriptor).

Value	Format specification	Output field	Remarks
.01367333	(SS, G13.3E6)	.137-000002	the same as E13.3E6
.594356	(SS, G13.3)	.594	F edit descriptor
75.332	(SS, G13.3)	75.33	F edit descriptor
450.2368	(SS, G13.3)	.450E+03	E edit descriptor
		1234567890123 ← Position	

When a G edit descriptor has the same effect as an E edit descriptor, a scale factor may be in effect.

12.3.12 I Edit Descriptors: Iw Iw.m

The I edit descriptor may be used to describe the editing of integer data. The corresponding input/output list item must be of type integer.

Input:

The input field must contain (apart from blank characters) a character string having the form of an integer literal constant. The plus sign may be omitted. The specification of **m** is ignored; that is, on input, an **Iw.m** edit descriptor is treated as an **Iw** edit descriptor. **w** must be positive.

```
READ (18, '(I3,I7,I5,I3,I2,I4)', END=100) i, j, k, l, m, n
539  -34     27  9    4 7      ← Input record
123456789112345678921234        ← Position
```

If unit 18 has not been connected by an OPEN statement specifying BLANK='ZERO', the execution of the READ statement defines the variables i, j, k, l, m, and n as follows: i = 539, j = -34, k = 27, l = 9, m = 0, and n = 47.

Output:

The form of the output field is as follows: leading blank characters (if necessary), possibly a sign, followed by a sequence of digits which has the form of an unsigned integer literal constant.

The form of the output field is processor-dependent, because the output of the plus sign is partly processor-dependent.

On output with an **Iw.m** edit descriptor, if the value to be transferred occupies fewer than **m** positions, the value is transferred right-justified with leading zeros as a decimal value with **m** digits. And the leading (**w** − **m**) positions of the output field are filled with blank characters.

On output of the value zero with an **Iw** edit descriptor, the digit zero is transferred right-justified with (**w** − 1) leading blank characters. On output of the value zero with an **Iw.m** edit descriptor, **m** zeros are transferred right-justified; if **m** is zero, no digits but **m** blank characters are transferred, giving a field of all blank characters. When **m** and **w** are both zero on output of the value zero, one blank character is transferred to the output field.

```
i = 57;  j = 7694327;  k = -3976
WRITE (16, '(I7.4,I10,I8)') i, j, k
```

The WRITE statement creates the following record:

```
   0057   7694327   -3976      ← Output record
1234567891123456789212345       ← Position
```

12.3.13 L Edit Descriptor: Lw

The L edit descriptor may be used to describe the editing of logical data. The corresponding input/output list item must be of type logical.

Input:

The form of the input field must be as follows: optional blank characters, optionally followed by a decimal point, followed by the letter T for the value *true* or by the letter F for the value *false*, optionally followed by any other characters,

12.3 Edit Descriptors

which are ignored. If the processor supports lower-case letters, the letters t and f are equivalent to the corresponding upper-case letters.

```
LOGICAL lx, ly, lz
READ (14, '(L5,L10,L7)', END=100) lx, ly, lz
   T    .FALSE.      TRU      ←— Input record
1234567891123456789212         ←— Position
```

The execution of the READ statement defines the variables lx, ly, and lz as follows: lx = .TRUE., ly = .FALSE., and lz = .TRUE..

Output:

The form of the output field is as follows: (**w** − 1) blank characters, followed by the letter T or F depending on whether the value of the corresponding output list item is *true* or *false*.

```
LOGICAL :: la = .FALSE., lb = .FALSE., lc = .TRUE.
WRITE (18, '(L5,L8,L3)') la, lb, lc
```

The WRITE statement creates the following record:

```
    F       F  T    ←— Output record
1234567891123456    ←— Position
```

12.3.14 O Edit Descriptors: Ow Ow.m

The O descriptor is used to describe the editing of octal data. The corresponding input/output list item must be of type integer.

Input:

The input field may contain the octal digits *0, 1, 2, 3, 4, 5, 6,* or *7*; and it may contain blank characters. It must not contain a sign, a decimal point, or an exponent. The specification of **m** has no effect on input. **w** must be positive.

```
INTEGER oi, oj, ok
READ (22, '(O3,O16,O5)', END=100) oi, oj, ok
 207         0131005 4000        ←— Input record
12345678911234567892 1234         ←— Position
```

The execution of the READ statement defines the variables oi, oj, and ok as follows: oi = 135, oj = 45573, and ok = 2048.

Output:

For an **Ow** edit descriptor, at least one octal digit without leading zeros is transferred to the output field. For an **Ow.m** edit descriptor, if the value to be transferred occupies fewer than **m** positions, the value is transferred right-justified with leading zeros as an octal number with **m** digits. And the leading (**w** − **m**) positions of the output field are filled with blank characters.

On output of the value zero with an **Ow** edit descriptor, the digit zero is transferred right-justified with (**w** − 1) leading blank characters. On output of the value zero with an **Ow.m** edit descriptor, **m** zeros are transferred right-justified; if **m** is zero, no digits but **m** blank characters are transferred, giving a field of all blank characters. When **m** and **w** are both zero on output of the value zero, one blank character is transferred to the output field.

```
INTEGER :: oi = 9, oj = 110, ok = 46
WRITE (34, '(O7.4,O10,O8)') oi, oj, ok
```

The WRITE statement creates the following record:

```
    0011       156      56       ←— Output record
12345678911234567789212345        ←— Position
```

12.3.15 P Edit Descriptor, Scale Factor

The P edit descriptor is used to change the scale factor for D, E, EN, ES, F, or G edit descriptors. During the conversion of a datum between its internal and its extenal representation, the scale factor may have effect on the position of the decimal point and on the value of the datum.

kP	**kPDw.d**	**kPEw.d**	**kPEw.dEe**	**kPENw.d**	**kPENw.dEe**
kPESw.d	**kPESw.dEe**	**kPFw.d**	**kPGw.d**	**kPGw.dEe**	

At the beginning of the execution of a formatted data transfer statement, the scale factor has always the value zero. That is, the real or double precision real data to be transferred are (theoretically) multiplied by 10^0 during the conversion between the internal and the external representation; if this scale factor is not changed, the position of the decimal point and the value of the data are not changed.

When format control encounters a P edit descriptor, the new scale factor affects subsequent D, E, EN, ES, F, and G edit descriptors until another P edit descriptor is encountered or until the execution of this data transfer statement terminates. To nullify the effect of a nonzero scale factor for subsequent edit descriptors, the scale factor must be reset to zero by specifying 0P.

Input:

If the input field contains an exponent, the scale factor has no effect.

If the input field contains no exponent, the input value is multiplied by the internal representation of 10^{-k}, and this new value is stored.

12.3 Edit Descriptors

```
READ (15, '(-2P,F10.2,E11.2)', END=100) a, b
2.7182818 2.7182818E0    ← Input record
123456789112345678921    ← Position
```

The execution of the READ statement defines the variables a and b as follows:
a = 271.82818 and b = 2.7182818.

Output:

D and E edit descriptor: The "normal" external representation of a real or double precision real value consisting of a significand and an exponent is changed such that the value of the significand is multiplied by 10^k and the value of the exponent is reduced by **k**.

If the scale factor **k** is negative with $-d < k \leq 0$, **d** digits of the significand following the decimal point are transferred; that is, the fractional part of the significand has |**k**| leading zeros, followed by (**d** − |**k**|) significant digits.

If the scale factor **k** is positive with $0 < k < d+2$, **k** digits of the significand are transferred preceding the decimal point and (**d** − **k** + 1) digits of the significand are transferred following the decimal point.

```
CHARACTER (LEN=*) fmt
PARAMETER (fmt='(-3PE12.4,-1PE12.4,0PE12.4,1PE12.4,3PE12.4)')
E = -2.7182818
WRITE (16, fmt) e, e, e, e, e
```

The WRITE statement creates the following record:

```
-0.0003E+04 -0.0272E+02 -0.2718E+01 -2.7183E+00 -271.83E-02
```

On output with a D or E edit descriptor, if a scale factor is in effect, the internal value of the output datum is the same as the actually transferred value. Only the position of the decimal point is shifted within the significand, and this is corrected by an appropriate adjustment of the exponent.

EN and ES edit descriptor: A P edit descriptor has no effect on output.

F edit descriptor: The internal value of the output datum is multiplied by 10^k during the conversion to the external character representation. The position of the decimal point within the output field, which is given by **d**, is not affected by the scale factor.

```
E = -2.7182818
WRITE (16, '(-1PF10.3,0PF10.3,1PF10.3,3PF10.3)') e, e, e, e
```

The WRITE statement creates the following record:

```
    -0.272    -2.718   -27.183 -2718.282      ← Output record
1234567891123456789212345678931234567894      ← Position
```

On output with an F edit descriptor, if a nonzero scale factor is in effect, the internal value of the output datum differs from the value of the external representation by one or more orders of magnitude.

G edit descriptor: A P edit descriptor has no effect when the G edit descriptor has the effect of an F edit descriptor. But when the G edit descriptor has the effect of an E edit descriptor, a P edit descriptor has the same effect on the G edit descriptor as on an E edit descriptor.

12.3.16 Sign Control Edit Descriptors: S SP SS

The S, SP, and SS edit descriptors may be used to control the output of plus signs in numeric output fields. These edit descriptors may affect the "normal" processor-dependent way of producing a plus sign in numeric output fields.

SP: Positive values in subsequent numeric output fields are represented with a leading plus sign.

SS: Positive values in subsequent numeric output fields are represented without a leading plus sign.

S: The normal processor-dependent way of producing a plus sign for positive numeric output values is reestablished.

At the beginning of the execution of a formatted output statement, sign control is processor-dependent. The effect is as if format control has encountered an S edit descriptor as the first edit descriptor in the format specification.

```
DOUBLE PRECISION d
DATA i, a, d /165, 1234.56, 67.4D-3/
WRITE (16, '(SP,I5,F10.2,SS,F10.2)') i, a, d
```

The WRITE statement creates the following record:

```
 +165  +1234.56       .07      ←— Output record
12345678911234567892123 45       ←— Position
```

12.3.17 Slash Edit Descriptor: /

The slash edit descriptor indicates the end of the data transfer in or from the current record.

Sequential input: When format control encounters a slash edit descriptor, the remaining portion of the input record is skipped without transferring data, and the file is positioned at the beginning of the next record. This record becomes the current record. That is, if there are more input list items, data

12.3 Edit Descriptors

may be transferred in these list items from this subsequent record; otherwise this subsequent record is skipped without transferring data.

```
READ (14, '(3I5//2I10)', END=100) i, j, k, l, m
   453   33 2876   112        ← 1st input record
 15432 2345    7  3418        ← 2nd input record
    44    2  378     4        ← 3rd input record
 12345678911234567892         ← Position
```

The execution of the READ statement defined the variables i, j, k, l, and m as follows: i = 453, j = 33, k = 2876, l = 44, and m = 2.

Direct input: When format control encounters a slash edit descriptor, the record number is increased by one and the file is positioned at the beginning of the next record. This record becomes the current record. That is, if there are more input list items, data may be transferred in these list items from this subsequent record; otherwise this subsequent record is skipped without transferring data.

Sequential output: When format control encounters a slash edit descriptor, the current record is terminated and a new record is created. This new record becomes the current record. It is the last record in the file (at this time). In the case of an external file, if the current record is empty when the slash edit descriptor is encountered, an empty record is created; in the case of an internal file, the record is filled with blank characters.

Direct output: When format control encounters a slash edit descriptor, the record number is increased by one and the file is positioned at the beginning of the record with this new record number. This record becomes the current record. That is, if there are more output list items, the output data may be transferred in this current record. If the current record is empty when the slash edit descriptor is encountered, it is filled with blank characters.

12.3.18 Tabulator Edit Descriptors: Tn TLn TRn

The T, TL, and TR edit descriptors may be used to specify the position at which the next character will be transferred in or from the current record.

The T and TL edit descriptors may be used to specify a position backward from the current position. Positioning backward to a position which has been processed already is allowed until the *left tab limit* is reached. The **left tab limit** is given by the position of the current record immediately at the beginning of the execution of the formatted data transfer statement. Note that this may be a position which is *not* at the beginning of the current record. If an input/output statement transfers more than one record, the left tab limit is given by the first character position of each current record (possibly except the first one).

Tn: The transfer of the next character is to occur at the character position **n**. In this case, the character positions are counted beginning at the left tab limit.

TLn: The transfer of the next character is to occur at the character position **n** characters backward from the current position. If this position is backward from the left tab limit, the transfer of the next character is to occurs at the left tab limit.

TRn: The transfer of the next character is to occur at the character position **n** characters forward from the current position. A TR edit descriptor has a similar effect as an X edit descriptor.

If any of the skipped positions are of type nondefault character, the effect of the T, TL, or TR edit descriptor is processor-dependent.

Input:

The characters in an input record may be transferred more than once (optionally with different edit descriptors). Even such a character position may be specified that is beyond the end of the record; in this case, no characters must be transferred from this position of the input record.

```
READ (18, '(F10.6,T1,F10.3)', END=100) a, b
   45824569    ←— Input record
   1234567890  ←— Position
```

The execution of the READ statement defines the variables a and b as follows: a = 45.824569 and b = 45824.569.

Output:

Suppose, format control has encountered a T, TL, or TR edit descriptor, when characters are transferred to or after the specified position, those character positions are filled with blank characters which are skipped without being previously filled.

```
WRITE (16, '(" railway",TR5,"racecourse",T2," run")')
```

The WRITE statement creates the following record:

```
   runway     racecource      ←— Output record
   12345678911234567892123    ←— Position
```

12.3.19 X Edit Descriptor: nX

The X edit descriptor has a similar effect as the TR edit descriptor. When format control encounters an X edit descriptor, the transfer of the next character is to occur at the character position **n** characters forward from the current position.

12.3 Edit Descriptors

If any of the skipped positions are of type nondefault character, the effect of the X edit descriptor is processor-dependent.

Input:

The next **n** characters in the input record are skipped without transferring data. Even a character position may be specified that is beyond the end of the record; in this case, no characters must be transferred from this position of the input record.

```
CHARACTER (LEN=*), PARAMETER :: fmt= &
                   &     '(I2,5X,F7.2,6X,I2,5X,F7.2,X15)'
READ (14, fmt, END=100) no1, pr1, no2, pr2

 1 pair   33.90 DM    5 pairs 29.90 DM/pair     ← Record (length 42)
123456789112345678921234567893123456789412      ← Position
```

The execution of the READ statement defines the variables no1, no2, pr1, and pr2 as follows: no1 = 1, no2 = 5, pr1 = 33.90, and pr2 = 29.90.

Output:

The next **n** character positions of the output record are skipped. Suppose, format control has encountered an X edit descriptor, when characters are transferred to or after the specified position, those character positions are filled with blank characters which are skipped without being previously filled.

```
DATA i, j, k /10101, 10010, 11001/
WRITE (14, '(3(I5,3X))') i, j, k
```

The WRITE statement creates the following record:

```
10101   10010   11001        ← Output record
123456789112345678921        ← Position
```

The output record has the length 21 (and not 24), because the last three skipped character positions are not filled with blank characters.

12.3.20 Z Edit Descriptors: Zw Zw.m

The Z edit descriptor is used to describe the editing of hexadecimal data. The corresponding input/output list item must be of type integer.

Input:

The input field may contain the hexadecimal digits *0, 1, 2, 3, 4, 5, 6, 7, A, B, C, D, E,* or *F*; and it may contain blank characters. If the **Fortran** processor supports lower-case letters, the letters *a* to *f* appearing in hexadecimal data are equivalent to corresponding upper-case letters. The input field must contain no sign, no decimal point, and no exponent. The specification of **m** has no effect on input. **w** must be positive.

```
INTEGER zi, zj, zk
READ (22, '(Z3,Z16,Z5)', END=100) zi, zj, zk

10A         000100F C000      ← Input record
12345678911234567892123 4     ← Position
```

The execution of the READ statement defines the variables z_i, z_j, and z_k as follows: $z_i = 266$, $z_j = 4111$, and $z_k = 49152$.

Output:

For a **Zw** edit descriptor, at least one octal digit without leading zeros is transferred to the output field. For a **Zw.m** edit descriptor, if the value to be transferred occupies fewer than **m** positions, the value is transferred right-justified with leading zeros as a hexadecimal number with **m** digits. And the leading (**w** − **m**) positions of the output field are filled with blank characters.

On output of the value zero with a **Zw** edit descriptor, the digit zero is transferred right-justified with (**w** − 1) leading blank characters. On output of the value zero with a **Zw.m** edit descriptor, **m** zeros are transferred right-justified; if **m** is zero, no digits but **m** blank characters are transferred, giving a field of all blank characters. When **m** and **w** are both zero on output of the value zero, one blank character is transferred to the output field.

```
INTEGER :: zi = 9, zj = 318, zk = 46
WRITE (34, '(Z7.4,Z10,Z8)') zi, zj, zk
```

The WRITE statement creates the following record:

```
   0009        13E      2E    ← Output record
1234567891123456789212345     ← Position
```

13 PROGRAM UNITS AND SUBPROGRAMS

A **program unit** is a *main program*, an *external subprogram*, a *module*, or a *block data program unit*. It is a physically complete sequence of program lines which ends with an END statement. A program unit may be processed independently of any other program unit by the processor.

Program units are those building blocks of a program which the programmer has to provide. A **program** consists of one main program and any number of other program units; that is, it may consist only of a main program. The program units of a program may appear in any order.

A **subprogram** is a *subroutine* or a *function*. Subprograms are program lines which belong together, which may be referenced in other parts of the executable program, and which solve particular parts of the complete problem solution of the program.

A subprogram may be an independent program unit such as an external subroutine or an external function. It may be embedded in a program unit as in the case of a *module subprogram*, an *internal subprogram*, or a *statement function*. In addition to these *user-defined subprograms*, there are *intrinsic subprograms*.

A *block data program unit* contains only specification statements. It may be used to initialize variables in named common blocks. A *module* may be used to put related declarations, specifications, and definitions (also subprogram definitions) together.

Block data program units and modules are not executable. Main programs, subroutines, and functions are executable. Note that modules may contain executable subprograms, the so called *module subprograms*.

During the execution of a program, at first the main program takes over control by executing its first executable statement. The main program may invoke subprograms. A subprogram may invoke other subprograms and, under certain circumstances, it may invoke itself. If no error condition occurs, the execution of the program is terminated when the END statement of the main program is executed or when a STOP statement (in the main program or in a subprogram) is executed.

13.1 Main Program

A **main program** is a program unit that has a first statement which is *no* BLOCK DATA, FUNCTION, MODULE, or SUBROUTINE statement.

PROGRAM name ⟵ PROGRAM statement; may be omitted
⋮ ⟵ Specification part; may be omitted
⋮ ⟵ Execution part; may be omitted
CONTAINS ⟵ May be omitted together with any subsequent internal subprograms
⋮ ⟵ Internal subprograms
END [PROGRAM [name]] ⟵ Executable END statement

The first statement in a main program may be (and should be) a PROGRAM statement. The last statement must be an END statement. Additional PROGRAM statements must not appear. The scoping unit of a main program must not contain BLOCK DATA, ENTRY, MODULE, or RETURN statements.

```
PROGRAM sample
!-------------------------------------------------------------------
  USE sample_module                       !    specification part
  :
  REAL, DIMENSION (4, 10) :: m1, m2, m3
!-------------------------------------------------------------------
  OPEN (15, FILE=..., FORM=...)           !    execution part
  CALL form(m1, m2)
  :
  IF (m1(3, 4) >= 78.95) PRINT *, m2(3, 4)
!-------------------------------------------------------------------
CONTAINS                                  !    internal subprograms
  SUBROUTINE form (fp1, fp2)
  :
  END SUBROUTINE form
  :
!-------------------------------------------------------------------
END PROGRAM sample
```

PROGRAM statement: The PROGRAM statement may be used to specify a program name. This **name** is a global name, which must not be used as a local name within the main program and even within the whole program.

Specification part: The specification part of a main program must not include INTENT, OPTIONAL, PRIVATE, and PUBLIC statements. And the corresponding attributes must not be specified in a type declaration statement within a main program.

Execution part: The execution part contains the executable statements and constructs of the main program.

Internal subprograms : A main program may contain subprograms. These *internal subprograms* appear subsequent to the execution part of the main program between the CONTAINS statement and the END statement of the main program.

END statement: If the END statement has the form END PROGRAM, the **name** of the main program may be specified in the END statement.

13.2 Modules

A **module** is a nonexecutable program unit which contains only declarations, specifications, and/or definitions (also subprogram definitions). The *public* entities in a module may be made *accessible* to other scoping units. An entity is **public** if it has the PUBLIC attribute.

A reference to a module is accomplished by a USE statement. The effect of such a module reference is that the public entities of the module become accessible in the scoping unit containing the USE statement. A module may reference other modules, but it must neither directly nor indirectly reference itself.

MODULE name	⟵ MODULE statement
⋮	⟵ Specification part; may be omitted
CONTAINS	⟵ May be omitted together with any subsequent module subprograms
⋮	⟵ Module subprograms
END [MODULE [name]]	⟵ Nonexecutable END statement

The first statement in a module must be a MODULE statement. The last statement must be an END statement. Additional MODULE statements must not appear in a module. A module must not include a PROGRAM or a BLOCK DATA statement.

MODULE statement: The **name** of the module is used in the program where public entities from the module are to be made accessible (by an appropriate USE statement). The name is a global name. It must not be used as a local name within the module. The name of the module may be specified in the END MODULE statement of the module.

```
MODULE list
  REAL, DIMENSION (10) :: a, e
  ⋮
  CHARACTER (LEN=17)   :: card
END MODULE list
```

Specification part: The specification part of a module must not include statement function definitions, ENTRY, FORMAT, INTENT, and OPTIONAL statements. And the corresponding attributes must not be specified in a type declaration statement within the specification part of a module. If an object of a data type for which default initialization is specified in its type definition appears in the specification part of a module and does not have the ALLOCATABLE or POINTER attribute, the object must have the SAVE attribute. The names of the module subprograms may appear within the specification part of a module only in PRIVATE or PUBLIC statements. Executable statements must not appear in the specification part of a module.

Module subprograms : A module may contain subprograms. These *module subprograms* appear subsequent to the specification part of the module between the CONTAINS statement and the END statement of the module.

END statement: If the END statement has the form END MODULE, the **name** of the module may be specified in the END statement.

13.2.1 USE Statement

The USE statement is used to reference a module. Where an appropriate USE statement appears, declarations, specifications, and definitions may become accessible from the referenced module; and we say that the scoping unit containing the USE statement has *access to the module*. The following entities can be made accessible from a module only if they are public: named variables, module subprograms, derived types, named constants, namelist groups, interface blocks, and generic identifiers defined by generic interface blocks (for instance, generic names, generic operators, and the generic assignment).

USE module

USE module, local_name => use-name [, local_name => use-name]...

USE module, ONLY :

USE module, ONLY : only [, only]...

module is the name of a module, and each **local_name** is a local name in the scoping unit containing the USE statement. Each **use-name** is the name of a public entity in the **module**. And each **only** in the *only-list* is one of the following:

use-name
local_name => use-name
generic_name
OPERATOR (operator)
ASSIGNMENT (=)

The last three specifications designate corresponding generic interface blocks.

Entities in the scoping unit with the USE statement and public entities in the referenced module may become associated. This is called *USE association*.

A scoping unit may include more than one USE statement. These may be USE statements referencing different modules, if necessary. But two or more USE statements may reference even the same module; in this case, if two or more USE statements appear with an only-list, a subsequent specification is interpreted as a continuation of the last specification.

If a USE statement *without* the keyword ONLY is processed, *all* public entities within the specified module are made accessible. A USE statement *with* the keyword ONLY makes *only* the specified public entities of the module accessible. In both cases, public named entities may be renamed for use in the scoping unit containing the USE statement by specifying a local name for them.

Module entities *must* be renamed if there are name conflicts between entities accessed from different modules or between entities accessed from the module and entities in the scoping unit containing the USE statement.

`USE graphic_lib`

All public entities in module `graphic_lib` are made accessible such that each named entity has the same name both in the scoping unit containing the USE statement and in the module.

`USE graphic_lib, cir => circle, ar => arc, li => line`

As above, except the module entities `circle`, `arc`, and `line` are accessible by the local names `cir`, `ar`, and `li`, respectively, in the scoping unit containing the USE statement.

Local names

If a **local_name** is specified in a USE statement, this is (in the scoping unit containing the USE statement) the local name for the module entity **use-name**.

An accessible entity in the referenced module has one or more local names. These names are: The name of the entity in the referenced module if that name appears as **use-name** in any only-list for that module; each of the **local_name**s the entity is given for that module; and the name of the entity in the referenced module if that name does not appear as a **use-name** for that module.

A local name of an entity made accessible by a USE statement must not be respecified in a type declaration statement or in another specification statement in the scoping unit containing the USE statement except as a member of a namelist group in a NAMELIST statement. But if a USE statement appears in a module, the names of the entities made accessible by the inner USE statement may be specified in PUBLIC or PRIVATE statements in the module containing the USE statement. They are treated as entites of the module; that is, they become public or private entities in the module containing the USE statement. If an entity made accessible to a module is not specified in a PUBLIC or PRIVATE statement in this module, it has the default accessibility attribute.

```
MODULE m1
  INTEGER, DIMENSION (100) :: measurements
  :
END MODULE m1
MODULE m2
  USE m1, mean => measurements
  :
END MODULE m2
PROGRAM mp
  USE m1
  USE m2
  :
END PROGRAM
```

In the main program **mp**, the local names **measurements** and **mean** identify the same integer array.

USE with ONLY

If a scoping unit needs only a part of a module, the keyword ONLY followed by an only-list may be specified in the USE statement. In this case, only those (public) entities are made accessible which are specified in the only-list. A USE statement which includes the keyword ONLY does not override a USE statement *without* the keyword ONLY.

```
USE statistic_lib, ONLY : gauss, hg => histogram
```

Only the module entities **gauss** and **histogram** are made accessible. These entities are identified by the local names **gauss** and **hg** in the scoping unit containing this USE statement.

13.3 Block Data Program Units

A **block data program unit** is a nonexecutable program unit that is used to initialize objects in named common blocks.

BLOCK DATA [name]	⟵ BLOCK DATA statement
:	⟵ Specification part; may be omitted
END [BLOCK DATA [name]]	⟵ Nonexecutable END statement

The first statement in a block data program unit must be a BLOCK DATA statement. The last statement must be an END statement. Additional BLOCK

13.3 Block Data Program Units

DATA statements must not appear in a block data program unit. A block data program unit must not contain PROGRAM, MODULE, FUNCTION, and SUBROUTINE statements, and executable statements.

A program may contain more than one block data program unit. The named block data program units must be unambiguously identified by different names. And one of the block data program units may remain unnamed.

BLOCK DATA statement: The **name** is a global name.

Specification part: The specification part of a block data program unit may include only USE statements, type declaration statements, IMPLICIT statements, derived type definitions, and the following additional specification statements: COMMON, DATA, DIMENSION, EQUIVALENCE, INTRINSIC, PARAMETER, POINTER, SAVE, and TARGET. Note that it must not contain an interface block.

The following attributes must not be included in a type declaration statement within a block data program unit: ALLOCATABLE, EXTERNAL, INTENT, OPTIONAL, PRIVATE, and PUBLIC.

Only variables in *named* common blocks may be initialized in a block data program unit. The objects to be initialized must not be pointers. If an object in a named common block is initialized, all storage units in the common block storage sequence must be specified even if they are not all initialized.

A block data program unit may initialize variables in several named common blocks. But a particular named common block may appear only within one block data program unit in a program.

END statement: If the END statement has the form END BLOCK DATA, the **name** of the block data program unit may be specified in the END statement.

```
BLOCKDATA
  CHARACTER type *12, range *15, pre *9, alloc *17
  INTEGER line_width, font, sl
  COMMON /head/ type, range, pre, alloc
  COMMON /page/ line_width, font, sl
  DATA range, pre /'range of values', 'precision'/
  DATA alloc /'memory allocation'/, line_width, sl /79, 60/
END BLOCKDATA

BLOCKDATA params
  PARAMETER (pi=3.14159)
  REAL angle (3, 9), measure (9)
  COMMON /three/ angle, /cm/ measure
  DATA angle /27*pi/
  DATA measure(1), measure(2), measure(4), measure(7) /4*1./
END BLOCKDATA params
```

The first block data program unit is unnamed. The variables **type** and **font** in the common blocks **head** and **page**, respectively, are not initialized. The second block data program unit is named. The array **measure** is not completely initialized.

13.4 Subprograms

A **subprogram** is either a *function* or a *subroutine*. Functions and subroutines are different from each other with regard to subprogram definition and subprogram reference. A subprogram is a user-defined independent program unit, a user-defined embedded subprogram, or an intrinsic subprogram.

A subprogram is not an independent part of a program. It will be executed only when it is referenced and (explicitly or implicitly) invoked. The referencing scoping unit and the referenced subprogram may exchange informations.

Classification of subprograms

Subprograms may be classified as follows:

- An **external subprogram** is a user-defined subprogram which is an independent program unit. The subprogram may be defined by another language than **Fortran**.

- An **internal subprogram** is a user-defined subprogram which looks like an external subprogram, but its subprogram definition is not a program unit. It is a part of a program unit or of another subprogram.

- A **module subprogram** is a user-defined subprogram which looks like an external subprogram, but its subprogram definition is not a program unit. It is a part of a module. If it is public, its subprogram definition my be made accessible to other program units or subprograms.

- A **statement function** is a user-defined function. Its function definition consists only of one statement. It is a part of a program unit or of a subprogram.

- An **intrinsic subprogram** is a predefined subprogram. It is provided by the **Fortran** processor.

An **operator function** is an external function or a module function which defines the interpretation of a defined operator or which defines the interpretation of an extension of an intrinsic operator.

An **assignment subroutine** is an external subroutine or a module subroutine which defines the interpretation of a defined assignment statement.

13.4 Subprograms

A **dummy subprogram** is a dummy argument which is specified as a subprogram or which appears in a subprogram reference as the name of the referenced subprogram.

A function may be referenced only in an expression. The function reference may appear as an operand in the expression or the function may be implicitly (that is, automatically) invoked when a defined operator or an extended intrinsic operator is encountered during expression evaluation. A subroutine is either explicitly referenced and invoked by the execution of a CALL statement or it is implicitly invoked when a defined assignment statement is executed.

Each external subprogram, internal subprogram, and module subprogram has a *main entry point*. The name of this main entry point is specified in the FUNCTION or SUBROUTINE statement, respectively. Normally, these subprograms are invoked by referencing the main entry point. But external subprograms and module subprograms may contain one or more additional entry points. The names of these additional entry points are specified in ENTRY statements, one ENTRY statement for each additional entry point. Each of these entry points of a subprogram defines a *subprogram interface*, which may be referenced.

A **subprogram interface** has the following characteristics:

- Classification of the interface (function or subroutine);
- Specific name of the entry point;
- Interface of a pure or elemental subprogram;
- Names of the dummy arguments;
- Kind of dummy arguments (datum, subprogram, or asterisk);
- Characteristics of the dummy arguments;
- Generic name of the subprogram (if any); and possibly
- Characteristics of the function result.

The names of the additional entry points of external subprograms and module subprograms are subject to the same rules as their main entry points. There is no difference between a subprogram reference via the main entry point and a subprogram reference via another entry point. Therefore, we do not distinguish the name of the main entry point from the names of the additional entry points. And each one is called a *subprogram name*.

The execution of an external subprogram, internal subprogram, or module subprogram begins with the execution of the first executable statement following the referenced entry point specified in the FUNCTION, SUBROUTINE, or ENTRY statement.

Pure subprograms

A **pure subprogram** cannot cause side effects because it does not include any operation that is able to modify the status of a variable outside the scoping unit of the subprogram (except in the case of a pure subroutine which is allowed to modify certain actual arguments associated with its dummy arguments) and because it does not include a STOP statement or any input/output statement for an external file.

A pure subprogram is a user defined subprogram with the keyword PURE in the prefix of its SUBROUTINE or FUNCTION statement, an intrinsic function or the intrinsic MVBITS subroutine, or a statement function referencing only pure functions.

Elemental subprograms

An **elemental subprogram** is defined with scalar dummy argument but under certain conditions is allowed to be referenced with array valued actual arguments. Each elemental subprogram is a pure subprogram.

An elemental subprogram is an intrinsic or a user defined elemental subprogram. A user defined elemental subprogram is defined with the keyword ELEMENTAL in the prefix of its SUBROUTINE or FUNCTION statement.

13.4.1 User-Defined Functions (except Statement Functions)

A function begins (apart from optional leading comment lines) with a FUNCTION statement, it ends with an END statement, and contains at least one additional executable statement between the FUNCTION and the END statement. A function may contain any statements except BLOCK DATA, MODULE, and PROGRAM statements.

A function is referenced and invoked during expression evaluation either explicitly when the function name is encountered as an operand or implicitly when a defined operator or an extended intrinsic operator is encountered. The invoked function returns a value to the place of the function reference, this value is the function result. Apart from this function result, the argument list and common blocks may be used to exchange informations between the function and the invoking scoping unit. When a RETURN statement or the END statement of the function is executed, the execution of the function is terminated, and control returns to the expression containing the function reference.

13.4.1.1 Function Definition

```
[...] FUNCTION name ([...]) [...]    ←— FUNCTION statement
    ⋮                                 ←— Specification part; may be omitted
    ⋮                                 ←— Execution part
CONTAINS                              ←— May be omitted together with the
                                         subsequent internal subprograms
    ⋮                                 ←— Internal subprograms
END [ FUNCTION [ name ]]              ←— Executable END statement
```

FUNCTION statement

The FUNCTION statement is used to specify the name of (the main entry point) of the function and may be used to specify a dummy argument list. The FUNCTION statement must be the first statement of a function.

[prefix] **FUNCTION name ( ) [RESULT (result)]**

[prefix] **FUNCTION name (dummy_argument [, dummy_argument]...) &**
                                                    [RESULT (result)]

The **prefix** contains the following optional specifications: a *type specification* for the function, the keyword RECURSIVE if the function is recursive, the keyword PURE if the function is pure, and/or the keyword ELEMENTAL if the function is elemental. The keywords may appear in any order at most once. RECURSIVE and ELEMENTAL must not appear within the same FUNCTION statement. When written in free source form, blanks are required to separate the keywords within the **prefix**.

name is the specific name of the function. For an external function, a dummy function, or a module function, an additional *generic name* may be specified by a generic interface block.

Each **dummy_argument** is the name of a variable or the name of a dummy subprogram. And **result** is the name of the *result variable*. This name *must* be specified for the directly recursive invocation of the function and *may* be specified for other invocations of the function.

Each entry point of a function has a type, a kind type parameter, and (if applicable) a character length. These attributes may be given by default implicit typing, by an IMPLICIT statement, or by an explicit type declaration in the specification part of the function. For the main entry point, these attributes may be specified in the FUNCTION statement. If an entry point is of derived type, this type must be defined within the specification part of the function or must be available by USE association or host association.

If the length specification in the FUNCTION statement is preceded by an asterisk *, the comma after the type specification must be omitted.

The entry points of a function may have additional attributes, such as the DIMENSION and the POINTER attribute. The DIMENSION attribute must be specified in the specification part of the function. For an array-valued non-pointer function, the function result defined by the *result variable* (see next page) must be an explicit-shape array. For a pointer function, the POINTER attribute must be specified in the specification part of the function.

```
REAL FUNCTION mean (x)
  DIMENSION x (100)
  mean = 0.
  DO, i=1,100
    mean = mean + i/100.*x(i)
  END DO
  mean = mean/100.
END FUNCTION
```

Function result

The attributes of the invoked entry point of a function determine the characteristics of the function result.

On return to the invoking expression, the function result is given by the value of the *result variable*. This **result variable** is a local variable in the function.

If RESULT is *not* specified, the result variable has the same name as the function. And all occurrences of the **name** in the execution part of the function are interpreted as references to this result variable. The characteristics of the function result are given by the attributes and the value of this result variable.

If RESULT is specified, a separate **result** variable is specified. And all occurrences of the **name** in the execution part of the function are interpreted as recursive function references. The characteristics of the function result are given by the attributes and the value of this **result** variable. In this case, there is no local variable having the same name as the function. And the name of the function must not appear in type declaration statements or other specification statements within the scoping unit of the function.

If an ENTRY statement is used to specify an additional entry point, there also is a result variable defined for this entry point. And in analogy to the main entry point, this result variable has either the same name as this additional entry point or a separate result variable must be specified together with the keyword RESULT in the ENTRY statement.

Nonpointer function: If the function result is *not* a pointer, the result variable of the invoked entry point must become defined with a valid value during the execution of the function. This value may be used or may be redefined during the execution of the function. Finally, when a RETURN statement or the END statement of the function is executed, the current value of this result variable is the function result, which is then reused as an operand in the invoking expression.

Pointer function: If the function result is a pointer, the pointer result variable must become associated with a target during the execution of the function, or the association status of this pointer must become disassociated. The result variable of the function may be an array, that is, an array pointer. The shape of the result variable is determined by the shape of the result variable when control returns to the invoking expression. The length specification of a pointer functions of type character must not be an asterisk.

Association of the result variables

The result variables of a function are associated with each other. The result variables of the main entry point and of the additional entry points identify the same variable if they have the same characteristics such as the type, the kind type parameter, the length, the POINTER attribute (yes or no), the shape (if they are nonpointers), the dependence on other entities (if the specification of the kind type parameter, of the character length, or of an array bound is an expression), and for character functions, if they have the same character length asterisk * or if they do not have assumed length.

Otherwise, the result variables of the main entry point and of the additional entry points are storage associated, and they must be all scalars, they must be all nonpointers, and every entry point must be of one of the types default integer, default real, double precision real, default complex, or default logical.

If a (nonpointer) function has entry points (that is, result variables) which do not identify the same variable, exactly that result variable must be defined on return to the invoking scoping unit which belongs to the invoked entry point.

If two or more entry points of a function identify the same variable, any of these result variables may be defined on return to the invoking scoping unit.

Dummy arguments

The names of the dummy arguments are local names which must not appear in EQUIVALENCE, DATA, INTRINSIC, PARAMETER, or SAVE statements. They may appear in subsequent type declaration statements or in other specification statements. In a COMMON statement, the name of a dummy argument may be used only as the name of a common block.

Dummy arguments being arrays must be specified as arrays in type declaration statements, by DIMENSION, POINTER, or TARGET statements in the specification part of the function.

Internal subprograms, CONTAINS statement

External functions and module functions may contain internal subprograms. If CONTAINS is written, at least one internal subprogram must follow.

END statement

If the END statement has the form END FUNCTION, the **name** of the function may be specified in the END statement.

Recursive functions

The keyword RECURSIVE must be specified in the FUNCTION statement if the function directly or indirectly invokes itself. This is independent on whether the function recursively references its main entry point, one of its additional entry points, or different entry points. The keyword RESULT and the **result** variable must be specified if the function invokes its main entry point directly recursively. RECURSIVE and ELEMENTAL must not appear within the same FUNCTION statement.

```
RECURSIVE FUNCTION fff (duar) RESULT (f)
  REAL, DIMENSION (:) :: duar;  REAL, DIMENSION (SIZE(duar)) :: f
  n = SIZE(duar)
  IF (n <= 1) THEN
    f = duar
  ELSE
    n = n/2
    f(:n)   = fff(duar(:n))
    f(n+1:) = fff(duar(n+1:)) + f(n)
  END IF
END FUNCTION fff
```

If the keyword RESULT followed by a result variable appears in the FUNCTION statement, the name of the result variable must be different from the name of the function; and the name of the function must not appear in type declaration statements or other specification statements in the specification part of the function. A recursive character function must not be specified to have the character length *, that is, it must not have assumed length.

Pure functions

A user defined pure function must not have any means to cause side effects. Therefore it must not include any operation that could modify the status of any variable outside its scoping unit, it must not include a STOP statement, and it must not include any input/output statement for an external file.

A user defined pure function is specified with PURE or ELEMENTAL in the prefix of its FUNCTION statement,

The following conditions apply to user defined pure functions:

- The specification part of a pure function must specify INTENT(IN) for each dummy argument which is a variable without the POINTER attribute.

- A local variable declared in the specification part or within an internal subprogram of a pure function must neither explicitly nor implicitly have the SAVE attribute.

- The specification part of a pure function must include for each dummy subprogram an interface block with an interface definition specifying that the dummy subprogram is pure.

- If a dummy function, a recursive function outside its own scoping unit, or a nonrecursive external function is used in a context that requires it to be pure, then an interface block for the function must be available within the scope of that use specifying PURE or (optionally for the nonrecursive case) ELEMENTAL within the prefix of the FUNCTION statement.

- All internal subprograms of a pure function must also be pure.

- Any subprogram referenced explicitly or implicitly within a pure function must be pure.

- A pure function must not contain a STOP statement and must not contain a PRINT, OPEN, CLOSE, BACKSPACE, ENDFILE, REWIND, and INQUIRE statement.

- A pure function must not contain a READ statement or WRITE statement for an external file.

Elemental functions

A user defined elemental function is specified with the keyword ELEMENTAL in the prefix of its FUNCTION statement.

The above rules for pure functions also apply to elemental functions. Additionally, each dummy argument and the result variable of an elemental function must be scalar non-pointer variables. Within a specification expression, a dummy argument or a subobject of a dummy argument must not appear accept as the argument to one of the intrinsic functions BIT_SIZE, KIND, LEN, or to one of the numeric inqiry functions DIGITS, EPSILON, HUGE, MAXEXPONENT, MINEXPONENT, PRECISION, RADIX, RANGE, and TINY. ELEMENTAL and RECURSIVE must not appear within the same FUNCTION statement.

13.4.1.2 Explicit Invocation of a Function, Function Reference

A function is explicitly invoked by a function reference during expression evaluation. A function invocation returns a value for use in the expression at the place of the function reference; this value is the function result.

name ( )

name (aa [, aa]...)

The **name** is the (specific or generic) name of the main entry point or of one of the additional entry points of a function. And each **aa** is an actual argument specification, which may be written as a *positional argument* or as a *keyword argument*.

For a positional argument, only the actual argument is specified. A keyword argument has the form **dummy_argument = actual_argument**, where **dummy_argument** is the name of a dummy argument from the interface of the referenced function. This name is used as a keyword for the **actual_argument** which will become associated with this dummy argument.

A function defined with an empty dummy argument list must be referenced with an empty actual argument list; that is, the parentheses must not be omitted. If the actual argument list includes a keyword argument, all subsequent actual arguments must be specified as keyword arguments.

On invocation of a function, the **name** used in the function reference and the actual arguments must have appropriate characteristics. In addition, it may be important whether or not an interface block containing an interface definition of the referenced entry point is available in the invoking scoping unit.

Characteristics of the name used in the function reference

In the scoping unit containing the function reference, the entity with this **name** must correspond to the referenced entry point with regard to name, type, kind type parameter, (if applicable) character length, and shape. In addition, the dependence on other entities is a characteristic if the kind type parameter specification, the character length or an array bound is not a constant expression.

13.4 Subprograms

Character function: For a character function, the character length of the entity **name** used in the function reference must correspond to the length specified in the function definition for the referenced entry point.

If the definition of an external function specifies the length * for a result variable of the function, this result variable assumes its length from the length specified for the referenced function name in the invoking scoping unit. This length * is called the **assumed length**. In this case, the invoking scoping unit must not specify the character length * for the name of the referenced function.

Assumed length must not be specified for internal functions, module functions, recursive functions, pure functions, pointer functions, operator functions, and array-valued functions.

Execution

The invocation of a function is executed as follows:

1. Evaluation of the actual arguments which are expressions with operators or which are enclosed in parentheses; and determination of the position of the actual arguments which are subobjects.
2. Association of the arguments.
3. Execution of the function beginning with the first executable statement following the invoked entry point.
4. Return to the expression containing the function reference, where the function result is now available as an operand.

```
REAL mean, min, jan (100), feb (100)
 :
min - mean(jan) * 100.
ref = mean(feb) * 10.
IF (ref .LT. min) min = ref
 :
END PROGRAM
```

The function **mean** is invoked twice. In the specification part of the main program, the entity **mean** is declared to be a real entity because **mean** is defined as a real function.

Recursive reference: For each invocation of a function, a separate **instance** of the function is created and executed. Each instance of a function has an independent set of dummy arguments and nonsaved local variables. An internal subprogram or a statement function which is directly invoked from an instance of a function has access to the entities of the instance of the host function by host association.

All other entities are shared by all instances of the function. For example, if a saved variable is defined (with a valid value) in one instance of the function, the variable may be reused with this value in any other instance of the function.

Elemental reference: A reference to an elemental function is an elemental reference if one ore more actual arguments are arrays and all these arrays have the same shape.

If an elemental function is referenced, all actual arguments must be conformable. If all actual arguments are scalar, the result is scalar. If there is one actual array argument or if there are several conformable actual array arguments, the result has the same shape as the array argument(s). And the values of the elements of the result are the same as if the scalar function were applied separately for corresponding elements of the actual argument array(s). The total result of the function is independent of the order of the element-wise invocations of the function.

13.4.1.3 Operator Functions

An *operator function* defines the interpretation of a defined operator or of an extended intrinsic operator. It is automatically invoked when the operator is encountered during expression evaluation.

A function may be used as an operator function only if it has one or two non-optional dummy arguments, if the dummy arguments are data objects, if the arguments are specified with the INTENT(IN) attribute, and if the result variable of the function has no assumed length.

Whether or not such a function eventually is implicitly invoked as an operator function depends on additional conditions.

13.4.2 User-Defined Subroutines

A subroutine begins (apart from optional leading comment lines) with a SUBROUTINE statement, it ends with an END statement, and may contain between the SUBROUTINE and the END statement additional executable statements. A subroutine may contain any statements except BLOCK DATA, MODULE, and PROGRAM statements.

A subroutine is referenced and invoked when a CALL statement is executed or when a defined assignment statement is executed. The argument list and common blocks may be used to exchange informations between the subroutine and the invoking scoping unit. When a RETURN statement or the END statement of the subroutine is executed, the execution of the subroutine is terminated, and control returns to the invoking scoping unit.

13.4.2.1 Subroutine Definition

```
[...] SUBROUTINE name [(...)] [...]     ← SUBROUTINE statement
   ⋮                                    ← Specification part; may be omitted
   ⋮                                    ← Execution part; may be omitted
CONTAINS                                ← May be omitted together with the
                                          subsequent internal subprograms
   ⋮                                    ← Internal subprograms
END [SUBROUTINE [name]]                 ← Executable END statement
```

SUBROUTINE statement

The SUBROUTINE statement is used to specify the name (of the main entry point) of the subroutine and may be used to specify a dummy argument list. The SUBROUTINE statement must be the first statement of a subroutine.

[**prefix**] **SUBROUTINE name**

[**prefix**] **SUBROUTINE name ( )**

[**prefix**] **SUBROUTINE name (dummy_argument** [**, dummy_argument**]... **)**

The **prefix** contains the following optional specifications: the keyword RECURSIVE if the subroutine is recursive, the keyword PURE if the subroutine is pure, and/or the keyword ELEMENTAL if the subroutine is elemental. The keywords may appear in any order at most once. RECURSIVE and ELEMENTAL must not appear within the same SUBROUTINE statement. When written in free source form, blanks are required to separate the keywords within the **prefix**.

name is the specific name of the subroutine. For an external subroutine, a dummy subroutine, or a module subroutine, an additional *generic name* may be specified by a generic interface block.

Each **dummy_argument** is the name of a variable, the name of a dummy subprogram, or an asterisk *.

```
SUBROUTINE summa (x, y, r)
  REAL, DIMENSION (100) :: x, y, r
  r = x + y
  WHERE (r > 775.) r = 775.
END
```

Dummy arguments

The names of the dummy arguments are local names which must not appear in EQUIVALENCE, DATA, INTRINSIC, PARAMETER, or SAVE statements.

They may appear in subsequent type declaration statements or in other specification statements. In a COMMON statement, the name of a dummy argument may be used only as the name of a common block.

Dummy array arguments must be specified as arrays in type declaration statements, by DIMENSION, POINTER, or TARGET statements in the specification part of the subroutine.

An asterisk specified as a dummy argument means that an alternate return specifier must be specified and associated as a corresponding actual argument in a subroutine reference.

Internal subprograms, CONTAINS statement

External subroutines and module subroutines may contain internal subprograms. If CONTAINS is written, at least one internal subprogram must follow.

END statement

If the END statement has the form END SUBROUTINE, the **name** of the subroutine may be specified in the END statement.

Recursive subroutines

The keyword RECURSIVE must be specified in the SUBROUTINE statement if the subroutine directly or indirectly invokes itself. This is independent on whether the subroutine recursively references its main entry point, one of its additional entry points, or different entry points. RECURSIVE and ELEMENTAL must not appear within the same SUBROUTINE statement.

Pure subroutines

A user defined pure subroutine must not have any means to cause side effects. Therefore it must not include any operation that could modify the status of any variable outside its scoping unit (except actual arguments associated with certain of its dummy arguments), it must not include a STOP statement, and it must not include any input/output statement for an external file.

A user defined pure subroutine is specified with the keyword PURE or ELEMENTAL in the prefix of its SUBROUTINE statement,

The following conditions apply to user defined pure subroutines:

- The specification part of a pure subroutine must specify INTENT attributes for each dummy argument which is a variable without the POINTER attribute.

- A local variable declared in the specification part or within an internal subprogram of a pure subroutine must neither explicitly nor implicitly have the SAVE attribute.
- The specification part of a pure subroutine must include for each dummy subprogram an interface block with an interface definition specifying that the dummy subprogram is pure.
- If a dummy subroutine, a recursive subroutine outside its own scoping unit, or a nonrecursive external subroutine is used in a context that requires it to be pure, then an interface block for the subroutine must be available within the scope of that use specifying PURE or (optionally for the nonrecursive case) ELEMENTAL within the prefix of the SUBROUTINE statement.
- All internal subprograms of a pure subroutine must also be pure.
- Any subprogram referenced explicitly or implicitly within a pure subroutine must be pure.
- A pure subroutine must not contain a STOP statement and must not contain a PRINT, OPEN, CLOSE, BACKSPACE, ENDFILE, REWIND, and INQUIRE statement.
- A pure subroutine must not contain a READ statement or WRITE statement for an external file.

Elemental subroutines

A user defined elemental subroutine is specified with the keyword ELEMENTAL in the prefix of its SUBROUTINE statement.

The above rules for pure subroutines also apply to elemental subroutines. Additionally, each dummy argument of an elemental subroutine must be a scalar non-pointer variable. ELEMENTAL and RECURSIVE must not appear within the same SUBROUTINE statement.

13.4.2.2   Explicit Invocation of a Subroutine, CALL Statement

A subroutine is explicitly invoked by the execution of a CALL statement.

CALL name

CALL name ( )

CALL name (aa [, aa]...)

The **name** is the (specific or generic) name of the main entry point or of one of the additional entry points of a subroutine. And each **aa** is an actual argument specification, which may be written as a *positional argument* or as a *keyword argument*.

For a positional argument, only the actual argument is specified. A keyword argument has the form **dummy_argument = actual_argument**, where **dummy_argument** is the name of a dummy argument from the interface of the referenced subroutine. This name is used as a keyword for the **actual_argument** which will become associated with this dummy argument.

In addition, if a user-defined subroutine is referenced, an *alternate return specifier* may be specified as an actual argument. An **alternate return specifier** is written as an asterisk * followed by a statement label of a branch target statement in the scoping unit containing the CALL statement; this branch target statement is the *alternate return target*.

A subroutine which is defined without a dummy argument list or with an empty argument list may be referenced by a CALL statement optionally either without an actual argument list or with an empty actual argument list. If the actual argument list includes a keyword argument, all subsequent actual arguments must be specified as keyword arguments. If alternate return specifiers are specified, they must appear before the keyword arguments.

On invocation of a subroutine, the actual arguments must have appropriate characteristics. In addition, it may be important whether or not an interface block containing an interface definition of the referenced entry point is available in the invoking scoping unit.

Execution

A CALL statement is executed as follows:

1. Evaluation of the actual arguments which are expressions with operators or which are enclosed in parentheses; and determination of the position of the actual arguments which are subobjects.

2. Association of the arguments.

3. Execution of the subroutine beginning with the first executable statement following the invoked entry point.

4. Return to the invoking scoping unit containing the CALL statement.

```
REAL, DIMENSION (100) :: jan, feb, y1993
READ (14, END=100) jan, feb
CALL terms(jan, feb, y1993)
WRITE (16) y1993
  ⋮
END
```

After return to the invoking scoping unit, the execution of the CALL statement terminates, and program execution is continued with the WRITE statement.

Recursive reference: For each invocation of a subroutine, a separate *instance* of the subroutine is created and executed. Each instance of a subroutine has an independent set of dummy arguments and nonsaved local variables. An internal subprogram or a statement function which is invoked directly from an instance of a subroutine has access to the entities of the instance of the host subroutine by host association.

All other entities are shared by all instances of the subroutine. For example, if a saved variable is defined (with a valid value) in one instance of the subroutine, the variable may be reused with this value in any other instance of the subroutine.

Elemental reference: A reference to an elemental subroutine is an elemental reference if all actual arguments are scalar or if all actual arguments being associated with INTENT(OUT) and INTENT(INOUT) dummy arguments are arrays that have the same shape and are conformable with the remaining actual argument.

If there is an actual array argument being associated with an INTENT(OUT) or INTENT(INOUT) dummy argument, the values of the elements of these actual arguments are the same as if the scalar subroutine were applied separately for corresponding elements of the actual argument array(s). The total result of the evaluations is independent of the order of the element-wise invocations of the subroutine.

13.4.2.3 Assignment Subroutines

An *assignment subroutine* defines the interpretation of a defined assignment statement. It is automatically invoked when the defined assignment statement is executed.

A subroutine may be used as an assignment subroutine only if it has two nonoptional dummy arguments, if the dummy arguments are data objects, if the first dummy argument is specified with the INTENT(OUT) or INTENT(INOUT) attribute, and if the second dummy argument is specified with the INTENT(IN) attribute.

Whether or not such a subroutine eventually is implicitly invoked as an assignment subroutine depends on additional conditions.

13.4.3 External Subprograms

An external subprogram is a user-defined subprogram which is not embedded in a module, in a main program, or in a subprogram. It has a main entry point and may have one or more additional entry points.

Subprogram name

The name of an external subprogram and the names of its additional entry points are global names. If the name of an external subprogram is the same as the name of an intrinsic subprogram, the external subprogram is used in the scoping unit containing the subprogram reference if the name is specified in an EXTERNAL statement or if an interface block for the external subprogram is available.

Function: Type, kind type parameter, and (if applicable) character length of an external function must be implicitly or explicitly specified in the invoking scoping unit if the reference is not directly recursive and, in addition, if no interface block for the function is available in the scoping unit with the function reference.

The character length * (i.e. assumed length) may be specified for an external character function which is no array-valued function, no recursive function, no pure function, and no pointer function.

```
PARAMETER (m=10, n=12)
REAL, DIMENSION (m, n) :: a, b, h
INTERFACE
  FUNCTION pytha (x, y, e, z)
    INTEGER e, z
    REAL, DIMENSION (e, z) :: pytha, x, y
  END FUNCTION pytha
END INTERFACE
  ⋮
h = 3.*pytha(x=a, y=b, e=m, z=n) - 1.
  ⋮
END

FUNKTION pytha (x, y, e, z)
  INTEGER e, z
  REAL, DIMENSION (e, z) :: pytha, x, y
  pytha = SQRT(x*x + y*y)
END FUNCTION pytha
```

The right-hand side of the assignment statement h =... is an array expression which has the shape (/m, n/).

Invocation of an external subprogram

The form of the subprogram reference and its possibilities depend on whether only FORTRAN 77 features are to be used or whether any of the extended Fortran 95 features are to be used.

Extended Fortran 95 features regarding subprogram reference and subprogram definition are:

- The subprogram reference may include *keyword (actual) arguments*;
- The subprogram reference need not include a corresponding actual argument for an optional dummy argument;
- A subroutine may be implicitly invoked as an *assignment subroutine* (under particular circumstances);
- A function may be implicitly invoked as an *operator function* (under particular circumstances);
- A subprogram may be invoked in a context that requires it to be pure;
- A subprogram may be invoked by an elemental subprogram reference; and
- A subprogram may be referenced not only by its specific name but also by its *generic name*.

- A generic subprogram name may be "overloaded"; that is, the generic name may identify more than one subprogram in the same scoping unit;
- A subprogram may have *optional* dummy arguments;
- A function result may be an *array*;
- A function result of a character function may have a character length which is neither constant nor assumed;
- A function result may be a *pointer*;
- A subprogram may be an elemental subprogram; and
- A dummy argument may be an *assumed-shape array* or it may have the POINTER or the TARGET attribute.

These extended Fortran 95 features may be used where the subprogram interface is fully known; otherwise, only the restricted FORTRAN 77 features are allowed. The interface of an external subprogram is fully known where the external subprogram invokes itself directly (recursively) or where an interface block for the referenced subprogram is available.

13.4.4 Internal Subprograms

An internal subprogram is embedded in a surrounding main program or subprogram which is the so-called host. An internal subprogram is a part of its host and has access to certain entities of its host by *host association*.

With the following exceptions, internal subprograms have the same form and the same properties as external subprograms. The exceptions are:

- The name of an internal subprogram is not a global name;
- The name of an internal subprogram must not be supplied as an actual argument in a subprogram reference;
- The length specification of an internal character function must not be an asterisk *;
- An internal subprogram must not contain an additional entry point; that is, it must not contain an ENTRY statement;
- An internal subprogram must not contain another internal subprogram;
- And the END statement of an internal subprogram must be an END FUNCTION statement or an END SUBROUTINE statement, respectively.

The subprogram definition of any internal subprogram within the main program or within a subprogram must appear after the execution part of its respective host between the CONTAINS statement and the END statement of its host.

Invocation of an internal subprogram

With the following exceptions, a reference to an internal subprogram has the same form and the same properties as a reference to an external subprogram:

1. A reference to an internal subprogram may appear only in this subprogram, in its host, or in another internal subprogram of its host.
2. An internal subprogram cannot be used as an operator function or as an assignment subroutine.
3. An internal subprogram cannot have a generic name.

Therefore, the interface of an internal subprogram is fully known wherever a reference to the internal subprogram is allowed to appear.

```
PROGRAM plane
   ⋮
   xpoint = r * rotate(angle=phi) ! reference to internal function
   CALL error(x=xpoint-10)        ! reference to internal subroutine
   ⋮
CONTAINS                          ! internal subprograms
```

```
  REAL FUNCTION rotate (angle)
    REAL angle, pi
    ⋮
    rotate = COS(angle + pi)
    CALL error(x=rotate)        ! reference to internal subroutine
  END FUNCTION rotate

  SUBROUTINE error (x)
    REAL x
    IF (x < 0) STOP 'X is negative'
  END SUBROUTINE error
END PROGRAM plane
```

13.4.5 Module Subprograms

A module subprogram is embedded in a module and has access to certain entities from the module by *host association*.

With the following exceptions, module subprograms have the same form and the same properties as external subprograms. The exceptions are:

- The name of a module subprogram is not a global name;
- The length specification of a character module function must not be an asterisk *; and
- The END statement of a module subprogram must be an END FUNCTION statement or an END SUBROUTINE statement.

The subprogram definitions of all module subprograms must appear after the specification part of the module between the CONTAINS statement and the END statement of the module.

Invocation of a module subprogram

A reference to a module subprogram has the same form and the same properties as a reference to an external subprogram with the following exceptions: a reference to a module subprogram may appear only in this subprogram, in another module subprogram of this module, or in a scoping unit which has access to the module (if the module subprogram is public).

Therefore, the interface of a module subprogram is fully known wherever the reference to the module subprogram is allowed to appear.

```
PROGRAM plane
  USE move_lib                          ! module reference
  ⋮
  xpoint = r * rotate(angle=phi)        ! module function reference
  ⋮
  CALL error(x=xpoint-10)               ! module subroutine reference
  ⋮
END PROGRAM plane

MODULE move_lib
  ⋮
CONTAINS                                ! module subprograms
  REAL FUNCTION rotate (angle)
    REAL angle, pi
    ⋮
    rotate = COS(angle + pi)
    CALL error(x=rotate)                ! module subroutine reference
  END FUNCTION rotate

  SUBROUTINE error (x)
    REAL x
    IF (x < 0) STOP 'X is negative'
  END SUBROUTINE error
END MODULE move_lib
```

13.4.6 Dummy Subprograms

A dummy argument identifying a subprogram is called a **dummy subprogram**. A dummy argument identifies a subprogram if the EXTERNAL attribute is specified for the name of the dummy argument by an EXTERNAL statement or in a type declaration statement, if the dummy argument is used in a subprogram reference as the name of the invoked subprogram, or if an interface block for a subprogram with this name is available. Except in the first case, it is determinable whether the dummy subprogram is a *dummy function* or a *dummy subroutine*.

Invocation of a dummy subprogram

A reference to a dummy subprogram has the same form and the same properties as a reference to an external subprogram.

If any of the extended **Fortran 95** features is used in the reference to a dummy subprogram, an interface block for this dummy subprogram must be available in the scoping unit containing the subprogram reference.

13.4.7 Statement Functions

A statement function consists only of one statement which is embedded in the specification part of a main program or of a subprogram. The statement function definition has the form of an intrinsic assignment statement. A statement function reference may appear only in the scoping unit containing the statement function definition.

13.4.7.1 Statement Function Definition

name ( ) = expression

name (dummy_argument [, dummy_argument]...) = expression

The **name** of the statement function identifies the *scalar* function result. Each **dummy_argument** identifies a scalar data object. The **expression** is a scalar expression which may include only intrinsic operators and the following *scalar* operands: literal and named constants, named and unnamed variables, function references, or subexpressions of this form enclosed in parentheses.

Statement function name: The **name** is local to the scoping unit containing the statement function definition. It must not have the EXTERNAL attribute and must not be used as an actual argument in a subprogram reference. And it must not appear in other specification statements except in a type declaration statement (for the specification of the type, kind type parameter, and (if applicable) character length of the statement function). It may be reused as the name of a common block.

If an entity with the same name is accessible by host association in the scoping unit containing the statement function definition, an explicit type declaration of the function name must appear before the statement function definition.

The type of a statement function is given by the type of the name of the statement function. If the statement function is a character function, the character length specification for the function name must be a constant specification expression.

Dummy arguments: Dummy arguments of a statement function must be scalar data objects. Each dummy argument has the same type, kind type parameter, and (if applicable) character length as the entity having the same name in the scoping unit containing the statement function definition. The length specification of a character dummy argument must be a constant specification expression.

The scope of each dummy argument is the statement function definition. A particular name must appear only once in the dummy argument list, but it may be reused as a dummy argument of the same type in another statement function definition within the same main program or subprogram.

The name of a dummy argument of a statement function may be reused in the scoping unit containing the statement function as a variable of the same type, kind type parameter, and (if applicable) character length. It may be the name of a dummy argument in a FUNCTION, SUBROUTINE, or ENTRY statement (if the ENTRY statement appears before the statement function definition). And finally, the name may be reused as the name of a common block.

Right-hand side: If the **expression** includes a function reference, neither the function definition nor the function reference must require any of the extended Fortran 95 features mentioned at the end of 13.4.3. The function reference must not be an elemental reference. The function result must be scalar and the function must not be an intrinsic transformational function. If an actual argument to such a function is array valued, it must be an array name.

If the **expression** includes a statement function reference, the definition of this (other) statement function must appear before the definition of the statement function just being defined; that is, a statement function reference may appear only after the definition of the statement function. A recursive invocation of a statement function is not allowed.

Normally, a type declaration statement or any other specification statement affecting one of the operands of the right-hand side must precede the statement function definition or must be accessible by USE association or host association. Note that an explicit type declaration following the statement function definition is allowed if it confirms the implicit typing of the operand.

A variable appearing in the **expression** must be either a dummy argument of the statement function or a scalar variable accessible in the host scoping unit containing the statement function definition. If the name of a dummy argument is the same as the name of a variable in the host scoping unit of the statement function, the dummy argument is used within the statement function.

Assignment: The evaluation of the **expression** and the assignment of its result to the left-hand side of the equals are governed by the rules for intrinsic assignment statements.

13.4.7.2 Invocation of a Statement Function

A statement function reference has the same form as a reference to a "normal" user-defined function.

name ( )

name (actual_argument [, actual_argument]...)

The type of the function result is given by the type of the **name** of the statement function.

An actual argument in a statement function reference is allowed to be the name of a dummy argument of the subprogram containing the statement function definition. Each actual argument must identify a data object. The actual arguments in a statement function reference are *positional arguments* and must agree with the corresponding dummy arguments with regard to order, number, type, and kind type parameter. And if a dummy argument is of type character, the length of the associated actual argument must be greater than or equal to the length of the dummy argument. All actual arguments must be scalar.

The invocation of a statement function is executed as follows:

1. Evaluation of the actual arguments which are expressions with operators or which are enclosed in parentheses; and determination of the position of the actual arguments which are array elements or character substrings.

2. Association of the arguments.

3. Evaluation of the expression on the right-hand side of the equals.

4. Conversion of the result of the right-hand side according to the type, kind type parameter, or character length of the statement function if those of the right-hand side are different. The type conversion, the kind type parameter conversion, and the adjustment of the character length of the result of the evaluation of the right-hand side are governed by the rules for intrinsic assignment statements.

A statement function reference may appear only in the main program or subprogram containing the statement function definition.

```
CHARACTER (LEN=1) :: upper_case, lower_case
upper_case (lower_case) = CHAR( ICHAR(lower_case) + &
                        &   ICHAR('A') - ICHAR('a'))
PRINT *, upper_case('t') // 'e' // upper_case('x')
```

The statement function upper_case converts an (ASCII or EBCDIC) lower-case letter into an upper-case letter. CHAR and ICHAR are intrinsic functions. The PRINT statement writes the character string **TeX**.

There are the following restrictions on statement function references: keyword arguments are not allowed, an interface block is not available, actual arguments must not be array objects, and actual arguments must not be omitted.

13.4.8 Interface Blocks

An **interface block** may be used to specify the characteristics of one or more subprogram interfaces. These may be the characteristics of main entry points or of additional entry points (as specified by ENTRY statements).

An interface block may contain the following specifications:

- One or more *interface definitions*, in which *all* characteristics of particular external subprograms or dummy subprograms are specified. This form of an interface block is called a *specific interface block*.

- A *generic name*, the names of those module subprograms, and the interface definitions of those external subprograms or dummy subprograms which are specified to have this generic name. This interface block is used to specify a (common) *generic name* for the subprograms whose *specific names* are specified in this interface block.

  This form of an interface block is a special form of a *generic interface block* and is called an *interface block with a generic name*.

- The designation of an operator, the names of those module functions, and the interface definitions of those external functions or dummy functions which are specified to be the operator functions for this operator. This form of an interface block is a special form of a *generic interface block* and is called an *operator interface block*.

- The assignment operator, the names of those module subroutines, and the interface definitions of those external subroutines or dummy subroutines which are specified to be the assignment subroutines for an extension of the assignment operator. This form of an interface block is a special form of a *generic interface block* and is called an *assignment interface block*.

Interface definition: As far as *external subprograms* and *dummy subprograms* are concerned, an interface block may contain for such a subprogram an interface definition, which begins with a FUNCTION or SUBROUTINE statement, respectively, which ends with an END statement, and which contains between these two statements additional specification statements that characterize the interface, that is, the function result (if any) and the dummy arguments.

An interface definition within an interface block must be consistent with the subprogram interface in the subprogram definition. That is, the characteristics of the interface definition within the interface block must agree with the subprogram definition. There are two exceptions: the interface definition may specify that the subprogram is not pure though the subprogram is defined to be pure, and the names of the dummy arguments within the interface definition and within the subprogram definition may be different. If the names of a particular dummy argument are different, the dummy argument name specified in the interface block overrides the name specified in the subprogram interface of the subprogram definition.

An interface definition for a pure subprogram must specify INTENT attributes for all dummy arguments except dummy subprograms, arguments with POINTER attribute, and asterisk dummy arguments.

An interface block appearing in the specification part of a subprogram must not contain an interface definition for the main entry point or any other entry point of the subprogram containing this interface block.

INTERFACE [...] ⟵ INTERFACE statement
⋮ ⟵ *Interface definition(s)*; may be omitted
MODULE PROCEDURE ... ⟵ MODULE PROCEDURE statement(s); may be or must be omitted
⋮
END INTERFACE [...] ⟵ END INTERFACE statement

An interface block must not contain an ENTRY statement. And the MODULE PROCEDURE statement must not appear in a *specific interface block*. An interface definition within an interface block is allowed to specify attributes and to define values for data entities except dummy arguments and function result variables, but such specifications and definitions are ineffective.

An interface block may be specified in the specification part of a subprogram or program unit except in a block data program unit.

In a generic interface block, the MODULE PROCEDURE statement is used to specify the specific names of those module subprograms which define the generic properties of the generic name, the operator, or the assignment, respectively.

MODULE PROCEDURE name [**, name**]...

A MODULE PROCEDURE statement may appear in a generic interface block only if the interface block is specified in the specification part of a module or if the interface block references a module. Each **name** identifies a module subprogram in the module containing this interface block or identifies a module subprogram accessible by USE association.

The form of a **specific interface block** is:

INTERFACE
⋮   ⟵ *Interface definition(s)*; may be omitted
END INTERFACE

Such a specific interface block enables the specified external or dummy subprograms to use the extended **Fortran 95** features described at the end of 13.4.3.

The form of an **interface block with a generic name** is:

INTERFACE generic_name
⋮   ⟵ *Interface definition(s)*; may be omitted
  MODULE PROCEDURE ...   ⟵ MODULE PROCEDURE statement(s);
                                may be or must be omitted
⋮
END INTERFACE [generic_name]

Such an interface block specifies a **generic_name** for all those external subprograms, dummy subprograms, and module subprograms which are specified in the interface block. These subprograms may then be referenced optionally by their specific names or by this generic name. If the generic name of a subprogram is available but the specific name is private, the subprogram may be referenced by its generic name but it cannot be referenced by its specific name.

If an interface block with a generic name specifies more than one specific subprogram name, the generic name becomes *overloaded* by these specific names. The generic name may be the same as any one of the specific names in the interface block. And the generic name may be the same as any other generic name of an interface block that is specified or accessible in the same scoping unit. A generic interface block must specify either only subroutines or only functions.

Overloading of a generic subprogram name by specific subprogram names is allowed if each of the affected interfaces unambiguously matches exactly one subprogram at the time of the reference to this generic name.

An interface block is a part of the specification part of its host scoping unit. The specifications of an interface block are accessible in the host scoping unit, but the specifications in the host scoping unit are *not* accessible in the interface block.

```
PROGRAM mp
  INTERFACE                        ! interface block in main pr.
    SUBROUTINE circle (x, y, r)    ! specific name
    ⋮
    END SUBROUTINE circle
  END INTERFACE
```

13.4 Subprograms

```
          ⋮                                     ← remaining spec. part of main pr.
          ⋮                                     ← execution part of main program
CONTAINS                                        ! internal subpr. of main pr.
  SUBROUTINE dot (x, y)
    INTERFACE                                   ! interface block in int. subr.
      REAL FUNCTION zoom (pro)                  ! specific name
        ⋮
      END FUNCTION zoom
    END INTERFACE
    ⋮                                           ← remaining spec. in int. subr.
    ⋮                                           ← execution part of int. subr.
  END SUBROUTINE dot
    ⋮                                           ← remaining internal subprograms
END PROGRAM mp
MODULE graph
  INTERFACE link                                ! generic name
    SUBROUTINE line (start, end)                ! specific name
      ⋮
    END SUBROUTINE line
    SUBROUTINE vector (xy, phi, l)              ! specific name
      ⋮
    END SUBROUTINE vector
    MODULE PROCEDURE segment
  END INTERFACE
CONTAINS                                        ! module subprograms
  REAL FUNCTION rotate (box)
    ⋮                                           ← specifications in module function
    INTERFACE                                   ! interface bl. in module function
      SUBROUTINE colour (c)                     ! specific name
        ⋮
      END SUBROUTINE colour
    END INTERFACE
    ⋮                                           ← execution part of module function
  END FUNCTION rotate
  SUBROUTINE segment (a, b)
    ⋮
  END SUBROUTINE segment
END MODULE graph
```

Note that two or more generic interface blocks accessible in a scoping unit are interpreted as a single generic interface if they have the same generic identifier (i. e. generic name, generic operator, or generic assignment).

Dummy subprograms

An interface block containing an interface definition of a dummy subprogram specifies that the dummy argument name identifies either a function or a subroutine having exactly the characteristics as specified in the interface block.

Outside the interface block, the name of the dummy subprogram must not appear in the specification part of the scoping unit containing the interface block except in an OPTIONAL statement.

Operator interface blocks, assignment interface blocks

Operator interface blocks and assignment interface blocks are presented in context with defined operators and defined assignment statements in 7.5.1 and 8.2, respectively.

13.4.9 Overloaded Generic Subprogram Names

Normally, a user-defined subprogram is identified in a scoping unit by exactly one name. This name is the *specific* name of the subprogram. But in addition to the specific name, a subprogram may have a *generic* name. Such a generic name of an external subprogram, a module subprogram, or a dummy subprogram is specified in the INTERFACE statement of a generic interface block.

If an interface block with a generic name contains names or interface definitions of two or more subprograms, the generic name is **overloaded** because it identifies more than one subprogram. Overloading of a generic subprogram name by specific subprogram names is allowed if each of the affected interfaces specified by the generic interface block unambiguously matches exactly one subprogram when the generic subprogram name is used in a subprogram reference.

Therefore, any two such functions or two such subroutines having the same generic name in a scoping unit must be distinguishable from each other when they are referenced by their common generic name. They are distinguishable if one of them has more nonoptional dummy arguments of a particular data type, kind type parameter, and rank than the other has dummy arguments (including optional dummy arguments) of that data type, kind type parameter, and rank. They also are distinguishable if one of them has a nonoptional dummy argument that corresponds by position in the argument list to a dummy argument with different properties in the other, and if it has a nonoptional dummy argument that corresponds by name in the argument list to a dummy argument with different properties in the other. These properties of the corresponding dummy argument in the other argument list are: presence, data type, kind type parameter, or rank. The dummy argument that disambiguates by position must either be the same as or occur earlier in the argument list than the one that disambiguates by keyword.

Intrinsic subprograms

If the generic name specified in the INTERFACE statement of a generic interface block is the name of an intrinsic subprogram, the user-defined specific subprograms specified in this generic interface block extend the predefined meaning of this intrinsic subprogram. In analogy to the above rule (concerning the unambiguity of the references to overloaded user-defined generic subprograms), the references to all subprograms having the same generic name must be unambiguous, as though the intrinsic subprogram would consist of a collection of specific intrinsic subprograms and as though all the interface definitions of these specific intrinsic subprograms were also specified in the generic interface block.

13.4.10 Additional Entry Points, ENTRY Statement

External subprograms and module subprograms may specify in addition to their main entry point one or more *additional entry points*. The name of such an additional entry point must be specified in an ENTRY statement. An additional entry point of a subprogram defines (in the same way as the main entry point) a *subprogram interface*, which may be referenced to invoke the subprogam.

Normally, the name of an additional entry point is a specific name. But the generic name of an additional entry point also is available where an interface block with this generic name is available.

When a subprogram is invoked by reference to an additional entry point, the execution of the subprogram begins at the first executable statement following the ENTRY statement with the referenced entry point name.

An ENTRY statement may appear in an external subprogram or in a module subprogram between the SUBROUTINE or FUNCTION statement, respectively, and the END statement of the subprogram; more precisely, it may appear between USE and CONTAINS, but must not appear within an executable construct.

ENTRY name [RESULT (result)]

ENTRY name ( ) [RESULT (result)]

ENTRY name (dummy_argument [, dummy_argument]...) [RESULT (result)]

The **name** identifies the additional entry point.

An additional entry point of a function may be specified without a dummy argument list and (in contrary to its main entry point) even without the parentheses. But in any case, the parentheses must appear in the function reference. Note that, apart from that, all the normal rules for the main entry point of a subprogram also hold for any additional entry point.

Within a program, an additional entry point of an external subprogram may be referenced in any other program unit. An additional entry point of a module subprogram may be referenced in any module subprogram of the same module. A *public* additional entry point of a module subprogram may be referenced in any scoping unit which has access to the module.

The name of an additional entry point must not appear in the same subprogram as a dummy argument in the FUNCTION statement, SUBROUTINE statement, or in another ENTRY statement. And the name must not be specified in the same subprogram in an EXTERNAL or INTRINSIC statement.

The normal rules for the result variable of the main entry point of a function also hold for the result variable of an additional entry point. Note that the result variables of a function are associated with each other.

A dummy argument name in an ENTRY statement must appear only in executable statements following this ENTRY statement, except the same dummy argument name also is specified earlier in the FUNCTION or SUBROUTINE statement or in another ENTRY statement within the same subprogram.

If a subprogram executes a statement which includes an array or a character entity and if a dummy argument appears in the declaration of such an entity for the specification of an array bound or a character length, then these dummy arguments must be specified in the dummy argument list of the invoked entry point. And if a subprogram executes a statement which includes a dummy argument as an operand, then this dummy argument must be specified in the dummy argument list of the invoked entry point. If such a permissible dummy argument has the OPTIONAL attribute, the dummy argument must be *present*.

```
FUNCTION hyp ()
  REAL cth, sch, csch, sinh2, cosh2
  !      statement function definitions
  sinh2 (y) = EXP(y) - EXP(-y)
  cosh2 (y) = EXP(y) + EXP(-y)
  !      end of statement function definitions
  PRINT *, ' Unsuitable reference!',' Use CTH, SCH, or CSCH.'
RETURN

ENTRY cth (x)
  cth = cosh2(x) / sinh2(x)
RETURN
```

```
ENTRY sch (x)
  sch = 2. / cosh2(x)
RETURN
ENTRY csch (x)
  csch = 2. / sinh2(x)
END
```

Recursive additional entry point: The keyword RECURSIVE must be specified in the FUNCTION or SUBROUTINE statement of the subprogram containing the ENTRY statement if the additional entry point is invoked directly or indirectly recursively.

The keyword RESULT and the **result** variable must be specified in the ENTRY statement of a function if this function is invoked directly recursively, that is, if the function invokes this entry point within its own execution part; otherwise, they may be omitted.

Pure additional entry point: Either the keyword PURE or the keyword ELEMENTAL must be specified in the FUNCTION or SUBROUTINE statement of the subprogram containing the ENTRY statement if the subprogram being given by the additional entry point is a pure subprogram which is used in a context that requires it to be pure.

Elemental additional entry point: The keyword ELEMENTAL must be specified in the FUNCTION or SUBROUTINE statement of the subprogram containing the ENTRY statement if the subprogram being given by the additional entry point is an elemental subprogram which is invoked elsewhere in the program by an elemental reference.

Function result: The attributes of the referenced additional entry point of a function determine the characteristics of the function result. After return to the expression containing the function reference, the function result is the value of the result variable which is given by the ENTRY statement with the referenced entry point name.

13.4.11 Return from the Invoked Subprogram

When the invoked subprogram executes a RETURN statement or its END, END FUNCTION, or END SUBROUTINE statement, the execution of the (instance of the) subprogram is terminated, control returns to the invoking main program or subprogram, respectively, and the arguments become disassociated.

If the invoked subprogram is a function, the function result is supplied for use in the expression at the place of the reference to the function.

If the invoked subprogram is a subroutine, the execution of the CALL statement is completed. The statement where program execution continues after completion of the execution of the CALL statement depends on the appearance or nonappearance of alternate return specifiers in the CALL statement and on the use of the corresponding dummy arguments within the invoked subroutine. If no alternate return specifiers are specified in the CALL statement or if the invoked subroutine does not use them, a "normal" return of control takes place.

RETURN

RETURN integer_expression

The **integer_expression** must be scalar.

A RETURN statement may appear in external, internal, or module subprograms. Such a subprogram may contain more than one RETURN statement. The second form of the RETURN statement (with the integer expression) is called an **alternate RETURN statement** and may appear only in a subroutine.

"Normal" return from a subroutine

When a subroutine executes its END or END SUBROUTINE statement or when it executes a RETURN statement without an integer expression, the execution of the subroutine is completed and the execution of the invoking main program or subprogram continues with the first executable statement following the CALL statement.

```
      INTEGER d
      READ (*, *) d
      CALL discriminant(d)
55    CONTINUE
      ⋮
      END PROGRAM

      SUBROUTINE discriminant (n)
        SELECT CASE (n)
          CASE (:-1); WRITE (*, *) ' Discriminant is negative'; RETURN
          CASE (0);   WRITE (*, *) ' Discriminant is zero';     RETURN
          CASE (1:);  WRITE (*, *) ' Discriminant is positive'
        END SELECT
      END SUBROUTINE
```

Regardless of whether the invoked subroutine **discriminant** executes one of the RETURN statements or the END statement, control returns to statement 55 within the main program after completion of the CALL statement.

Alternate RETURN statement

When an alternate RETURN statement is executed in a subroutine, control is transferred to an executable statement in the invoking scoping unit; this alternate return target is specified by an alternate return specifier in the CALL statement.

The following rules govern the execution of an alternate RETURN statement:

- The actual argument list in the CALL statement must specify one or more alternate return specifiers.
- The dummy argument list of the referenced entry point of the subroutine must specify one or more asterisks.
- The number of asterisks in the dummy argument list must be equal to the number of alternate return specifiers in the actual argument list of the CALL statement.

When the invoked subroutine executes an alternate RETURN statement, the value of the **integer_expression** determines the branch target statement where program execution continues after return from the subroutine. If the value of the integer expression is n such that $1 \leq n \leq$ (number of asterisks in the dummy argument list), then the nth alternate return specifier of the actual argument list identifies the branch target statement which will be executed after the completion of the execution of the CALL statement. Otherwise, the alternate RETURN statement has the same effect as a normal RETURN statement.

```
      READ (*, *) nn;  CALL discriminant (nn, *60, *70)
   50 WRITE (*, *) ' Discriminant is negative'; STOP
   60 WRITE (*, *) ' Discriminant is zero';    STOP
   70 WRITE (*, *) ' Discriminant is positive'
      END PROGRAM
      SUBROUTINE discriminant (n, *, *)
        SELECT CASE (n)
          CASE (:-1); RETURN
          CASE (0);   RETURN 1
          CASE (1:);  RETURN 2
        END SELECT
      END SUBROUTINE
```

When n is negative, control is transferred to statement 50 within the main program. When n is zero, control is transferred to statement 60, and when n is positive, control is transferred to statement 70 within the main program.

13.5 Internal Program Communication

Program units and subprograms may exchange informations about data, subprogram names, and alternate return targets. This communication may take place via argument lists, common blocks, result variables of functions, shared local variables, and external files. External files which allow programs to communicate with the outside world are presented in chapter 11.

13.5.1 Argument Lists

An *actual argument list* may appear in a subprogram reference. A *dummy argument list* may appear in a subprogram interface (within a subprogram definition) or in an interface definition (within an interface block). The actual arguments in the subprogram reference and the corresponding dummy arguments of the subprogram become associated during the invocation of the subprogram.

The communication between these argument lists is called the *argument association*. Arguments may be used to exchange values. They may also be used to transfer particular subprogram names to the invoked subprogram. And in the case of a subroutine, informations about alternate return targets may be transferred to the invoked subroutine.

A dummy argument is called an **input argument** if it can receive informations from the invoking scoping unit but cannot transfer informations back to the invoking scoping unit. A dummy argument is called an **output argument** if it can transfer informations to the invoking scoping unit but cannot receive informations from the invoking scoping unit. And a dummy argument is called an **input/output argument** if it can both receive informations from the invoking scoping unit and transfer informations to the invoking scoping unit.

Though a dummy argument is an output argument or an input/output argument, there may be circumstances such that this dummy argument is not allowed to transfer data back to the invoking scoping unit.

13.5.1.1 Dummy Argument List

Dummy arguments are used in place of actual arguments in the subprogram definition. When the subprogram is invoked, the actual arguments become associated with the dummy arguments such that they are used instead of the dummy arguments during the execution of the subprogram.

Dummy arguments of a *user-defined* subprogram (except a statement function) may be specified in the dummy argument list of a FUNCTION, SUBROUTINE, or ENTRY statement of a subprogram definition. The dummy arguments of

13.5 Internal Program Communication

an external subprogram may be additionally specified in the dummy argument list of a FUNCTION or SUBROUTINE statement of an interface definition within an interface block. The dummy arguments of a dummy subprogram may be specified only in an interface definition within an interface block. And the dummy arguments of a statement function may be specified on the left-hand side of the equals in the statement function definition. The dummy arguments of the *intrinsic* subprograms are predefined.

Each dummy argument is either the name of a variable, the name of a subprogram, or an asterisk.

Characteristics of dummy arguments

An actual argument which is associated with a dummy argument during subprogram invocation must have certain properties. These properties depend on the characteristics of the dummy argument.

Dummy data object: The characteristics of a dummy argument identifying a data object are: the type, the kind type parameter, the character length, the shape, the INTENT attribute, the OPTIONAL attribute, the POINTER attribute, the TARGET attribute, the dependence of the kind type parameter, of the character length, or of the array bounds on the value and properties of other data entities, and assumed properties such as assumed size (of an array), assumed-shape (of an array), or assumed length (of a character dummy argument).

Dummy subprogram: The characteristics of a dummy argument identifying a subprogram are: the availability of an interface block containing an interface definition for the dummy subprogram, the classification of the dummy subprogram as a function or subroutine, whether it is pure, whether it is elemental, the OPTIONAL attribute, the characteristics of its dummy arguments, and (in case of a dummy function) the characisctics of the function result.

Asterisk dummy argument: An asterisk dummy argument does not have any characteristic.

Dummy argument names

The name of a dummy argument must not be specified in an ALLOCATABLE, DATA, EQUIVALENCE, INTRINSIC, PARAMETER, SAVE, or COMMON statement (except as the name of a common block). And the name of a dummy argument must be different from the name of a statement function and from the name of an entry point in an ENTRY, FUNCTION, or SUBROUTINE statement in the same scoping unit.

13.5.1.2 Actual Argument List

Actual arguments may be specified in the actual argument list of a subprogram reference. The actual argument list may be empty or may be entirely absent. The form of the actual argument list is described in detail in connection with the description of subprogram references.

Actual arguments may be constants, variables, function references, expressions with operators and/or parentheses, subprogram names, and alternate return specifiers.

Data entities: If the actual argument is a data entity, the corresponding dummy argument must be the name of a variable.

Subprogram name: If the actual argument is a subprogram name, the corresponding dummy argument must identify a dummy subprogram. If the actual argument is the specific name of an intrinsic subprogram, the INTRINSIC attribute must be specified for this name. If the actual argument identifies a dummy subprogram or is the specific name of an external subprogram and if there is no interface block for this subprogram available in the scoping unit containing the actual argument list, the EXTERNAL attribute must be specified for this name. If a name with EXTERNAL attribute is available (i.e. specified or via USE association accessible) in a scoping unit, then the name must not appear as specific subprogram name within an interface block being available in the scoping unit.

Such an actual argument must have the EXTERNAL or INTRINSIC attribute. Note that names of internal subprograms, of intrinsic subroutines, of generic subprograms, of some specific intrinsic functions (14.1), and of statement functions must *not* be supplied as actual arguments.

Alternate return specifiers: If the actual argument is an alternate return specifier, the corresponding dummy argument must be an asterisk. Alternate return specifiers may be specified only in CALL statements.

Keyword (actual) arguments

The name of a dummy argument of an internal subprogram, of a module subprogram, and of an intrinsic subprogram may be used as an argument keyword to specify the actual argument in the subprogram reference.

The name of a dummy argument of a statement function must not be used as a keyword to specify the actual argument in the statement function reference.

The name of a dummy argument of an external subprogram may be used as an argument keyword in the actual argument list of a subprogram reference only when an interface block for the external subprogram is available in the scoping unit containing the subprogram reference, or when the reference to the external subprogram is directly recursive.

If a keyword argument is specified in the actual argument list, all subsequent actual arguments must be specified as keyword arguments. And conversely, if an actual argument is specified as positional argument, all preceding actual arguments must be specified without an argument keyword. If an actual argument is specified corresponding to an optional dummy argument, all subsequent actual arguments must be specified as keyword arguments.

13.5.2 Argument Association

During the execution of a CALL statement and during the execution of a function reference, the actual arguments (in the actual argument list of the subprogram reference) become associated with the corresponding dummy arguments (in the dummy argument list of the referenced subprogram). **Association** means here that a particular data entity, subprogram, or alternate return target is identified both (in the scoping unit with the subprogram reference) by the actual argument and (in the referenced subprogram) by the dummy argument.

If the actual argument list does not include keyword arguments, the actual arguments and the corresponding dummy arguments become *positionally* associated; that is, the first actual argument becomes associated with the first dummy argument, the second actual argument with the second dummy argument, and so on. If a keyword argument is specified in the actual argument list, the actual argument is associated with the dummy argument which is identified by this keyword.

If there are *optional* dummy arguments in the subprogram interface, the number of actual arguments in the actual argument list of the subprogram reference may be less than or equal to the number of dummy arguments in the corresponding dummy argument list. An actual argument *must* be specified and associated for each nonoptional dummy argument, and an actual argument *may* be specified and associated for each optional dummy argument.

Normally, wherever the name of a dummy argument from the dummy argument list of the referenced entry point is used in the invoked subprogram, the corresponding actual argument is associated with this dummy argument during the subprogram invocation. This association remains in existence during the execution of the invoked subprogram, even if one or more actual arguments depend on variables which are redefined during the execution of the subprogram.

When a subprogram is recursively referenced, another instance of the subprogram is executed, which has its own set of dummy arguments. And the association of the actual arguments with the dummy arguments of this instance of the subprogram remains in existence only for the time of the execution of this instance of the subprogram.

Except when the dummy argument is an asterisk, both the dummy argument and the associated actual argument must either be data entities or identify subprograms.

```
MODULE oppar
  :
CONTAINS
SUBROUTINE compute (fct, *, solution, method, strategy, print)
  INTERFACE
    FUNCTION fct (x)
      REAL fct, x
    END FUNCTION fct
  END INTERFACE
  REAL solution
  INTEGER, OPTIONAL :: method, strategy, print
  :
END SUBROUTINE compute
END MODULE oppar
```

This module subroutine may be referenced as follows:

```
CALL compute(f, *75, 11)
CALL compute(f, *75, solution=11)
CALL compute(f, *75, solution=11, method=nine, strategy=long)
CALL compute(f, *75, 11, nine, strategy=long, print=2)
CALL compute(fct=f, *75, solution=11)              ! <-- illegal
```

Arguments may become associated across several levels of subprogram references; that is, a dummy argument which identifies a data object or a subprogram may be supplied within the same subprogram as an actual argument in a subprogram reference. This is not allowed if the dummy argument is an asterisk.

```
FUNCTION fu (x)
  :
  CALL sub(x)
  :
END FUNCTION
```

The dummy argument x is used as an actual argument in the CALL statement.

The arguments become disassociated when a RETURN statement or the END statement of the invoked subprogram is executed. The dummy arguments are undefined between two invocations of the same subprogram.

13.5.2.1 Data Objects as Dummy Arguments

If the actual argument is a data entity, the corresponding dummy argument must be a named variable, and

- The associated arguments must be of the same type;
- The associated arguments must have the same kind type parameter;
- The character lengths of the associated arguments must be suitable (see 13.5.2.3). Note that the lengths automatically agree if the dummy argument is specified with assumed length *;
- The associated arguments must have the same rank if the dummy argument is a scalar and the subprogram is not elemental, if the dummy argument is a pointer, if the subprogram is invoked by a reference to its generic name, if the subprogram is implicitly invoked as an operator function, if the subprogram is implicitly invoked as an assignment subroutine, if the dummy argument is an array except an explicit-shape array or an assumed-size array, or if the dummy argument is an array and the arguments are not sequence associated; and
- The associated arguments must have the same shape if the dummy argument is a pointer, if the dummy argument is an array except an explicit-shape array or an assumed-size array, or if the dummy argument is an array and the arguments are not sequence associated.

Expressions, array elements, array sections, and character substrings are not associated with the corresponding dummy arguments until they are preprocessed internally to determine their values and their position, respectively. If such an actual argument includes expressions needed to determine the position, its position remains fixed for the time of the execution of the invoked subprogram, even if the expressions contain variables which are redefined during the execution of the subprogram.

If the dummy argument does not have a POINTER or TARGET attribute, any pointers associated with the actual argument do *not* become associated with the dummy argument. They remain associated with the actual argument. If the dummy argument is not a pointer and the corresponding actual argument is a pointer, the actual argument must be currently associated with a target and the dummy argument becomes argument associated with that target. If the dummy argument has the TARGET attribute and the corresponding actual argument has no TARGET attribute or is an array section with a vector subscript, pointers associated with the dummy argument become undefined when control returns to the invoking scoping unit. If the dummy argument is an explicit-shape array or an assumed-size array and has the TARGET attribute, the pointer association status of each pointer being associated with the dummy argument when the subprogram terminates is processor-dependent. If the dummy argument has the TARGET attribute and is a scalar or an assumed-size array and its associated actual argument has the TARGET attribute and is no array section with vector subscript, then the pointers associated with the actual argument become associated with the corresponding dummy argument; and after termination of the subprogram, the pointers being associated with the dummy argument remain associated with the actual argument.

Derived type arguments

An actual argument of derived type may be associated with a dummy argument of the same type or of an equal type. This is guaranteed if the *same* derived type definition is used to declare the type of both the actual argument and the corresponding dummy argument. This is possible if the derived type definition is accessible by host association or by USE association in the different scoping units of the arguments.

But if the corresponding actual and dummy arguments are of a derived type with SEQUENCE attribute, they need not be of the same derived type, because they are allowed to be of *equal* derived types. Such sequenced types are equal if they have the same name and have type components which agree in order, name, rank, shape, type, kind type parameter, and (if applicable) character length.

13.5.2.2 Implicit Association of Two Dummy Arguments

The reference to a subprogram may cause one dummy argument of the referenced subprogram to become associated with another dummy argument of the referenced subprogram.

```
CALL sub(a, a)
  ⋮
END

SUBROUTINE sub (x, y)
  ⋮
END SUBROUTINE
```

The same actual argument a is associated with both dummy arguments. Thus these dummy arguments become implicitly associated with each other.

Such a subprogram invocation must not cause one of the implicitly associated dummy arguments to be defined, redefined, or undefined during the execution of the invoked subprogram (including any other subprogram invoked from the first subprogram).

13.5 Internal Program Communication

If two actual arguments partially overlap, those portions of the corresponding dummy arguments which correspond to the overlapped portions of the associated actual arguments must not be defined, redefined, or become undefined during the execution of the subprogram.

The invocation of a subprogram may cause a dummy argument to become associated with a variable in a common block:

```
COMMON a
CALL sub(a)
  :
END

SUBROUTINE sub (x)
  COMMON b
  :
END SUBROUTINE
```

The actual argument a is associated with the dummy argument x. The variable b in the invoked subroutine and the variable a in the common block in the referencing program are storage associated via the COMMON statement. Therefore, the dummy argument x is implicitly associated with the variable b.

Such a subprogram invocation must not cause the dummy argument or the variable in the common block to be defined, redefined, or to become undefined during the execution of the subprogram.

13.5.2.3 Length of Character Dummy Arguments

If the dummy argument is of type character, the associated actual argument also must be of type character. And their kind type parameters must agree. If the dummy argument is of type *nondefault* character, the lengths of the actual argument and of the corresponding dummy argument must agree.

If the dummy argument is of type *default* character, the lengths of the arguments need not agree. If the dummy argument is scalar, its length must be not greater than the length of the associated actual argument. In this case, the leading characters of the actual argument become associated with the dummy argument, beginning with the first character of the actual argument and associating as many characters as correspond to the length of the dummy argument. If assumed length * is specified for the dummy argument, it automatically has the correct length (see below).

If the dummy argument is an array of type default character which is not an assumed-shape array, there is a similar rule for the total length of the dummy argument. In this case, the total length of the dummy argument must not be greater than the number of characters of the actual argument, beginning with the first character of the associated actual argument up to the last character of the actual argument or of the remaining portion of the parent array of the actual argument. During argument association, characters of the actual argument become associated with the dummy argument, beginning with the first character of the actual argument and associating as many characters as correspond to the total length of the dummy argument.

If the dummy argument is an assumed-shape array of type default character, its character length must agree with the length of the associated actual argument.

If the actual argument is a whole array, an array section, or an array expression, the character length of the actual argument is given by the length of a single array element. If the actual argument is a character substring, the length of the actual argument is given by the length of the character substring. If the actual argument is a character expression with concatenation operator(s), the length of the actual argument is given by the sum of the lengths of the operands of the character expression.

Dummy arguments with assumed length

If the length * is specified for a dummy argument, this dummy argument automatically assumes its length from the associated actual argument during the invocation of the subprogram.

```
    CHARACTER actarg *4, y *15, chrfct *8
    actarg = 'rail'
10  y = chrfct(actarg) // 'station'
    ⋮
    END

    SUBROUTINE sub
      CHARACTER chrfct *11, par *6, z *13
20    z = chrfct(par) // ' engine'
    ⋮
    END SUBROUTINE sub

    CHARACTER *(*) FUNCTION chrfct (darg)
      CHARACTER (LEN=*) :: darg
    ⋮
      chrfct = darg // 'way '
    END FUNCTION chrfct
```

When the function chrfct is referenced in statement 10, its dummy argument darg has the length 4; and when referenced in statement 20, it has the length 6.

13.5.2.4 Scalar Arguments

If a dummy argument of a non-elemental subprogram is a *scalar*, at most a *scalar* actual argument may become associated with this dummy argument.

If the actual argument is scalar, the corresponding dummy argument must be scalar, except the actual argument is an array element or is a character substring of an array element and the corresponding dummy argument is an explicit-shape array or an assumed-size array.

13.5.2.5 Dummy (Argument) Arrays

A dummy argument being an array is a **dummy argument array** or **dummy array**. The associated actual argument must be an array of the same rank
- If the dummy argument is an assumed-shape array,
- If the dummy argument is an array pointer,
- If the subprogram is referenced by its generic name,
- If the subprogram is implicitly invoked as an operator function, or
- If the subprogram is implicitly invoked as an assignment subroutine.

If the dummy argument and the associated actual argument are array entities, the array elements of the actual argument become associated with the array elements of the dummy argument in array element order. Therefore, if the dummy argument has the same shape as the associated actual argument, corresponding array elements (which have the same position) become associated.

Normally, the dummy argument and the associated actual argument must have the same shape. But there are exceptions if the reference to the subprogram is elemental or if the dummy argument is an explicit-shape array or assumed-size array. In the last case, the actual argument may be *sequence associated* with the dummy argument.

Variable dummy arrays

If an array bound specification of a dummy array is a nonconstant expression, the array is a **variable (dummy) array**. A variable array may be an explicit-shape array, an assumed-shape array, or an assumed-size array.

Depending on the kind of the array, an array bound may be a specification expression that also includes variables and function references, the upper bound of the last dimension may be an asterisk, or all upper bounds may be omitted. The variables may be scalar integer dummy arguments, variables in a common block, or scalar integer variables which are accessible by host association or by USE association. The referenced functions must be scalar integer intrinsic functions and their actual arguments also must be such specification expressions as described here.

```
REAL FUNCTION xmean (x, n)
  DIMENSION x (n)
  xmean = 0.                  ! \
  DO i=1,n                    ! |
    xmean = xmean + x(i)      ! >  xmean = SUM(x) / n
  END DO                      ! |
  xmean = xmean / n           ! /
END FUNCTION xmean
```

The size of array x is given through the second dummy argument.

13.5.2.6 Dummy (Argument) Pointers

If the dummy argument is a pointer, the associated actual argument also must be a pointer. And types, kind type parameters, (if applicable) character lengths, and ranks of the corresponding arguments must agree.

During argument association, the dummy argument receives the pointer association status of the actual argument. If the actual argument is currently associated, the dummy argument pointer becomes associated with the target which is associated with the actual argument. The association with this target may be modified within the subprogram by execution of an ALLOCATE statement or pointer assignment statement. In such a case, if the target of the dummy argument is not a variable that becomes undefined when control returns from the subprogram, the actual argument receives the pointer association status of the dummy argument at termination of the subprogram.

A pointer with an undefined association status must not be reused until it has got a defined association status by the execution of an ALLOCATE statement, pointer assignment statement, or NULLIFY statement.

13.5.2.7 Sequence Association

Normally, the shapes of a dummy argument and its associated actual argument must agree. But if the dummy argument is an explicit-shape array or an assumed-size array, shapes and even ranks need not agree if the actual argument specifies an *element sequence*. In this case, a particular kind of argument association applies that is based on the linear memory model which is given by the "array element order" described in 4.4.1.

Element sequence: An actual argument defines an **element sequence** if it is a whole array, an array section, an array expression, an array element, or a character substring of an array element.

If the actual argument is an array entity, the element sequence is the sequence of all array elements of the actual argument in array element order. If the actual argument is an array element, the element sequence consists of this array element and all subsequent array elements of the parent array in array element order.

If the actual argument is of type default character, the element sequence begins with the first character position of the actual argument and ends with the last character position of (the parent array of) the actual argument. If the actual argument is a substring of an array element, the element sequence is interpreted as a sequence of array elements, each of which has the length of the dummy argument.

Sequence association means that not only the specified actual argument but possibly additional elements of the element sequence are associated with the dummy argument. In this case, the ranks and shapes of the corresponding arguments need not agree, but the number of elements in the dummy array must not exceed the number of elements in the element sequence of the actual argument. If the dummy argument is an assumed-size array, the number of elements in the dummy array agrees always with the number of elements in the element sequence of the actual argument.

Whole array, array section, array expression: If the associated actual argument is an array entity of type noncharacter, the size of the dummy array must not be greater than the size of the actual argument array. Each array element of such an actual argument becomes associated with the corresponding array element of the dummy array at the same position (in array element order); that is, the first array element of the actual argument becomes associated with the first element of the dummy array, the second element of the actual argument array with the second element of the dummy array, and so on.

If the associated actual argument is an array of type character and has the same character length as the dummy array, the single corresponding array elements of the argument arrays become associated in the same way as the array elements of noncharacter arrays. If the lengths of the actual argument and the corresponding dummy argument are different, there is no direct association between the array elements at the same position.

Array element: If the associated actual argument is an array element of type noncharacter, the corresponding dummy argument array must not be greater than the parent array of the actual argument plus one less the position of the actual argument within its parent array. The dummy argument determines how many elements of the element sequence defined by the actual argument become actually associated.

```
DIMENSION aa_array (25)
CALL sub(aa_array(7))
  ⋮
END

SUBROUTINE sub (da_array)
  DIMENSION da_array(19)
  ⋮
END SUBROUTINE sub
```

The 7th element of the actual argument array `aa_array` is associated with the first element of the dummy array `da_array`. This results in the following additional associations: the 8th element of `aa_array` is associated with the 2nd element of `da_array`, the 9th element of `aa_array` with the 3rd element of `da_array`, and so on. The dummy array `da_array` must not be greater than $(25 + 1 - 7 =)$ 19.

Character entity: If the actual argument is an array of type character, an array element of type character, or a character substring of an array element, the element sequence defined by this actual argument is associated with the dummy argument character-wise, beginning with the first character position of the actual argument.

The first character position of an array element of type character or of a substring of such an array element may be the nth character position of the parent array. Starting from the total length of the parent array (in characters) and the position of the first character of the actual argument within its parent array, the possible maximum length (in characters) of the corresponding dummy argument can be determined. The dummy argument determines how many characters of the element sequence defined by the actual argument become actually associated.

```
CHARACTER aa_char (15) *3
CALL sub(aa_char(4)(3:3))
END

SUBROUTINE sub (da_char)
  CHARACTER da_char (3) *2
  ⋮
END SUBROUTINE
```

13.5 Internal Program Communication

The execution of the CALL statement causes the following argument association:

$$
\begin{aligned}
\text{aa\_char}\,(4)\,(3{:}3) &\iff \text{da\_char}\,(1)\,(1{:}1) \\
\text{aa\_char}\,(5)\,(1{:}1) &\iff \text{da\_char}\,(1)\,(2{:}2) \\
\text{aa\_char}\,(5)\,(2{:}2) &\iff \text{da\_char}\,(2)\,(1{:}1) \\
\text{aa\_char}\,(5)\,(3{:}3) &\iff \text{da\_char}\,(2)\,(2{:}2) \\
\text{aa\_char}\,(6)\,(1{:}1) &\iff \text{da\_char}\,(3)\,(1{:}1) \\
\text{aa\_char}\,(6)\,(2{:}2) &\iff \text{da\_char}\,(3)\,(2{:}2)
\end{aligned}
$$

13.5.2.8 Assumed-Size Arrays

An assumed-size array must not be of a derived type for which default initialization is specified in its type definition.

There is no explicit value available for the upper bound in the last dimension of an assumed-size array. This means that the size of the dummy array is not known in the subprogram. Therefore, this kind of dummy array has always the maximum size which results from the size of the associated actual argument.

It is the task of the programmer to ensure that only those elements of the assumed-size array are referenced within the invoked subprogram which are actually associated with an array element of the actual argument.

```
SUBROUTINE sub (a, b, c, n)
  DIMENSION a (5, 3, *), b (n, 3, *)
  CHARACTER (LEN=*), DIMENSION (10, *) :: c
  :
END SUBROUTINE sub
```

The subroutine sub has three assumed-size arrays, the dummy arrays a, b, and c. Array b also is a variable array, because one of its array bounds is a variable.

13.5.2.9 Assumed-Shape Arrays

Each dimension of an assumed-shape array automatically has the same extent as the corresponding dimension of the associated actual argument array. If a dummy argument is an assumed-shape array, the associated actual argument must not be an assumed-size array or a scalar.

13.5.2.10 Restrictions on the Association of Data Entities

For the time of the execution of a subprogram, there are certain restrictions on arguments which are data entities. Such a dummy argument must *not* be redefined

- If the associated actual argument is a constant, a function reference, or another expression with operator(s) or enclosed in parentheses;
- If the associated actual argument appears more than once in the actual argument list;
- If the actual argument is in a common block which also is specified in the invoked subprogram;
- If the associated actual argument of a function reference would cause side effects in the invoking scoping unit; or
- If the associated actual argument is an array section with a vector subscript.

While a data entity is associated with a dummy argument, its allocation status must not be modified. While a data entity is associated with a dummy argument, any action, which affects the value of the data entity or of a part of it, must be taken only through the dummy argument, except the dummy argument has the POINTER attribute, the part is a pointer object or a part of a pointer object, or the dummy argument has the TARGET attribute but no INTENT(IN) attribute, the dummy argument is scalar or an assumed-shape array, and the actual argument is a target that is no array section with a vector subscript. If any part of the data entity is defined as a result of a definition of the dummy argument, this part must be referenced during the execution of the subprogram only through the dummy argument; this rule applies to the whole scoping unit of the subprogram.

13.5.2.11 Dummy Subprograms

An actual argument to be associated with a dummy subprogram may be: the specific name of an intrinsic function, an external subprogram, a module subprogram, or a dummy subprogram (of the invoking scoping unit).

There are exceptions for some intrinsic functions: the specific names of the functions for type conversion and lexical comparisons and the specific names of the minimum/maximum functions must not be specified as actual arguments.

If a specific name which is supplied as an actual argument is equal to a generic name, only the specific name is associated. If the name of an intrinsic function is associated with a dummy function, the dummy function does not receive the generic properties (if any) of the actual argument. Therefore, the type of a dummy function must agree with the types of all specific functions which are associated as actual arguments during the invocations of the subprogram.

If an interface block for a dummy subprogram is available within the subprogram definition, the characteristics of the dummy subprogram and its associated actual subprogram must agree, except that a pure actual subprogram may be associated with a dummy subprogram which is not pure.

Interface block: If an interface block for a dummy subprogram is available in the invoking scoping unit, the characteristics of the actual subprogram and of the corresponding dummy subprogram must agree.

13.5.2.12 Asterisk Dummy Arguments

A dummy argument which is an asterisk * is called an **asterisk (dummy) argument**. An asterisk argument may appear in the dummy argument list of a SUBROUTINE or ENTRY statement of a subroutine. In the subroutine reference, an alternate return specifier must be supplied as an actual argument corresponding to such a dummy argument.

13.5.3 Optional Dummy Arguments

A dummy argument which has the OPTIONAL attribute is an **optional** dummy argument. In the subprogram reference, no actual argument need be specified corresponding to an optional dummy argument; in this case, no corresponding actual argument is associated with the dummy argument during the execution of the subprogram.

But if an actual argument is specified corresponding to an optional dummy argument, this actual argument may itself be a dummy argument of the subprogram which contains the subprogram reference. If this dummy argument of the outer subprogram is also an optional argument, no actual argument need be specified in the reference to this subprogram. But if an actual argument is specified corresponding to this optional dummy argument, this actual argument may itself be a dummy argument of the subprogram which contains the subprogram reference. — And so on, for any depth of nested references to subprograms with optional dummy arguments.

An optional dummy argument is **not present** during subprogram execution if no actual argument is associated with it. The following restrictions apply to an optional dummy argument that is not present:

- If it is a dummy data object, it must not be referenced or defined.
- If it is a dummy data object of a derived type with default initilization, the initialization is ineffective.

- If it is a dummy subprogram, it must not be invoked during program execution.
- It must not be specified as an actual argument corresponding to a nonoptional dummy argument except in a reference to the PRESENT intrinsic function.
- A subobject of it must not be specified as an actual argument corresponding to an optional dummy argument.
- If it is an array, it must not be specified as an actual argument in a reference to an elemental subprogram if the actual argument list does not contain another array argument with the same shape corresponding to a nonoptional dummy argument.
- If it is a pointer, it must not be specified as an actual argument corresponding to a nonpointer dummy argument except in a reference to the PRESENT intrinsic function.

Except as noted in the above list, an optional dummy argument which is not present may be specified as an actual argument corresponding to an optional dummy argument, and this optional dummy argument is then also considered to be not associated with an actual argument. That is, the property of presence may be inherited from the actual argument.

A dummy argument is **present** during subprogram execution if an actual argument is associated with it which is no dummy argument in the referencing scoping unit. A present dummy argument may be specified as an actual argument corresponding to a dummy argument, then this last dummy argument also is considered to be present. A nonoptional dummy argument must be present.

External subprograms (which are referenced not directly recursively) and dummy subprograms may have optional dummy arguments only if there is an interface block for the subprogram available in the scoping unit containing the subprogram reference. Statement functions do not have optional arguments.

13.5.4 Dummy Argument with INTENT Attribute

A dummy argument which is a data object may be specified to have an INTENT attribute specifying the intended use of the dummy argument.

A dummy argument having the INTENT(IN) attribute is an *input argument* which may be used but must not be defined or become undefined within the subprogram. A dummy argument having the INTENT(OUT) attribute is an *output argument*. Such a dummy argument is initially undefined except it is of a derived type with default initialization specified in its type definition. And a dummy argument having the INTENT(INOUT) attribute is an *input/output argument*.

A dummy argument with INTENT(IN) or a subobject of such a dummy argument must not appear as the left-hand side of an assignment statement or pointer assignment statement, as an input list item in a READ statement, as the name of a variable in a NAMELIST statement if the belonging group-name appears in a NML= specifier in a READ statement, as an internal file in a WRITE statement, as the integer variable in an IOSTAT= or SIZE= specifier in an input/output statement, as a definable variable in a specifier in an INQUIRE statement, as a status variable in an ALLOCATE or DEALLOCATE statement, or as an actual argument in a reference to a subprogram that is able to redefine the actual argument.

If the INTENT(OUT) or INTENT(INOUT) attribute is specified for a dummy argument, it must be definable. If the INTENT(OUT) attribute is specified for a dummy argument, the actual argument automatically becomes undefined when it becomes associated with the dummy argument except the dummy argument is of a derived type with default initialization specified in its type definition.

13.5.5 Common Blocks

Normally, argument lists are used to exchange data between program units and subprograms of a program. But in the case of user-defined program units and subprograms, common blocks may also be used to perform this kind of communication.

The data in a common block may be referenced and defined in (nearly) all program units and subprograms where the common block is specified. Because variables in a particular common block appearing in different scoping units are not associated by their names but by their storage sequences, these storage associated variables may have (but need not have) different names in each scoping unit.

A reference to the value of a variable in a common block is valid only if the variable is defined; that is, the value must have the same type, the same kind type parameter, and (if applicable) the same character length as the local variable which is used to reference the value. There is an exception: the real part and the imaginary part of a complex variable may be separately referenced in another scoping unit as real variables.

Within a subprogram, variables in a named common block which is also specified in the main program and variables in the blank common preserve their definition status when the END statement or a RETURN statement of the subprogram is executed. Under certain circumstances, variables in a named common block which is specified in a subprogram but not in the main program also preserve their definition status.

14 INTRINSIC SUBPROGRAMS

An intrinsic subprogram is a predefined subprogram supplied by the Fortran processor. A *standard-conforming* Fortran processor must include at least the intrinsic functions and subroutines described in this chapter. Even if a Fortran processor supplies more intrinsic subprograms, a *standard-conforming* program must not make use of these additional intrinsic subprograms.

14.1 Intrinsic Functions

Intrinsic functions may be classified as *inquiry functions*, *elemental functions*, or *transformational functions*.

Inquiry functions return informations about those properties of the argument(s) which do not depend on the value(s) of the argument(s). If an actual argument inquired about consists of a single variable, it may be undefined.

Elemental functions are defined for scalar arguments, but they also can process array arguments element-by-element in order to return a corresponding array result.

Transformational functions are all other (standard-conforming) intrinsic functions. Nearly all of them are defined for array arguments. The rank of the function result may be different from the rank of the argument(s).

Names of intrinsic functions

Nearly every intrinsic function has more than one name. An intrinsic function may have a **generic name** and none, one, or more **specific names**. For example, the intrinsic function which returns the square root of its argument has the generic name SQRT and the specific names SQRT, DSQRT, and CSQRT. Therefore, an intrinsic function may be classified according to its name as a *generic function* or *specific function*.

If an intrinsic function which has a generic name and a specific name is referenced by its specific name, the type of the actual argument(s) must agree with the type of the referenced function; exceptions see below. For example, if a specific function is to be used to calculate the square root of a double precision real (actual) argument, then the double precision real name DSQRT must be used in the function reference.

If an intrinsic function (except INT, REAL, NINT, and ABS) which has a generic name and more than one specific name is referenced by its generic name, the result automatically has the same type as the actual argument(s).

```
PRINT *, SQRT(2.)
PRINT *, SQRT(2.D0)
```

The first PRINT statement prints the value $\sqrt{2}$ as a (default) real value, such as 1.414214; and the second PRINT statement prints $\sqrt{2}$ as a double precision real value, such as 1.414213562373096.

Generic names simplify the reference to intrinsic functions, because the same generic function name may be used regardless of what the type of the actual argument(s) is.

If the name of an intrinsic function is supplied as an actual argument in a subprogram reference, this name must be specified in an INTRINSIC statement within the invoking scoping unit.

Only a *specific* name of an intrinsic function may be specified as an actual argument in a subprogram reference; exceptions see below. During the execution of the referenced subprogram, the dummy function corresponding to the (actual) specific function may be referenced only with scalar arguments. In the case of an elemental function, the dummy function does not receive any elemental properties from the actual argument.

Exceptions: without regard to whether or not the INTRINSIC attribute is specified, the names of the following specific functions must *not* be specified as actual arguments:

- Type conversion: CHAR   FLOAT   ICHAR   IDINT
                     IFIX   INT   REAL   SNGL

- Lexical comparison: LGE   LGT   LLE   LLT

- Minimum/Maximum: DMAX1   DMIN1
                     MAX0   AMAX0   MIN0   AMIN0
                     AMAX1   MAX1   AMIN1 MIN1

The type of an intrinsic function is predefined and need not be specified in a type declaration statement. If the generic name of an intrinsic function is specified in a type declaration statement, the generic function does *not* lose its generic properties.

Masked actual arguments: The function definitions of some intrinsic functions contain an *optional* argument MASK, which may be used to specify a logical mask. This logical mask is useful if not all but only selected array elements of another array argument are to be processed by the function. If this optional argument MASK is present, those array elements of the argument need *not* be defined which are *not* selected by the mask. Note that some intrinsic functions contain a *nonoptional* argument MASK which also defines a logical mask.

14.1 Intrinsic Functions

14.1.1 Table of Intrinsic Functions

Generic name	Specific name	No. of arg.	Type of 1st arg.	Type of result	Description
ABS	IABS	1	integer	integer	Absolute value
	ABS		real	real	$y = |x|$
	DABS		double	double	
	CABS		complex	real	$y = \sqrt{(\Re(x))^2 + ((\Im(x))^2}$
ACHAR	–	1	integer	char.	Char. in given position of ASCII collating sequence
ACOS	ACOS	1	real	real	Arccosine
	DACOS		double	double	$y = \arccos x$
					y in radians
ADJUSTL	–	1	char.	char.	Adjust left
ADJUSTR	–	1	char.	char.	Adjust right
AIMAG	AIMAG	1	complex	real	Imag. part of compl. arg. $y = \Im(x)$
AINT	AINT	1 [,2]	real	real	Truncate
	DINT		double	double	$y = \text{int } x$
–	AMAX0	2 [,3,...]	integer	real	Maximum value
–	MAX1		real	integer	$y = \max(x1, x2, ...)$
–	AMIN0	2 [,3,...]	integer	real	Minimum value
–	MIN1		real	integer	$y = \min(x1, x2, ...)$
ALL	–	1 [,2]	logical	logical	True, if all values true
ALLOCATED	–	1	any	logical	Allocation status
ANINT	ANINT	1 [,2]	real	real	Nearest whole number:
	DNINT		double	double	$y = \text{int}(x + 0.5), x \geq 0$
					$y = \text{int}(x - 0.5), x < 0$
ANY	–	1 [,2]	logical	logical	True, if one value true
ASIN	ASIN	1	real	real	Arcsine
	DASIN		double	double	$y = \arcsin x$
					y in radians
ASSOCIATED	–	1 [,2]	any	logical	Pointer association status
ATAN	ATAN	1	real	real	Arctangent
	DATAN		double	double	$y = \arctan x$
					y in radians

Generic name	Specific name	No. of arg.	Type of 1st arg.	Type of result	Description
ATAN2	ATAN2 DATAN2	2	real double	real double	Arctangent $y = \arctan x1/x2$ y in radians
BIT_SIZE	–	1	integer	integer	Number of Bits
BTEST	–	2	integer	logical	Tests a bit
CEILING	–	1 [,2]	real double	integer integer	Smallest integer $\geq x$
CHAR	CHAR	1 [,2]	integer	char.	Type conversion: integer to character
CMPLX	– – – –	1 [,2,3]	integer real double complex	complex complex complex complex	Type conversion: numeric to complex
CONJG	CONJG	1	complex	complex	Conjugate $y = \Re(x) - i\,\Im(x)$
COS	COS DCOS CCOS	1	real double complex	real double complex	Cosine $y = \cos x$ x in radians
COSH	COSH DCOSH	1	real double	real double	Hyperbolic cosine $y = \cosh x$
COUNT	–	1 [,2]	logical	integer	Number of true elem.
CSHIFT	–	3	any	as 1. arg.	Circular shift
DBLE	– – – –	1	integer real double complex	double double double double	Type conversion: numeric to double precision real
DIGITS	– – –	1 1 1	integer real double	integer integer integer	Number of significant digits in the model of x
DIM	IDIM DIM DDIM	2	integer real double	integer real double	Positive difference $y = x1 - x2,\ x1 > x2$ $y = 0,\ x1 \leq x2$
DOT_PRODUCT	– –	2	numeric logical	numeric logical	Dot product
DPROD	DPROD	2	real	double	Double prec. product $y = x1 * x2$

14.1 Intrinsic Functions

Generic name	Specific name	No. of arg.	Type of 1st arg.	Type of result	Description
EOSHIFT	–	3 [,4]	any	as 1. arg.	End-off shift
EPSILON	–	1	real double	real double	Smallest number with exponent 0
EXP	EXP DEXP CEXP	1	real double complex	real double complex	Exponential function $y = e^x$
EXPONENT	– –	1	real double	integer integer	Exponent part of x in the model
FLOOR	– –	1 [,2]	real double	integer integer	Largest integer $\leq x$
FRACTION	– –	1	real double	real double	Fractional part of x in the model
HUGE	– – –	1	integer real double	integer real double	Largest integer in the model
IACHAR	–	1	char.	integer	Position of given char. in ASCII collating sequence
IAND	–	2	integer	integer	Logical AND
IBCLR	–	2	integer	integer	Clear bit
IBITS	–	3	integer	integer	Bit extraction
IBSET	–	2	integer	integer	Set bit
ICHAR	ICHAR	1	char.	integer	Type conversion: character to integer
IEOR	–	2	integer	integer	Exclusive OR
INDEX	INDEX	2 [,3]	char.	integer	Index of a substring in character string
INT	– INT IFIX IDINT –	1 [,2]	integer real real double complex	integer integer integer integer integer	Type conversion: numeric to integer
IOR	–	2	integer	integer	Inclusive OR
ISHFT	–	2	integer	integer	Logical shift
ISHFTC	–	2 [,3]	integer	integer	Circular shift
KIND	–	1	any	integer	Kind type parameter
LBOUND	–	1 [,2]	any	integer	Lower array bounds

Generic name	Specific name	No. of arg.	Type of 1st arg.	Type of result	Description
LEN	LEN	1	char.	integer	Length of char. str.
LEN_TRIM	–	1	char.	integer	Length without trailing blanks
LGE	LGE	2	char.	logical	Lexical comparison $y = (x1 \geq x2)$
LGT	LGT	2	char.	logical	Lexical comparison $y = (x1 > x2)$
LLE	LLE	2	char.	logical	Lexical comparison $y = (x1 \leq x2)$
LLT	LLT	2	char.	logical	Lexical comparison $y = (x1 < x2)$
LOG	ALOG DLOG CLOG	1	real double complex	real double complex	Natural logarithm $y = \ln x$ base e
LOGICAL	–	1 [,2]	logical	logical	Kind par. convers.
LOG10	ALOG10 DLOG10	1	real double	real double	Common logarithm $y = \log x$ Base 10
MATMUL	– –	2	numeric logical	numeric logical	Matrix multiplication
MAX	MAX0 AMAX1 DMAX1	2 [,3,...]	integer real double	integer real double	Largest value $y = \max(x1, x2, ...)$
MAXEXPONENT	–	1	real double	integer integer	Largest exponent in the model
MAXLOC	– –	1 [,2,3]	integer real double	integer integer integer	Location of largest value of array
MAXVAL	– –	1 [,2,3]	integer real double	integer real double	Largest value of array
MERGE	–	3	any	as 1. arg.	Merge 2 arrays
MIN	MIN0 AMIN1 DMIN1	2 [,3,...]	integer real double	integer real double	Smallest value $y = \min(x1, x2, ...)$

14.1 Intrinsic Functions

Generic name	Specific name	No. of arg.	Type of 1st arg.	Type of result	Description
MINEXPONENT	–	1	real	integer	Smallest exponent in
	–		double	integer	the model
MINLOC	–	1 [,2,3]	integer	integer	Location of smallest
	–		real	integer	value of array
	–		double	integer	
MINVAL	–	1 [,2,3]	integer	integer	Smallest value of
	–		real	real	array
	–		double	double	
MOD	MOD	2	integer	integer	Remainder:
	AMOD		real	real	$y =$
	DMOD		double	double	$x1 - (\text{int } x1/x2) * x2$
MODULO	–	2	integer	integer	Modulo function
	–		real	real	
	–		double	double	
NEAREST	–	2	real	real	Nearest different
			double	double	representable number
NINT	NINT	1 [,2]	real	integer	Nearest integer:
	IDNINT		double	integer	$y = \text{int}(x + 0.5), x \geq 0$
					$y = \text{int}(x - 0.5), x < 0$
NOT	–	1	integer	integer	Logical complement
NULL	–	[1]	any	any	Disassociated pointer
PACK	–	2 [,3]	any	as 1. arg.	Pack array 1-dimens.
PRECISION	–	1	real	integer	Decimal precision
	–		double	integer	
	–		complex	integer	
PRESENT	–	1	any	logical	Argument presence
PRODUCT	–	1 [,2,3]	numeric	as 1. arg.	Product of array elem.
RADIX	–	1	integer	integer	Base of model
	–		real	integer	
	–		double	integer	
RANGE	–	1	numeric	integer	Dec. exponent range
REAL	REAL	1 [,2]	integer	real	Type conversion:
	FLOAT		integer	real	numeric to real
	–		real	real	
	SNGL		double	real	
	–		complex	real	

Generic name	Specific name	No. of arg.	Type of 1st arg.	Type of result	Description
REPEAT	–	2	char.	char.	Repeated concatenat.
RESHAPE	–	2 [,3,4]	integer	as 2. arg.	Reshape array
RRSPACING	–	1	real double	real double	Recipr. of rel. spacing of numbers near x
SCALE	–	2	real double	real double	Scaling: $y = x1 * b^{x2}$
SCAN	–	2 [,3]	char.	integer	Position of a char.
SELECTED_INT_KIND	–	1	integer	integer	Smallest kind type param. of an int. type
SELECTED_REAL_KIND	–	1 [,2]	integer	integer	Smallest kind type param. of a real type
SET_EXPONENT	–	2	real double	real double	Set exponent part of a number
SHAPE	–	1	any	integer	Shape of array/scalar
SIGN	ISIGN SIGN DSIGN	2	integer real double	integer real double	Transfer of sign: $y = \|x1\|,\ x2 \geq 0$ $y = -\|x1\|,\ x2 < 0$
SIN	SIN DSIN CSIN	1	real double complex	real double complex	Sine $y = \sin x$ x in radians
SINH	SINH DSINH	1	real double	real double	Hyperbolic sine $y = \sinh x$
SIZE	–	1 [,2]	any	integer	Number of array elem.
SPACING	–	1	real double	real double	Absolute spacing near x
SPREAD	–	3	any	as 1. arg.	Adding a dimension to array
SQRT	SQRT DSQRT CSQRT	1	real double complex	real double complex	Square root $y = \sqrt{x}$
SUM	–	1 [,2,3]	numeric	as 1. arg.	Sum of array elements
TAN	TAN DTAN	1	real double	real double	Tangent $y = \tan x,\ x$ in rad.
TANH	TANH DTANH	1	real double	real double	Hyperbolic tangent $y = \tanh x$

Generic name	Specific name	No. of arg.	Type of 1st arg.	Type of result	Description
TINY	–	1	real double	real double	Smallest number in the model
TRANSFER	–	2 [,3]	any	as 2. arg.	Treat $x1$ as if of type of $x2$
TRANSPOSE	–	1	any	as arg.	Transpose of array
TRIM	–	1	char.	char.	Remove trailing blanks
UBOUND	–	1 [,2]	any	integer	Upper array bounds
UNPACK	–	3	any	as 1. arg.	Unpack 1-dim. array
VERIFY	–	2 [,3]	char.	integer	Verify set of characters in a string

14.2 Intrinsic Subroutines

Name	No. of arg.	Description
CPU_TIME	1	Returns processor time
DATE_AND_TIME	1 [,2,3,4]	Returns real time clock and date
MVBITS	5	Copies bits between integer objects
RANDOM_NUMBER	1	Returns pseudorandom number
RANDOM_SEED	0 [,1]	Queries or (re)starts pseudorandom number generator
SYSTEM_CLOCK	1 [,2,3]	Returns real-time clock informations

MVBITS is an **elemental subroutine**. The name of an intrinsic subroutine must *not* be specified as an actual argument in a subprogram reference.

14.3 Intrinsic Subprogram Reference

A reference to an intrinsic subprogram has the same form as a reference to a user-defined subprogram. On invocation, the actual arguments in the reference to an intrinsic subprogram become associated with the corresponding dummy arguments in the subprogram definition. An actual argument in a reference to a specific function must be of a *default* intrinsic data type, whereas generic functions accept all kinds of an intrinsic data type.

Usually, the type of the function result of a generic intrinsic function is determined by the type of the first actual argument in the function reference. The type of the result of a specific intrinsic function is determined by the type of the name of the function.

Keyword (actual) arguments: In any scoping unit of a program, actual arguments in references to intrinsic subprograms may be specified as keyword arguments because the names of the dummy arguments of the intrinsic subprograms are predefined. The specification of a keyword (actual) argument in a reference to an intrinsic subprogram must satisfy the same rules as a keyword argument in a reference to a user-defined subprogram.

i = INDEX(z, SUBSTRING='Kap.');   y = EXP(X=ee)

KIND dummy argument: An actual argument associated with the intrinsic function dummy argument KIND must be a scalar initialization expression with a result that specifies a representation method that is supported by the Fortran processor.

Elemental intrinsic subprogram reference: A reference to an elemental intrinsic subprogram may be an elemental reference. The same rules apply to such an elemental intrinsic subprogram reference as to elemental references to user-defined elemental functions (13.4.1.2) and subroutines (13.4.2.2).

In a reference to the intrinsic subroutine MVBITS, the actual arguments corresponding to the TO and FROM dummy arguments may be the same variable.

Inquiry functions: If the actual argument being inquired about in a reference to BIT_SIZE, DIGITS, EPSILON, HUGE, MAXEXPONENT, MINEXPONENT, PRECISION, RADIX, RANGE, or TINY is a pointer, it may have undefined or disassociated association status, and if it is allocatable, it may not be allocated.

14.4 Intrinsic Subprogram Definitions

Optional arguments appearing in headings and tables within this book are enclosed in square brackets. This bracket notation cannot describe all legal forms for argument usage. Note that the actual argument list in a reference to an intrinsic subprogram must not contain a leading comma, that [X, Y] and [X] [, Y] have the same interpretation, and that there are some intrinsic subprograms with more than one optional argument requiring at least one of its optional arguments to be present.

The **definition of a specific function** having an additional generic name may be obtained from the definition of the generic function in this way: leave out all optional dummy arguments (except for the generic functions MAX and MIN) and allow arguments only of default intrinsic types or double precision real type.

14.4 Intrinsic Subprogram Definitions

Specific Name	Generic Name	Argument Type
ABS (A)	ABS (A)	default real
ACOS (X)	ACOS (X)	default real
AIMAG (Z)	AIMAG (Z)	default complex
AINT (A)	AINT (A)	default real
ALOG (X)	LOG (X)	default real
ALOG10 (X)	LOG10 (X)	default real
AMAX0 (A1,A2 [,A3,...])	–	default integer
AMAX1 (A1,A2 [,A3,...])	MAX (A1,A2 [,A3,...])	default real
AMIN0 (A1,A2 [,A3,...])	–	default integer
AMIN1 (A1,A2 [,A3,...])	MIN (A1,A2 [,A3,...])	default real
AMOD (A,P)	MOD (A,P)	default real
ANINT (A)	ANINT (A)	default real
ASIN (X)	ASIN (X)	default real
ATAN (X)	ATAN (X)	default real
ATAN2 (Y,X)	ATAN2 (Y,X)	default real
CABS (A)	ABS (A)	default complex
CCOS (X)	COS (X)	default complex
CEXP (X)	EXP (X)	default complex
CHAR (I)	CHAR (I)	default integer
CLOG (X)	LOG (X)	default complex
CONJG (Z)	CONJG (Z)	default complex
COS (X)	COS (X)	default real
COSH (X)	COSH (X)	default real
CSIN (X)	SIN (X)	default complex
CSQRT (X)	SQRT (X)	default complex
DABS (A)	ABS (A)	double precision real
DACOS (X)	ACOS (X)	double precision real
DASIN (X)	ASIN (X)	double precision real
DATAN (X)	ATAN (X)	double precision real
DATAN2 (X)	ATAN2 (X)	double precision real
DCOS (X)	COS (X)	double precision real
DCOSH (X)	COSH (X)	double precision real
DDIM (X,Y)	DIM (X,Y)	double precision real
DEXP (X)	EXP (X)	double precision real
DIM (X,Y)	DIM (X,Y)	default real
DINT (A)	AINT (A)	double precision real
DLOG (X)	LOG (X)	double precision real
DLOG10 (X)	LOG10 (X)	double precision real
DMAX1 (A1,A2 [,A3,...])	MAX (A1,A2 [,A3,...])	double precision real
DMIN1 (A1,A2 [,A3,...])	MIN (A1,A2 [,A3,...])	double precision real
DMOD (A,P)	MOD (A,P)	double precision real

DNINT (A)	ANINT (A)	double precision real
DPROD (X,Y)	DPROD (X,Y)	default real
DSIGN (A,B)	SIGN (A,B)	double precision real
DSIN (X)	SIN (X)	double precision real
DSINH (X)	SINH (X)	double precision real
DSQRT (X)	SQRT (X)	double precision real
DTAN (X)	TAN (X)	double precision real
DTANH (X)	TANH (X)	double precision real
EXP (X)	EXP (X)	default real
FLOAT (A)	REAL (A)	default integer
IABS (A)	ABS (A)	default integer
ICHAR (C)	ICHAR (C)	default character
IDIM (X,Y)	DIM (X,Y)	default integer
IDINT (A)	INT (A)	double precision real
IDNINT (A)	NINT (A)	double precision real
IFIX (A)	INT (A)	default real
INDEX (STRING, SUBSTRING)	INDEX (STRING, SUBSTRING)	default character
INT (A)	INT (A)	default real
ISIGN (A,B)	SIGN (A,B)	default integer
LEN (STRING)	LEN (STRING)	default character
LGE (STRING_A, STRING_B)	LGE (STRING_A, STRING_B)	default character
LGT (STRING_A, STRING_B)	LGT (STRING_A, STRING_B)	default character
LLE (STRING_A, STRING_B)	LLE (STRING_A, STRING_B)	default character
LLT (STRING_A, STRING_B)	LLT (STRING_A, STRING_B)	default character
MAX0 (A1,A2[,A3,...])	MAX (A1,A2[,A3,...])	default integer
MAX1 (A1,A2[,A3,...])	–	default real
MIN0 (A1,A2[,A3,...])	MIN (A1,A2[,A3,...])	default integer
MIN1 (A1,A2[,A3,...])	–	default real
MOD (A,P)	MOD (A,P)	default integer
NINT (A)	NINT (A)	default real
REAL (A)	REAL (A)	default integer
SIGN (A,B)	SIGN (A,B)	default real
SIN (X)	SIN (X)	default real
SINH (X)	SINH (X)	default real
SNGL (A)	REAL (A)	double precision real
SQRT (X)	SQRT (X)	default real
TAN (X)	TAN (X)	default real
TANH (X)	TANH (X)	default real

14.4 Intrinsic Subprogram Definitions

14.4.1 Descriptions

The following are the descriptions of the generic intrinsic functions, of the specific intrinsic functions without a generic name, and of the intrinsic subroutines.

Note that the dummy arguments of the *specific* intrinsic functions have INTENT(IN). The nonpointer dummy arguments of the *generic* intrinsic functions also have INTENT(IN) if another intent is not stated explicitly.

ABS (A)

ABS is a generic elemental function, which returns the absolute value of the argument. The argument A must be of numeric type. If A is of type integer, real, or double precision real, the result is of the same type and kind type parameter as A. If A is of type complex, the result is of type real and is a processor-dependent approximation to $\sqrt{(\Re(A))^2 + (\Im(A))^2}$. Specific names are IABS, ABS, DABS, and CABS.

ACHAR (I)

ACHAR is a generic elemental function, which returns the character in the given position I of the ASCII collating sequence. The argument I must be of type integer. If $I > 127$, the result is processor-dependent. The result is of type default character and of length 1.

ACOS (X)

ACOS is a generic elemental function, which returns the arccosine expressed in radians. The argument X must be of type real or double precision real with $|X| \leq 1$. The result is of the same type and kind type parameter as X. The range of the result is $0 \leq ACOS(X) \leq \pi$. Specific names are ACOS and DACOS.

ADJUSTL (STRING)

ADJUSTL is a generic elemental function, which adjusts a given character string left by removing all leading blanks and inserting the same number of trailing blanks. The argument STRING must be of type character. The result is of the same type, kind type parameter, and length as STRING.

ADJUSTR (STRING)

ADJUSTR is a generic elemental function, which adjusts a given character string right by removing all trailing blanks and inserting the same number of leading blanks. The argument STRING must be of type character. The result is of the same type, kind type parameter, and length as STRING.

AIMAG (Z)

AIMAG is a generic elemental function, which returns the imaginary part of the complex argument Z. The result is of type real and has the same kind type parameter as the argument Z. Specific name is AIMAG.

AINT (A [, KIND])

AINT is a generic elemental function, which truncates the argument towards zero. The argument A must be of type real or double precision real. The result is of type real. If KIND is present, the result has the kind type parameter KIND; otherwise it is of type default real. Specific names are AINT and DINT.

ALL (MASK [, DIM])

ALL is a generic transformational function, which returns the value *true* if all elements in the argument MASK are *true*, or if DIM is present, if all elements in MASK along dimension DIM are *true*. The argument MASK must be an array of type logical. The optional argument DIM must be scalar and of type integer. The result is of type logical and has the same kind type parameter as MASK.

ALLOCATED (ARRAY)

ALLOCATED is a generic inquiry function, which returns the value *true* if the argument ARRAY is currently allocated. The argument ARRAY must be an allocatable array (of any type). The result is scalar of type default logical.

AMAX0 (A1, A2 [, A3, ...])

AMAX0 is a specific elemental function, which returns the real maximum of two or more given integer values. The arguments must be of type default integer. The result is of type default real.

AMIN0 (A1, A2 [, A3, ...])

AMIN0 is a specific elemental function, which returns the real minimum of two or more given integer values. The arguments must be of type default integer. The result is of type default real.

ANINT (A [, KIND])

ANINT is a generic elemental function, which returns the nearest whole number to the argument A. The argument A must be of type real or double precision real. If KIND is present, the result is of type real with the kind type parameter KIND; otherwise it has the same kind type parameter as the argument A. Specific names are ANINT and DNINT.

ANY (MASK [, DIM])

ANY is a generic transformational function, which returns the value *true* if any of the elements in the argument MASK are *true*, or if DIM is present, if any of the elements in MASK along dimension DIM are *true*. The argument MASK must be an array of type logical. The optional argument DIM must be scalar and of type integer. The result is of type logical and has the same kind type parameter as MASK.

ASIN (X)

ASIN is a generic elemental function, which returns the arcsine expressed in radians. The argument X must be of type real or double precision real with $|X| \leq 1$. The result is of the same type and kind type parameter as X. The range of the result is $-\pi/2 \leq \text{ASIN}(X) \leq \pi/2$. Specific names are ASIN and DASIN.

ASSOCIATED (POINTER [, TARGET])

ASSOCIATED is a generic inquiry function, which indicates whether the given POINTER is currently associated with a target or whether the given POINTER is currently associated with the given TARGET. The argument POINTER must be a pointer (of any type); its pointer association status must not be undefined. The optional argument TARGET must be a pointer or a target. It must have the same type, kind type parameter, character length (if applicable) and rank as POINTER. The result is scalar and of type default logical.

If TARGET is not present, the result is *true* if POINTER is currently associated with a target; otherwise the result is *false*.

If TARGET is present and is a scalar target, the result is *true* if TARGET is not a zero-size storage sequence and the target associated with POINTER occupies the same storage units as TARGET; otherwise or if the POINTER is disassociated the result is *false*.

If TARGET is present and is an array target, the result is *true* if the targets associated with POINTER and TARGET have the same shape, are neither of size zero nor arrays whose elements are zero-size storage sequences, and occupy the same storage units in array element order; otherwise or if POINTER is disassociated, the result is *false*.

If TARGET is present and is a scalar pointer, the result is *true* if the target associated with POINTER and the target associated with TARGET are not zero-size storage sequences and occupy the same storage units; otherwise or if either POINTER or TARGET is disassociated, the result is *false*.

If TARGET is present and is an array pointer, the result is *true* if the target associated with POINTER and the target associated with TARGET have the same shape, are neither of size zero nor arrays whose elements are zero-size storage sequences, and occupy the same storage units in array element order; otherwise or if either POINTER or TARGET is disassociated, the result is *false*.

ATAN (X)

ATAN is a generic elemental function, which returns the arctangent expressed in radians. The argument X must be of type real or double precision real. The result is of the same type and kind type parameter as X. The range of the result is $-\pi/2 \leq \text{ATAN}(X) \leq \pi/2$. Specific names are ATAN and DATAN.

ATAN2 (Y, X)

ATAN2 is a generic elemental function, which returns the arctangent for a pair of arguments. The arguments must be of type real or double precision real; their types and kind type parameters must agree. The result is of the same type and kind type parameter as the arguments. The result is a processor-dependent approximation to the principal value of the complex number (X, Y) and is expressed in radians. If $X \neq 0$, the result is $\arctan Y/X$, and the range of the result is $-\pi < \text{ATAN2}(Y, X) \leq \pi$. The arguments Y and X must not both be zero at the same time. Specific names are ATAN2 and DATAN2.

BIT_SIZE (I)

BIT_SIZE is a generic inquiry function, which returns the number of bits s in the model for bits within an integer of the same kind type parameter as I. The argument I may be scalar or array valued and must be of type integer. If the actual argument for I is a variable name, the variable need not be defined, if it is a pointer its association status is irrelevant, and if it is an allocatable array its allocation status is irrelevant. The result is a scalar integer with the same kind type parameter as I.

BTEST (I, POS)

BTEST is a generic elemental function, which tests a bit in the integer argument I. POS must be of type integer with $0 \leq \text{POS} < \text{BIT\_SIZE}(I)$. The result is of type default logical. The result has the value *true* if bit POS in I has the value 1, and it has the value *false* if bit POS in I has the value 0.

CEILING (A [, KIND])

CEILING is a generic elemental function, which returns the least default integer greater than or equal to its argument. The argument A must be of type real or double precision real. The result is of type integer. If the argument KIND is present, the result has the kind type parameter value KIND; otherwise it is of type default integer.

CHAR (I [, KIND])

CHAR is a generic elemental function, which returns the character in position I in the processor-dependent collating sequence. The argument I must be of type integer with $0 \leq I \leq n-1$, where n is the number of characters in the processor-dependent collating sequence having possibly the specified kind type parameter KIND. The optional argument KIND must be a scalar integer initialization expression. The result is of type character and of length 1. If KIND is present, the result has the kind type parameter KIND; otherwise it is of type default character. Note:

$ICHAR(CHAR(i, KIND(z))) = i$ for $0 \leq i \leq n-1$.
$CHAR(ICHAR(z), KIND(z)) = z$ for any character z (with the (specified) kind type parameter) which the Fortran processor can represent.

Specific name is CHAR.

CMPLX (X [, Y] [, KIND])

CMPLX is a generic elemental function, which converts X or (X, Y), respectively, to complex type. The argument X must be of numeric type. The optional argument Y must be of type integer, real, or double precision real. The optional argument KIND must be a scalar integer initialization expression. The result is of type complex. If the argument KIND is present, the result has the kind type parameter KIND; otherwise it is of type default complex.

If Y is absent and X is not of type complex, the result has the same value as CMPLX(X, 0) or CMPLX(X, 0, KIND), respectively.

If Y is absent and X is of type complex, the result has the same value as CMPLX(X, AIMAG(X)) or CMPLX(X, AIMAG(X), KIND), respectively.

CMPLX(X, Y, KIND) has the complex value (REAL(X, KIND), REAL(Y, KIND)).

CONJG (Z)

CONJG is a generic elemental function, which returns the conjugate of the complex argument Z. The result is of the same type and kind type parameter as Z. Specific name is CONJG.

COS (X)

COS is a generic elemental function, which returns the cosine. The argument X must be of type real, double precision real, or complex and must be specified in radians. The result is of the same type and kind type parameter as X. Specific names are COS, CCOS, and DCOS.

COSH (X)

COSH is a generic elemental function, which returns the hyperbolic cosine. The argument X must be of type real or double precision real. The result is of the same type and kind type parameter as X. Specific names are COSH and DCOSH.

COUNT (MASK [, DIM])

COUNT is a generic transformational function, which returns the number of elements in the logical array MASK that have the value *true*, or if DIM is present, which returns the number of *true* elements in MASK along dimension DIM. The optional argument DIM must be a scalar of type integer. The result is of type default integer.

CPU_TIME (TIME)

CPU_TIME is an intrinsic subroutine, which returns through its output argument a processor-dependent aproximation to the processor time (in seconds). If the **Fortran** processor does not support a meaningfull processor time, TIME is set to a processor-dependent negative value. TIME is an INTENT(OUT) argument and must be a real or double precision real scalar variable. The subroutine is not intended for comparisons of different **Fortran** processors but may be used to compare different algorithms on the same computer.

CSHIFT (ARRAY, SHIFT [, DIM])

CSHIFT is a generic tranformational function, which performs a circular shift on the 1-dimensional array ARRAY or performs a circular shifts on all complete 1-dimensional sections along a given dimension of the 2- or multi-dimensional array ARRAY. Array elements shifted out at one end of an array section are shifted in at the other end. Different array sections may be shifted by different amounts and in different directions. The argument ARRAY must be an array (of any type). The optional argument DIM must be a scalar of type integer. If DIM is omitted, the default DIM=1 is used. The argument SHIFT must be of type integer, it must be scalar if ARRAY is a 1-dimensional array. The result is of the same type, kind type parameter, and possibly character length as ARRAY. If SHIFT is scalar, the result is obtained by shifting every 1-dimensional array section that extends across the dimension DIM circularly SHIFT times. If SHIFT=1 and ARRAY is 1-dimensional, the array is shifted circularly to the left such that the first element of ARRAY becomes the last element of the result. If SHIFT is an array, it must have the shape of ARRAY with dimension DIM omitted. In this case, SHIFT must supply a separate value for each shift.

DATE_AND_TIME ([DATE] [, TIME] [, ZONE] [, VALUES])

DATE_AND_TIME is an intrinsic subroutine, which returns through its output arguments informations from the real-time clock. The form of the date and time informations is compatible with the representations defined in ISO 8601:1988.

DATE is an optional INTENT(OUT) argument and must be a scalar variable of type default character and of length ≥ 8. After return from the subroutine, the leftmost 8 characters contain a string of the form $CCYYMMDD$, with CC being the century, YY the year within the century, MM the month within the year, and DD the day within the month. If the Fortran processor does not support a calendar, DATE is set to blank.

TIME is an optional INTENT(OUT) argument and must be a scalar variable of type default character and of length ≥ 10. After return from the subroutine, the leftmost 10 characters contain a string of the form $hhmmss.sss$, with hh being the hour of the day, mm the minutes of the hour, and $ss.sss$ the seconds and the milliseconds of the minute. If the Fortran processor does not support a clock, TIME is set to blank.

ZONE is an optional INTENT(OUT) argument and must be a scalar variable of type default character and of length ≥ 5. After return from the subroutine, the leftmost 5 characters contain a string of the form $\pm hhmm$, where hh and mm are the time difference with respect to UTC. If the Fortran processor does not support a clock, ZONE is set ot blank.

VALUES is an optional INTENT(OUT) argument and must be a 1-dimensional array variable of type default integer and size ≥ 8. After return from the subroutine, the first 8 elements contain the following values: the year (for example, 1993), the month of the year, the day of the month, the time difference with respect to UTC in minutes, the hour of the day in the range of 0 to 23, the minutes of the hour in the range of 0 to 59, the seconds of the minute in the range of 0 to 60, and the milliseconds of the second in the range of 0 to 999. If the Fortran processor does not support a calendar or a clock, the corresponding elements of VALUES are set to $-$HUGE(0).

DBLE (A)

DBLE is a generic elemental function, which performs a conversion of the argument to type double precision real, namely REAL(A, KIND(0.0D0)). The argument A must be of numeric type. The result is of type double precision real and has as much precision of A or of the real part of A as a double precision real datum may contain.

DIGITS (X)

DIGITS is a generic inquiry function, which returns the number of significant digits in the model that includes the argument X; this is q if X is of type integer;

otherwise it is p (see appendix B). X must be a scalar or an array of type integer, real, or double precision real. The result is of type default integer.

DIM (X, Y)

DIM is a generic elemental function, which returns max(X–Y, 0). The arguments X and Y must be of type integer, real, or double precision real; their types and kind type parameters must agree. The result is of the same type and kind type parameter as the arguments. If X > Y, the result has the value (X − Y). If X ≤ Y, the result is zero.

DOT_PRODUCT (VECTOR_A, VECTOR_B)

DOT_PRODUCT is a generic transformational function, which performs dot-product multiplication of two numeric or two logical 1-dimensional arrays. The argument arrays must have the same size. If VECTOR_A is of numeric type, VECTOR_B may be of the same or any other numeric type. If the arguments are of numeric type, the type and kind type parameter of the result are those of the expression (VECTOR_A * VECTOR_B). If VECTOR_A is of type integer, real, or double precision real, the result is scalar and has the value SUM(VECTOR_A * VECTOR_B). And if VECTOR_A is of type complex, the result is scalar and has the value SUM(CONJG(VECTOR_A) * VECTOR_B). If the arguments are of type logical, the result also is of type logical; the kind type parameter is that of (VECTOR_A .AND. VECTOR_B). And the result is scalar and has the value ANY(VECTOR_A .AND. VECTOR_B).

DPROD (X, Y)

DPROD is a generic elemental function, which returns the product of the arguments as a double precision real result. The arguments must be of type default real. Specific name is DPROD.

EOSHIFT (ARRAY, SHIFT [, BOUNDARY] [, DIM])

EOSHIFT is a generic transformational function, which returns a result that is an end-off shift on the 1-dimensional array ARRAY or is an end-off shift on all complete 1-dimensional sections along a given dimension of the 2- or multi-dimensional array ARRAY. Array elements are shifted out at one end of a section and copies of a boundary value BOUNDARY are shifted in at the other end. BOUNDARY may be omitted for intrinsic types of ARRAY, in which case the default boundary value of zero, *false*, or blank, respectively, is inserted. BOUNDARY must be of the same type and kind type parameter as ARRAY. The optional argument DIM must be scalar and of type integer. Different sections may have different boundary values and may be shifted by

14.4 Intrinsic Subprogram Definitions

different amounts of SHIFT and in different directions. ARRAY must be an array of any type. The argument SHIFT must be of type integer. SHIFT must be scalar, if ARRAY is 1-dimensional; otherwise SHIFT may be scalar, or SHIFT may be an array whose shape is that of ARRAY with dimension DIM omitted, and must supply a separate value for each shift. The result is of the same type, kind type parameter, and, if applicable, character length as ARRAY.

EPSILON (X)

EPSILON is a generic inquiry function, which returns a positive model number that is almost negligible compared with the value one in the model that includes X. The argument must be of type real or double precision real. The result is scalar and is of the same type and kind type parameter as X. The result has the value b^{1-p} (see appendix B).

EXP (X)

EXP is a generic elemental function, which returns the value of the exponential function e^X. The argument X must be of type real, double precision real, or complex. The result is of the same type and kind type parameter as X. Specific names are EXP, CEXP, and DEXP.

EXPONENT (X)

EXPONENT is a generic elemental function, which returns the exponent part e of the argument X when represented as a model number. The argument X must be of type real or double precision real. The result is of type default integer. If X = 0, the result has the value zero.

FLOOR (A)

FLOOR is a generic elemental function, which returns the greatest integer less than or equal to the argument. The argument must be of type real or double precision real. The result is of type integer. If the argument KIND is present, the result has the kind type parameter value KIND; otherwise it is of type default integer.

FRACTION (X)

FRACTION is a generic elemental function, which returns the fractional part of the model representation of the argument. The argument X must be of type real or double precision real. The result is of the same type and kind type parameter as X. The result has the value $X \times b^{-e}$ (see appendix B).

HUGE (X)

HUGE is a generic inquiry function, which returns the largest number in the model that includes X. The argument X may be a scalar or an array and must be of type integer, real, or double precision real. The result is scalar and is of the same type and kind type parameter as X. If X is of type integer, the result has the value $(r^q - 1)$; and if X is of type real or double precision real, the result has the value $(1 - b^{-p})b^{e_{max}}$ (see appendix B).

IACHAR (C)

IACHAR is a generic elemental function, which returns the position of a given character in the ASCII collating sequence. The argument C must be of type default character and of length 1. The result is of type default integer. Note: IACHAR(ACHAR(i)) = i and ACHAR(IACHAR(z)) = z.

IAND (I, J)

IAND is a generic elemental function, which returns the logical AND of all bits in I and corresponding bits in J. The arguments I and J must be of type integer with the same kind type parameter. The result is of type integer and has the same kind type parameter as the arguments. Integers which are sequences of bits are interpreted according to the model for bit manipulation. The following truth table shows the effect of the function:

I	1	1	0	0
J	1	0	1	0
IAND(I, J)	1	0	0	0

IBCLR (I, POS)

IBCLR is a generic elemental function, which returns an integer with the same sequence of bits as the argument I except that bit position POS is set to 0. The argument I must be of type integer and is interpreted according to the model for bit manipulation as a sequence of bits. POS must be of type integer with $0 \leq$ POS $<$ BIT_SIZE(I). The result is of the same type and kind type parameter as I.

IBITS (I, POS, LEN)

IBITS is a generic elemental function, which returns a subsequence of the given sequence of bits in I. The argument I must be of type integer. POS must be of type integer with POS ≥ 0 and POS + LEN $\leq$ BIT_SIZE(I). LEN must be a nonnegative integer. The result is of type integer with the same kind type parameter as I. The result contains right adjusted the same sequence of bits as the LEN bits of I starting at bit POS; any remaining leading bits of the result are set to 0. Integers which are sequences of bits are interpreted according to the model for bit manipulation.

IBSET (I, POS)

IBSET is a generic elemental function, which returns an integer with the same sequence of bits as the argument I except that bit position POS is set to 1. The argument I must be of type integer and is interpreted according to the model for bit manipulation as a sequence of bits. POS must be of type integer with $0 \leq POS < BIT\_SIZE(I)$. The result is of the same type and kind type parameter as I.

ICHAR (C)

ICHAR is a generic function, which returns the position of a given character in the processor-dependent collating sequence of this character. The argument is of type character and of length 1. The result is of type default integer. Note: $ICHAR(CHAR(i, KIND(z))) = i$ and $CHAR(ICHAR(z), KIND(z)) = z$. Specific name is ICHAR.

IEOR (I, J)

IEOR is a generic elemental function, which returns the logical exclusive OR of all bits in I and corresponding bits in J. The arguments I and J must be of type integer with the same kind type parameter. The result is of type integer and has the same kind type parameter as the arguments. Integers which are sequences of bits are interpreted according to the model for bit manipulation. The following truth table shows the effect of the function:

I	1	1	0	0
J	1	0	1	0
IEOR(I, J)	0	1	1	0

INDEX (STRING, SUBSTRING [, BACK])

INDEX is a generic elemental function, which returns the starting position of SUBSTRING as a substring of STRING. The arguments STRING and SUBSTRING must be of type character with the same kind type parameter. The optional argument BACK must be of type logical. The result is of type default integer. The result is 0 if the substring does not occur in STRING or if the length of STRING is less than the length of SUBSTRING. If BACK is absent or present with the value *false*, the result is the position of the first occurrence of the substring SUBSTRING; if the length of SUBSTRING is zero, the result is 1. If BACK is present with the value *true*, the result is the position of the last occurrence of the substring SUBSTRING; if the length of SUBSTRING is zero, the result is LEN(STRING) + 1. Specific name is INDEX.

INT (A [, KIND])

INT is a generic elemental function, which performs a type conversion. The argument A must be of numeric type. The optional argument KIND must be a scalar integer initialization expression. If KIND is present, the result is of type integer and has the kind type parameter KIND; otherwise the result is of type default integer. If A is of type integer, the result has the value A. If A is of type real or double precision real, the result is A truncated towards zero. And if A is of type complex, the result is the real part of A truncated towards zero. Specific names are IDINT, IFIX, and INT.

IOR (I, J)

IOR is a generic elemental function, which returns the logical inclusive OR of all bits in I and corresponding bits in J. The arguments I and J must be of type integer with the same kind type parameter. The result is of type integer and has the same kind type parameter as the arguments. Integers which are sequences of bits are interpreted according to the model for bit manipulation. The following truth table shows the effect of the function:

I	1	1	0	0
J	1	0	1	0
IOR(I, J)	1	1	1	0

ISHFT (I, SHIFT)

ISHFT is a generic elemental function, which returns a sequence of bits equal to that of I except that the bits are shifted end-off SHIFT positions. The argument I must be of type integer. The argument SHIFT must be of type integer with |SHIFT| < BIT_SIZE(I). The result is of type integer and has the same kind type parameter as I. If SHIFT > 0, the sequence of bits is shifted to the left. If SHIFT < 0, the sequence of bits is shifted to the right. Bits shifted out to the left or to the right are lost and zeros are inserted into the gaps created. The integers which are sequences of bits are interpreted according to the model for bit manipulation.

ISHFTC (I, SHIFT [, SIZE])

ISHFTC is a generic elemental function, which returns a sequence of bits equal to that of I except that the SIZE rightmost bits or all bits are shifted circularly SHIFT positions. The argument I must be of type integer. The argument SHIFT

14.4 Intrinsic Subprogram Definitions

must be of type integer with $|\text{SHIFT}| < \text{BIT\_SIZE}(I)$. The optional argument SIZE must be of type integer with $0 < \text{SIZE} \leq \text{BIT\_SIZE}(I)$. If SIZE is absent, the default $\text{SIZE} = \text{BIT\_SIZE}(I)$ is used. The result is of type integer and has the same kind type parameter as I. If $\text{SHIFT} > 0$, the SIZE rightmost bits in I are shifted circularly to the left. If $\text{SHIFT} < 0$, the SIZE rightmost bits in I are shifted circularly to the right. The integers which are sequences of bits are interpreted according to the model for bit manipulation.

KIND (X)

KIND is a generic inquiry function, which returns the kind type parameter of the argument. The argument X may be scalar or array valued and of any intrinsic type. The result is scalar and of type default integer.

LBOUND (ARRAY [, DIM])

LBOUND is a generic inquiry function, which returns all lower bounds of array ARRAY or which returns the lower bound in dimension DIM. The argument ARRAY must be an array of any type, no scalar. The optional argument DIM must be scalar and of type integer. The result is of type default integer.

If DIM is present, the result is scalar. If ARRAY is an array section or another array expression other than a whole array and a structure component, LBOUND(ARRAY, DIM) returns the value 1. For a whole array or an array structure component, LBOUND(ARRAY, DIM) returns the value of the lower bound for dimension DIM of ARRAY if dimension DIM of ARRAY does not have extent zero or if ARRAY is an assumed-size array of rank DIM; otherwise the result has the value 1.

If DIM is absent, the result is a 1-dimensional array with a size which is equal to the rank of ARRAY. And the values of the array elements of the result are the lower bounds of ARRAY.

LEN (STRING)

LEN is a generic inquiry function, which returns the character length of the argument. The argument STRING must be of type character. The result is of type default integer. Specific name is LEN.

LEN_TRIM (STRING)

LEN_TRIM is a generic elemental function, which returns the character length of the argument without its trailing blanks. The argument STRING must be of type character. The result is of type default integer.

LGE (STRING_A, STRING_B)

LGE is a generic elemental function, which performs a lexical comparison of corresponding characters from the left to the right based on the ASCII collating sequence. Both arguments must be of type default character. The result is of type default logical. The result has the value *true* if string STRING_A follows string STRING_B lexically or if both strings are equal. The result has the value *false* if string STRING_A precedes string STRING_B lexically. If the strings have different lengths, the shorter one is padded (only for the comparison) with trailing blanks such that internally two strings of the same length are compared lexically. Specific name is LGE.

LGT (STRING_A, STRING_B)

LGT is a generic elemental function, which performs a lexical comparison of corresponding characters from the left to the right based on the ASCII collating sequence. Both arguments must be of type default character. The result is of type default logical. The result has the value *true* if string STRING_A follows string STRING_B lexically. The result has the value *false* if string STRING_A precedes string STRING_B lexically or if both strings are equal. If the strings have different lengths, the shorter one is padded (only for the comparison) with trailing blanks such that internally two strings of the same length are compared lexically. Specific name is LGT.

LLE (STRING_A, STRING_B)

LLE is a generic elemental function, which performs a lexical comparison of corresponding characters from the left to the right based on the ASCII collating sequence. Both arguments must be of type default character. The result is of type default logical. The result has the value *true* if string STRING_A precedes string STRING_B lexically or if both strings are equal. The result has the value *false* if string STRING_A follows string STRING_B lexically. If the strings have different lengths, the shorter one is padded (only for the comparison) with trailing blanks such that internally two strings of the same length are compared lexically. Specific name is LLE.

LLT (STRING_A, STRING_B)

LLT is a generic elemental function, which performs a lexical comparison of corresponding characters from the left to the right based on the ASCII collating sequence. Both arguments must be of type default character. The result is of type default logical. The result has the value *true* if string STRING_A precedes string STRING_B lexically. The result has the value *false* if string STRING_A follows string STRING_B lexically or if both strings are equal. If the strings have different lengths, the shorter one is padded (only for the comparison) with trailing blanks such that internally two strings of the same length are compared lexically. Specific name is LLT.

14.4 Intrinsic Subprogram Definitions

LOG (X)

LOG is a generic elemental function, which returns the natural logarithm (base e). The argument X must be of type real, double precision real, or complex. If X is of type real or double precision real, X must be positive. If X is of type complex, X must not be zero. The result is of the same type and kind type parameter as X. A result of type complex is the principal value with the imaginary part ω in the range $-\pi < \omega \leq \pi$. Specific names are ALOG, CLOG, and DLOG.

LOGICAL (L [, KIND])

LOGICAL is a generic elemental function, which performs a type (parameter) conversion. The argument L must be of type logical. The optional argument KIND must be a scalar integer initialization expression. If KIND is present, the result is of type logical and has the kind type parameter KIND; otherwise the result is of type default logical. The result has the value L.

LOG10 (X)

LOG10 is a generic elemental function, which returns the common logarithm (base 10). The argument X must be of type real or double precision real with $X > 0$. The result is of the same type and kind type parameter as X. Specific names are ALOG10 and DLOG10.

MATMUL (MATRIX_A, MATRIX_B)

MATMUL is a generic transformational function, which performs matrix multiplication of two numeric or two logical 1-dimensional or 2-dimensional arrays. The argument MATRIX_A must be an array of type numeric or logical. If MATRIX_A is of numeric type, MATRIX_B must be an array of numeric type; and if MATRIX_A is of type logical, MATRIX_B must be an array of type logical. The numeric type of MATRIX_A may be different from the numeric type of MATRIX_B. If MATRIX_A is 1-dimensional, MATRIX_B must be 2-dimensional; and if MATRIX_B is 1-dimensional, MATRIX_A must be 2-dimensional. The extent of the first (or single) dimension of MATRIX_B must be equal to the extent of the last (or single) dimension of MATRIX_A. If the arguments are of numeric type, the type and kind type parameter of the result are those of the expression (MATRIX_A * MATRIX_B). If the arguments are of type logical, the result also is of type logical, and the kind type parameter is that of (MATRIX_A .AND. MATRIX_B). The result is an array. The shape of the result depends on the shapes of the arguments, as follows:

If MATRIX_A has the shape (n, m) and MATRIX_B the shape (m, k), the result has the shape (n, k). If the arguments are of numeric type, the array element

(i, j) of the result has the value SUM(MATRIX_A$(i, :)$ * MATRIX_B$(:, j)$). And if both arguments are of type logical, the array element (i, j) of the result has the value ANY(MATRIX_A$(i, :)$ .AND. MATRIX_B$(:, j)$).

If MATRIX_A has the shape (m) and MATRIX_B has the shape (m, k), the result has the shape (k). If the arguments are of numeric type, the array element (j) of the result has the value SUM(MATRIX_A$(:)$ * MATRIX_B$(:, j)$). And if both arguments are of type logical, the array element (j) of the result has the value ANY(MATRIX_A$(:)$ .AND. MATRIX_B$(:, j)$).

If MATRIX_A has the shape (n, m) and MATRIX_B has the shape (m), the result has the shape (n). If the arguments are of numeric type, the array element (i) of the result has the value SUM(MATRIX_A$(i, :)$ * MATRIX_B$(:)$). And if both arguments are of type logical, the array element (i) of the result has the value ANY(MATRIX_A$(i, :)$ .AND. MATRIX_B$(:)$).

MAX (A1, A2 [, A3, ...])

MAX is a generic elemental function, which returns the maximum of two or more given values. The arguments must be of type integer, real, or double precision real; their types and kind type parameters must agree. The result is of the same type and kind type parameter as the arguments. Specific names are AMAX1, DMAX1, and MAX0.

MAXEXPONENT (X)

MAXEXPONENT is a generic inquiry function, which returns the maximum exponent in the model that includes the argument X. The argument X must be an array or a scalar of type real or double precision real. The result is scalar and of type default integer.

MAXLOC (ARRAY [, MASK]) and
MAXLOC (ARRAY, DIM [, MASK])

MAXLOC is a generic transformational function, which determines the first array element containing the maximum value of all elements in array ARRAY or of a given subset of elements in array ARRAY. The argument ARRAY must be an array of type integer, real, or double precision real. The argument DIM must be scalar and of type integer. The optional argument MASK must be of type logical and must have the same shape as ARRAY. DIM and MASK may be specified in any order. The result is of type default integer.

The result of MAXLOC(ARRAY) is a 1-dimensional array with a size which is equal to the rank of ARRAY and whose element values are the values of the subscripts of an element of ARRAY containing the maximum value of all elements of ARRAY. The result of MAXLOC(ARRAY, MASK=MASK) determines the position of the maximum value of those array elements in ARRAY which correspond to *true* elements in MASK.

If DIM is present, the result is an array of the rank of ARRAY reduced by one and with the shape of ARRAY without the dimension DIM. The function is applied to all 1-dimensional array sections which span right through dimension DIM.

Note that in either case not the declared array bounds of ARRAY are used, but value 1 is used as the lower bound in all dimensions of ARRAY.

MAXVAL (ARRAY [, MASK]) and
MAXVAL (ARRAY, DIM [, MASK])

MAXVAL is a generic transformational function, which returns the maximum value of all elements in array ARRAY or of a subset of elements in ARRAY. The argument ARRAY must be an array of type integer, real, or double precision real. The argument DIM must be scalar and of type integer. The optional argument MASK must be of type logical and must have the same shape as ARRAY. DIM and MASK may be specified in any order. The result is of the same type and kind type parameter as ARRAY.

If DIM is absent or if ARRAY is a 1-dimensional array, the result is scalar. The value of MAXVAL(ARRAY) is equal to the maximum value of all array elements in ARRAY. The value of MAXVAL(ARRAY, MASK=MASK) is equal to the maximum value of those array elements in ARRAY which correspond to *true* elements in MASK.

If DIM is present, the result is an array of the rank of ARRAY reduced by one and with the shape of ARRAY without the dimension DIM. The function is applied to all 1-dimensional array sections which span right through dimension DIM.

MAX1 (A1, A2 [, A3, ...])

MAX1 is a specific elemental function, which returns the integer maximum of two or more given real values. The arguments must be of type default real. The result is of type default integer.

MERGE (TSOURCE, FSOURCE, MASK)

MERGE is a generic elemental function, which returns a result composed of two given values. The argument TSOURCE may be of any type. FSOURCE must be of the same type, kind type parameter, and, if applicable, character length as TSOURCE. The argument MASK must be of type logical. The result is of the same type, kind type parameter, and possibly character length as TSOURCE. The result is equal to TSOURCE where MASK is *true*; otherwise the result is equal to FSOURCE.

MIN (A1, A2 [, A3, ...])

MIN is a generic elemental function, which returns the minimum of two or more given values. The arguments must be of type integer, real, or double precision real; their types and kind type parameters must agree. The result is of the same type and kind type parameter as the arguments. Specific names are AMIN1, DMIN1, and MIN0.

MINEXPONENT (X)

MINEXPONENT is a generic inquiry function, which returns the minimum exponent in the model that includes the argument X. The argument X must be an array or a scalar of type real or double precision real. The result is scalar and of type default integer.

MINLOC (ARRAY [, MASK]) and
MINLOC (ARRAY, DIM [, MASK])

MINLOC is a generic transformational function, which determines the first array element containing the minimum value of all elements in array ARRAY or of a given subset of elements in array ARRAY. The argument ARRAY must be an array of type integer, real, or double precision real. The argument DIM must be scalar and of type integer. The optional argument MASK must be of type logical and must have the same shape as ARRAY. DIM and MASK may be specified in any order. The result is of type default integer.

The result of MINLOC(ARRAY) is a 1-dimensional array with a size which is equal to the rank of ARRAY and whose element values are the values of the subscripts of an element of ARRAY containing the minimum value of all elements of ARRAY. The result of MINLOC(ARRAY, MASK=MASK) determines the position of the minimum value of those array elements in ARRAY which correspond to *true* elements in MASK.

If DIM is present, the result is an array of the rank of ARRAY reduced by one and with the shape of ARRAY without the dimension DIM. The function is applied to all 1-dimensional array sections which span right through dimension DIM.

Note that in either case not the declared array bounds of ARRAY are used, but value 1 is used as the lower bound in all dimensions of ARRAY.

14.4 Intrinsic Subprogram Definitions

MINVAL (ARRAY [, MASK]) and
MINVAL (ARRAY, DIM [, MASK])

MINVAL is a generic transformational function, which returns the minimum value of all elements in array ARRAY or of a subset of elements in ARRAY. The argument ARRAY must be an array of type integer, real, or double precision real. The argument DIM must be scalar and of type integer. The optional argument MASK must be of type logical and must have the same shape as ARRAY. DIM and MASK may be specified in any order. The result is of the same type and kind type parameter as ARRAY.

If DIM is absent or if ARRAY is an 1-dimensional array, the result is scalar. The value of MAXVAL(ARRAY) is equal to the minimum value of all array elements in ARRAY. The value of MAXVAL(ARRAY, MASK=MASK) is equal to the minimum value of those array elements in ARRAY which correspond to *true* elements in MASK.

If DIM is present, the result is an array of the rank of ARRAY reduced by one and with the shape of ARRAY without the dimension DIM. The function is applied to all 1-dimensional array sections which span right through dimension DIM.

MIN1 (A1, A2 [, A3, ...])

MIN1 is a specific elemental function, which returns the integer minimum of two or more given real values. The arguments must be of type default real. The result is of type default integer.

MOD (A, P)

MOD is a generic elemental function, which returns the remainder of A modulo P. The argument A must be of type integer, real, or double precision real. The types and kind type parameters of the arguments must agree. The result is of the same type and kind type parameter as A. If $P \neq 0$, the result has the value $(A - INT(A/P) * P)$. And if P=0, the result is processor-dependent. Specific names are AMOD, DMOD, and MOD.

MODULO (A, P)

MODULO is a generic elemental function, which returns A modulo P. The argument A must be of type integer, real, or double precision real. The types and kind type parameters of the arguments must agree. The result is of the same type and kind type parameter as A. If the arguments are of type integer, the result has the value $(A - FLOOR(REAL(A)/REAL(P)) * P)$. And if the arguments are of type real or double precision real, the result has the value $(A - FLOOR(A/P) * P)$. If P=0, the result is processor-dependent.

MVBITS (FROM, FROMPOS, LEN, TO, TOPOS)

MVBITS is a generic elemental subroutine, which copies the sequence of bits in FROM that begins at position FROMPOS and has the length LEN to TO beginning at position TOPOS. The arguments FROM, FROMPOS, LEN, and TOPOS have the INTENT(IN) attribute and must be of type integer. FROMPOS must be ≥ 0 with FROMPOS + LEN $\leq$ BIT_SIZE(FROM). LEN must be ≥ 0. TOPOS must be ≥ 0 with TOPOS + LEN $\leq$ BIT_SIZE(TO). The argument TO has the INTENT(INOUT) attribute and must be a variable of type integer and must have the same kind type parameter as FROM. Integers which are sequences of bits are interpreted according to the model for bit manipulation.

NEAREST (X, S)

NEAREST is a generic elemental function, which returns the nearest different representable machine number in the direction given by the sign of argument S. The argument X must be of type real or double precision real. The argument S must be of type real or double precision real; its value must not be zero. The result is of the same type and kind type parameter as X. If S is positive, the result is the next representable number which is greater than X. And if S is negative, the result is the next representable number which is smaller than X.

NINT (A [, KIND])

NINT is a generic elemental function, which returns the nearest integer to the argument A. The argument A must be of type real or double precision real. The result is of type integer. If KIND is present, the result has the kind type parameter KIND; otherwise it is of type default integer. Specific names are IDNINT and NINT.

NOT (I)

NOT is a generic elemental function, which returns the logical complement of all bits in the argument I. The argument I must be of type integer. The result is of type integer and has the same kind type parameter as I. Integers which are sequences of bits are interpreted according to the model for bit manipulation. The following truth table shows the effect of the function:

I	0	1
NOT(I)	1	0

NULL ([MOLD])

NULL is a generic transformational function, which returns a disassociated pointer. The optional argument MOLD may be of any type and must be a pointer

14.4 Intrinsic Subprogram Definitions

with an association status which may be undefined, disassociated, or associated. If MOLD is associated, its target need not be defined (with a valid value). If MOLD is present, the result has the same type, kind type parameter value, character length (if applicable), and rank as MOLD. If MOLD is absent, type, kind type parameter value, character length (if applicable), and rank of the result are those of the data entity being initialized or defined or being associated with NULL( ).

PACK (ARRAY, MASK [, VECTOR])

PACK is a generic transformational function, which packs into a 1-dimensional array those elements in the given array ARRAY that are selected by the conformable logical mask MASK. The argument ARRAY may be an array of any type. The argument MASK must be of type logical; it may be scalar or an array, but it must be conformable with ARRAY. The optional argument VECTOR must be a 1-dimensional array of the same type, kind type parameter, and, if applicable, character length as ARRAY. Its size must be greater than or equal to the number of *true* elements in MASK. If MASK is scalar and has the value *true*, VECTOR must have at least the same size as ARRAY.

The result is of the same type, kind type parameter, and possibly character length as ARRAY. If VECTOR is present, the 1-dimensional result has the same size as VECTOR. If VECTOR is absent and MASK is an array, the size of the result is equal to the number of *true* elements in MASK. If VECTOR is absent and MASK is scalar, the size of the result is equal to the size of ARRAY.

The i-th element of the result is equal to that element in ARRAY which corresponds to the i-th *true* element in MASK. The array elements are counted in array element order.

If VECTOR is present and if its size is greater than the number of *true* elements in MASK, the remaining elements of the result have the same value as the remaining elements in VECTOR.

PRECISION (X)

PRECISION is a generic inquiry function, which returns the decimal precision in the model that includes the argument X. The argument X must be of type real, double precision real, or complex. The result is of type default integer.

PRESENT (A)

PRESENT is a generic inquiry function, which determines whether argument A, which must be the name of an optional dummy argument that is accessible in the subprogram containing the reference to the PRESENT function, is present in that subprogram. Argument A may be scalar or array valued, of any type, a pointer, or a dummy subprogram. Dummy argument A has no INTENT attribute. The result is of type default logical. The result has the value *true*, if A is present; otherwise the result has the value *false*.

PRODUCT (ARRAY [, MASK]) and
PRODUCT (ARRAY, DIM [, MASK])

PRODUCT is a generic transformational function, which returns the product of all elements in array ARRAY or of a subset of elements in ARRAY. The argument ARRAY must be an array of numeric type. The argument DIM must be scalar and of type integer. The optional argument MASK must be of type logical and must have the same shape as ARRAY. The result is of the same type and kind type parameter as ARRAY.

If MASK and DIM are absent, the result is scalar and its value is the product of all array elements in ARRAY. If MASK is absent and DIM is present, the result is an array (or a scalar if ARRAY is 1-dimensional) of the rank of ARRAY reduced by one and with the shape of ARRAY without the dimension DIM. The function is applied to all 1-dimensional array sections which span right through dimension DIM.

If MASK is present, the function is applied at most to those elements in ARRAY which correspond to *true* elements in MASK.

RADIX (X)

RADIX is a generic inquiry function, which returns the base of the model that includes the argument X. The argument X must be of type integer, real, or double precision real and may be scalar or an array. The result is of type default integer. The result has the value r if X is of type integer and the value b if X is of type real or double precision real (see appendix B).

RANDOM_NUMBER (HARVEST)

RANDOM_NUMBER is subroutine, which returns through the argument HARVEST one pseudorandom number or an array of pseudorandom numbers from the uniform distribution over the range $0 \leq x < 1$. The argument HARVEST must be a variable of type real or double precision real, it may be scalar or an array, and it has the INTENT(OUT) attribute.

RANDOM_SEED ([SIZE] [, PUT] [, GET])

RANDOM_SEED is a subroutine, which initializes and restarts the pseudorandom number generator and allows inquiries. The pseudorandom numbers may be generated by the invocation of the subroutine RANDOM_NUMBER. The generation of pseudorandom numbers may be suspended at any point of this

14.4 Intrinsic Subprogram Definitions

sequence of pseudorandom numbers and may be resumed later exactly at this point. Either one argument may be present or no argument may be present. If no argument is present, the **Fortran** processor initializes the seed with a processor-dependent value. SIZE is an optional argument with INTENT(OUT) attribute; it must be a scalar variable of type default integer, which returns the size N of the seed array. PUT is an optional argument with INTENT(IN) attribute; it must be a 1-dimensional array of type default integer and size $\geq N$, which is used to reset the seed. GET is an optional argument with INTENT(OUT) attribute; it must be a 1-dimensional array variable of type default integer and size $\geq N$, which returns the current value of the seed. The value of GET may be later used to restart the seed.

Initialization of the pseuderandom number generator by a particular PUT argument is performed in a processor-dependent manner. The value returned by GET need not be the same as the value of PUT in an immediately preceding invocation of RANDOM_SEED. Though these values differ, when used as PUT argument in RANDOM_SEED the pseudorandom number sequences generated by RANDOM_NUMBER are equal.

RANGE (X)

RANGE is a generic inquiry function, which returns the decimal exponent range in the model that includes the argument X. The argument X must be of type numeric; it may be scalar or an array. X need not be defined. The result is scalar and of type default integer. If X is of type integer, the result has the value INT(LOG10(HUGE(X))). If X is of type real, double precision real, or complex, the result has the value INT(MIN(LOG10(*huge*)), –LOG10(*tiny*))), where *huge* is the largest positive number and *tiny* the smallest number in the model representing real numbers with the same kind type parameter as X.

REAL (A [, KIND])

REAL is a generic elemental function, which performs a type conversion. The argument must be of numeric type. The optional argument KIND must be a scalar integer initialization expression. If KIND is present, the result is of type real and has the kind type parameter KIND. If KIND is not present and if A is of type integer, real, or double precision real, the result is of type default real and its value is a processor-dependent approximation to A. If KIND is not present and if A is of type complex, the result is of type real and has the same kind type parameter as A and its value is a processor-dependent approximation to the real part of A. Specific names are FLOAT, REAL, and SNGL.

REPEAT (STRING, NCOPIES)

REPEAT is a generic transformational function, which concatenates several given copies of a string. The argument STRING must be scalar and of type character. The argument NCOPIES must be nonnegative, scalar, and of type integer. The result is scalar and of type character and has the same kind type parameter as STRING. The result is the string which consists of NCOPIES concatenated copies of STRING.

RESHAPE (SOURCE, SHAPE [, PAD] [, ORDER])

RESHAPE is a generic transformational function, which returns an array with shape SHAPE, which is of the same type, kind type parameter, and, if applicable, character length as SOURCE. SOURCE must be an array of any type. SHAPE must be a 1-dimensional array of type integer with at most 7 elements. The array elements in SHAPE must all be ≥ 0. If PAD is absent or of size zero, the size of SOURCE must be greater than or equal to the product of the elements in SHAPE (i.e., PRODUCT(SHAPE)). The size of the result is the product of the elements in SHAPE. The optional argument PAD must be an array and of the same type, kind type parameter, and, if applicable, character length as SOURCE. The optional argument ORDER must be a 1-dimensional array of type integer with the same shape as SHAPE. The value of ORDER must be a permutation of $(1, 2, \ldots, n)$, where n is the size of SHAPE.

If ORDER is absent, the values of the elements of the result are, in array element order, equal to the values of the elements in SOURCE in array element order followed by copies of PAD in array element order. If ORDER is present, the values of the elements of the result correspond in the permuted subscript order (ORDER(1), ORDER(2), ..., ORDER(n)) to the values of the elements in SOURCE in array element order followed by copies of PAD in array element order.

RRSPACING (X)

RRSPACING is a generic elemental function, which returns the reciprocal of the relative spacing of model numbers that include X near the argument value. The argument X must be of type real or double precision real. The result is of the same type and kind type parameter as X. The result has the value $|X \times b^{-e}| \times b^p$ (see appendix B).

SCALE (X, I)

SCALE is a generic elemental function, which returns the value of the argument X scaled by b^I, where b is the base in the model that includes X. The argument X must be of type real or double precision real. The result is of the same type and kind type parameter as X. The result has the value $X \times b^I$ (see app. B).

SCAN (STRING, SET [, BACK])

SCAN is a generic elemental function, which returns the position of a character in STRING that is in SET. The arguments STRING and SET must be of type character and their kind type parameters must agree. The optional argument BACK must be of type logical. The result is of type default integer. The result is 0 if no character in STRING is in SET or if the length of STRING or SET is zero. Otherwise: if BACK is absent or present with the value *false*, the result is the position of the leftmost character in STRING that is in SET; and if BACK is present with the value *true*, the result is the position of the rightmost character in STRING that is in SET.

SELECTED_INT_KIND (R)

SELECTED_INT_KIND is a generic transformational function, which returns a kind type parameter value of an integer type for the representation of all integers n in the range $-10^R < n < 10^R$. The argument R must be scalar and of type integer. The result is scalar and of type default integer. If the **Fortran** processor does not support such integer type, the result has the value -1. If the **Fortran** processor supports more than one integer type for the given range, the result has the value of the kind type parameter with the smallest decimal exponent range. If the **Fortran** processor supports more than one integer type for the given range with the same smallest decimal exponent range, the result has the value of the smallest value of their kind type parameters.

SELECTED_REAL_KIND ([P] [, R])

SELECTED_REAL_KIND is a generic transformational function, which returns a kind type parameter value of a real type for the representation of all reals with a decimal precision of at least P digits (as returned by the function PRECISION) and a decimal exponent range of at least R (as returned by the function RANGE). If the **Fortran** processor does not support such real type, the result has the value -1 if the minimum precision P is not supported, the value -2 if the minimum exponent range R is not supported, and the value -3 if neither the given P nor the given R are supported. If the **Fortran** processor supports more than one real type for the given arguments, the result has the value of the kind type parameter with the smallest decimal precision. If the **Fortran** processor supports more than one real type for the given arguments with the same smallest decimal precision, the result has the value of the smallest value of their kind type parameters.

SET_EXPONENT (X, I)

SET_EXPONENT is a generic elemental function, which returns the number in the model that includes X whose fractional part is the fractional part of the model representation of argument X and whose exponent part is the argument I. The argument X must be of type real or double precision real. The argument I must be of type integer. The result is of the same type and kind type parameter as X. The result has the value $(X \times b^{I-e})$ (see appendix B).

SHAPE (SOURCE)

SHAPE is a generic inquiry function, which returns the shape of the argument SOURCE. The argument SOURCE may be scalar or an array of any type; it must not be an assumed-size array, a currently disassociated pointer, or a currently deallocated allocatable array. The result is a 1-dimensional array of type default integer, whose size is equal to the rank of source. The value of the result is the shape of SOURCE. If SOURCE is scalar, the result has size zero.

SIGN (A, B)

SIGN is a generic elemental function, which returns the absolute value of the argument A times the sign of argument B. The arguments A and B must be of type integer, real, or double precision real; their types and kind type parameters must agree. Special case B = 0: normally, the result would have the value |A|; but if B is of type real or double precision real and the Fortran processor is able to distinguish between positive and negative zero, then the result is |A| if B = +0.0, and the result is −|A| if B = −0.0. The result is of the same type and kind type parameter as A. Specific names are DSIGN, ISIGN, and SIGN.

SIN (X)

SIN is a generic elemental function, which returns the sine. The argument X must be of type real, double precision real, or complex and must be specified in radians. The result is of the same type and kind type parameter as X. Specific names are CSIN, DSIN, and SIN.

SINH (X)

SINH is a generic elemental function, which returns the hyperbolic sine. The argument X must be of type real or double precision. The result is of the same type and kind type parameter as X. Specific names are SINH and DSINH.

SIZE (ARRAY [, DIM])

SIZE is a generic inquiry function, which returns the size of the array argument ARRAY or the size of ARRAY along dimension DIM. The argument ARRAY may be an array of any type, but no scalar. The optional argument DIM must be scalar and of type integer. The result is scalar and of type default integer. If DIM is absent, the result is the size of ARRAY. If DIM is present, the result is the extent of dimension DIM of ARRAY.

SPACING (X)

SPACING is a generic elemental function, which returns the absolute spacing of the model numbers that include X near the argument value. The argument must be of type real or double precision real. The result is of the same type and kind type parameter as X. If X is not zero, the result has the value b^{e-p}, provided this result is within range (see appendix B). Otherwise, the result is the same as that of TINY(X).

SPREAD (SOURCE, DIM, NCOPIES)

SPREAD is a generic transformational function, which returns an array with the rank of SOURCE increased by one that consists along its given dimension DIM of NCOPIES copies of SOURCE. The argument SOURCE may be scalar or an array of any type. The arguments DIM and NCOPIES must be scalar and of type integer. The result is of the same type, kind type parameter, and, if applicable, character length as SOURCE.

If SOURCE is scalar, the result is a 1-dimensional array with MAX(NCOPIES, 0) elements that have the same value as SOURCE. If SOURCE is an array with the shape $(d_1, d_2, \ldots, d_n)$, the shape of the result is $(d_1, d_2, \ldots, d_{DIM-1}, \text{MAX(NCOPIES, 0)}, d_{DIM}, \ldots, d_n)$. And the array element of the result with the subscript $(r_1, r_2, \ldots, r_{n+1})$ has the value SOURCE$(r_1, r_2, \ldots, r_{DIM-1}, r_{DIM+1}, \ldots, r_{n+1})$.

SQRT (X)

SQRT is a generic elemental function, which returns the square root $\sqrt{X}$. The argument X must be of type real, double precision real, or complex. If X is of type real or double precision real, it must not be negative. The result is of the same type and kind type parameter as X. If X is of type complex, the result is the principal value with the real part greater than or equal to zero. When the real part of the result is zero, the imaginary part of the result is not negative. Specific names are CSQRT, DSQRT, and SQRT.

SUM (ARRAY [, MASK]) and
SUM (ARRAY, DIM [, MASK])

SUM is a generic transformational function, which returns the sum of all elements in array ARRAY or of a subset of elements in ARRAY. The argument ARRAY must be an array of numeric type. The argument DIM must be scalar and of type integer. The optional argument MASK must be of type logical and must have the same shape as ARRAY. The result is of the same type and kind type parameter as ARRAY.

If MASK and DIM are absent, the result is scalar and its value is the sum of all array elements in ARRAY. If MASK is absent and DIM is present, the result is an array (or a scalar if ARRAY is 1-dimensional) of the rank of ARRAY reduced by one and with the shape of ARRAY without the dimension DIM. The function is applied to all 1-dimensional array sections which span right through dimension DIM.

If MASK is present, the function is applied at most to those elements in ARRAY which correspond to *true* elements of MASK.

SYSTEM_CLOCK([COUNT][,COUNT_RATE][,COUNT_MAX])

SYSTEM_CLOCK is an intrinsic subroutine, which returns through its output arguments informations from the real-time clock. All arguments are optional arguments with INTENT(OUT) attribute and must be scalar variables of type default integer.

COUNT is set to the processor-dependent value of the system clock. This system clock is incremented by one at each clock count until COUNT_MAX is reached and is set to zero on the next count. If the **Fortran** processor does not support a system clock, COUNT is set to −HUGE(0).

COUNT_RATE is set to the processor-dependent approximation to the number of processor clock counts per second. If the **Fortran** processor does not support a system clock, COUNT_RATE is set to zero.

COUNT_MAX is set to the maximum value that COUNT may have. If the **Fortran** processor does not support a system clock, COUNT_MAX is set to zero.

TAN (X)

TAN is a generic elemental function, which returns the tangent. The argument X must be of type real or double precision real and must be specified in radians. The result is of the same type and kind type parameter as X. Specific names are DTAN and TAN.

TANH (X)

TANH is a generic elemental function, which returns the hyperbolic tangent. The argument X must be of type real or double precision real. The result is of the same type and kind type parameter as X. Specific names are DTANH and TANH.

TINY (X)

TINY is a generic inquiry function, which returns the smallest number in the model that includes X. The argument X may be a scalar or an array and must be of type real or double precision real. The result is scalar and is of the same type and kind type parameter as X. The result has the value $b^{e_{min}-1}$ (see appendix B).

TRANSFER (SOURCE, MOLD [, SIZE])

TRANSFER is a generic transformational function, which returns a result that has the internal representation of SOURCE and the type, kind type parameter and, if applicable, character length of MOLD. This function allows internal transfers of data (that is, assignments) between data objects of different types without changing the internal representation. The arguments SOURCE and MOLD may be scalars or arrays of any type. The optional argument SIZE must be scalar and of type integer.

If MOLD is scalar and SIZE is absent, the result is scalar. And if MOLD is an array and SIZE is absent, the result is a 1-dimensional array; and the size of the result is just sufficient to hold the processor-dependent internal representation of SOURCE. If SIZE is present, the result is a 1-dimensional array of size SIZE.

If the physical representation of the result is as long as that of SOURCE or is longer than that of SOURCE, the leading part of the result has the same physical representation as SOURCE and the rest of the result is undefined; otherwise the result has the same physical representation as the leading part of SOURCE.

TRANSPOSE (MATRIX)

TRANSPOSE is a generic transformational function, which performs a matrix transpose for a given 2-dimensional array. The argument MATRIX may be a 2-dimensional array of any type. The result is of the same type, kind type parameter and, if applicable, character length as MATRIX. The result is a 2-dimensional array with the shape (n, m), where (m, n) is the shape of MATRIX. The array element (i, j) of the result has the value MATRIX(j, i) for $i = 1, 2, ..., n$ and $j = 1, 2, ..., m$.

TRIM (STRING)

TRIM is a generic transformational function, which returns the string STRING without its trailing blanks. The argument STRING must be scalar and of type character. The result is of the same type and kind type parameter as STRING. The length of the result is equal to the length of STRING reduced by the number of trailing blanks in STRING.

UBOUND (ARRAY [, DIM])

UBOUND is a generic inquiry function, which returns all upper bounds of array ARRAY or which returns the upper bound in dimension DIM. The argument

ARRAY must be an array of any type, no scalar. The optional argument DIM must be scalar and of type integer. The result is of type default integer.

If DIM is present, the result is scalar. If ARRAY is an array section or another array expression which is no whole array and no structure component, the result is equal to the extent of the dimension DIM. If the extent of the dimension DIM of ARRAY is not zero, the result is the value of the upper bound in dimension DIM. And if the extent of the dimension DIM of ARRAY is zero, the result is 0.

If DIM is absent, the result is a 1-dimensional array with a size which is equal to the rank of ARRAY. And the values of the array elements of the result are the upper bounds of ARRAY.

UNPACK (VECTOR, MASK, FIELD)

UNPACK is a generic transformational function, which unpacks a 1-dimensional array into an array of a given type and a given shape. The argument VECTOR may be a 1-dimensional array of any type. Its size must be greater than or equal to the number of *true* elements in MASK. The argument MASK must be an array of type logical. The argument FIELD must be of the same type, kind type parameter, and, if applicable, character length as VECTOR. FIELD may be scalar or an array, but it must be conformable with MASK.

The result is an array and is of the same type, kind type parameter, and possibly character length as VECTOR. It has the same shape as MASK. The array element of the result which corresponds to the ith *true* element in MASK has the value VECTOR(i) for $i = 1, 2, \ldots, t$, where t is the number of *true* elements in MASK. The array elements are counted in array element order.

All other elements of the result are equal to the value of FIELD if FIELD is scalar, or equal to the corresponding element in FIELD if FIELD is an array.

VERIFY (STRING, SET [, BACK])

VERIFY is a generic elemental function, which returns the value zero if each character in STRING is in SET, or which returns the position of a character of STRING which is *not* in SET. If the optional argument BACK is absent or if it is present with the value *false*, the result is the position of the leftmost such character which is not in SET. And if BACK is present with the value *true*, the result is the position of the rightmost such character which is not in SET.

A CHARACTER SETS AND COLLATING SEQUENCES

A.1 Processor-Dependent Character Sets

Default character type: The character set of the default character type contains at least all characters of the Fortran character set. In addition to the Fortran character set, a Fortran processor may support other representable characters, which may appear only in character literal constants, in character string edit descriptors, in comments, and in formatted records. The processor character set may contain control characters.

Nondefault character type: If a Fortran processor supports in addition to the default character type one or more nondefault character types, the characters of such a nondefault character type may appear only in character literal constants of that type or in formatted records. In the last case, the types of the datum in the record and of the corresponding input/output list item must agree. One character in the character set of a nondefault character type is designated as the blank character; this blank character of the nondefault character type may be used as the padding character for formatted input.

Collating sequence

Each processor-dependent character set has its own collating sequence. The first character has position 0, the second has position 1, and so on. The ICHAR intrinsic function returns for a given character the position of this character in the processor-dependent collating sequence. For example, the function reference ICHAR('X') returns the integer value 231 if the Fortran processor uses the EBCDIC character set. And conversely, the CHAR intrinsic function returns for a given nonnegative integer value the character that corresponds to this given position in the processor-dependent collating sequence. For example, CHAR(231) returns the character 'X' if the EBCDIC character set is used.

A.2 ASCII Character Set

The ASCII character set is a standard character set. The International Fortran Standard refers to ISO/IEC 646:1991. With regard to letters, digits, and special characters of the Fortran character set, ISO/IEC 646:1991 and ANSI X3.4-1986, which is the US national version of the ISO/IEC standard, are identical. An ASCII character is encoded by 7 bits. Therefore, a standard ASCII character set contains a total number of 128 characters. The first 32 characters (hex 00 to hex 1F) of the ASCII character set are *control characters* and *graphic characters*. Nearly all other characters (hex 20 to hex 7E) are *printable* characters.

Character Sets and Collating Sequences

b7 b6 b5 Bits b4 b3 b2 b1	0 0 0 Control Characters	0 0 1	0 1 0 Symbols, Digits	0 1 1	1 0 0 Upper-case Letters	1 0 1	1 1 0 Lower-case Letters	1 1 1
0 0 0 0	NUL (0, 0)	DLE (16, 10, 20)	SP (32, 20, 40)	0 (48, 30, 60)	@ (64, 40, 100)	P (80, 50, 120)	` (96, 60, 140)	p (112, 70, 160)
0 0 0 1	SOH (1, 1, 1)	DC1 (17, 11, 21)	! (33, 21, 41)	1 (49, 31, 61)	A (65, 41, 101)	Q (81, 51, 121)	a (97, 61, 141)	q (113, 71, 161)
0 0 1 0	STX (2, 2, 2)	DC2 (18, 12, 22)	" (34, 22, 42)	2 (50, 32, 62)	B (66, 42, 102)	R (82, 52, 122)	b (98, 62, 142)	r (114, 72, 162)
0 0 1 1	ETX (3, 3, 3)	DC3 (19, 13, 23)	# (35, 23, 43)	3 (51, 33, 63)	C (67, 43, 103)	S (83, 53, 123)	c (99, 63, 143)	s (115, 73, 163)
0 1 0 0	EOT (4, 4, 4)	DC4 (20, 14, 24)	$ (36, 24, 44)	4 (52, 34, 64)	D (68, 44, 104)	T (84, 54, 124)	d (100, 64, 144)	t (116, 74, 164)
0 1 0 1	ENQ (5, 5, 5)	NAK (21, 15, 25)	% (37, 25, 45)	5 (53, 35, 65)	E (69, 45, 105)	U (85, 55, 125)	e (101, 65, 145)	u (117, 75, 165)
0 1 1 0	ACK (6, 6, 6)	SYN (22, 16, 26)	& (38, 26, 46)	6 (54, 36, 66)	F (70, 46, 106)	V (86, 56, 126)	f (102, 66, 146)	v (118, 76, 166)
0 1 1 1	BEL (7, 7, 7)	ETB (23, 17, 27)	' (39, 27, 47)	7 (55, 37, 67)	G (71, 47, 107)	W (87, 57, 127)	g (103, 67, 147)	w (119, 77, 167)
1 0 0 0	BS (8, 8, 10)	CAN (24, 18, 30)	((40, 28, 50)	8 (56, 38, 70)	H (72, 48, 110)	X (88, 58, 130)	h (104, 68, 150)	x (120, 78, 170)
1 0 0 1	HT (9, 9, 11)	EM (25, 19, 31)	) (41, 29, 51)	9 (57, 39, 71)	I (73, 49, 111)	Y (89, 59, 131)	i (105, 69, 151)	y (121, 79, 171)
1 0 1 0	LF (10, A, 12)	SUB (26, 1A, 32)	* (42, 2A, 52)	: (58, 3A, 72)	J (74, 4A, 112)	Z (90, 5A, 132)	j (106, 6A, 152)	z (122, 7A, 172)
1 0 1 1	VT (11, B, 13)	ESC (27, 1B, 33)	+ (43, 2B, 53)	; (59, 3B, 73)	K (75, 4B, 113)	[(91, 5B, 133)	k (107, 6B, 153)	{ (123, 7B, 173)
1 1 0 0	FF (12, C, 14)	FS (28, 1C, 34)	, (44, 2C, 54)	< (60, 3C, 74)	L (76, 4C, 114)	\ (92, 5C, 134)	l (108, 6C, 154)	| (124, 7C, 174)
1 1 0 1	CR (13, D, 15)	GS (29, 1D, 35)	- (45, 2D, 55)	= (61, 3D, 75)	M (77, 4D, 115)	] (93, 5D, 135)	m (109, 6D, 155)	} (125, 7D, 175)
1 1 1 0	SO (14, E, 16)	RS (30, 1E, 36)	. (46, 2E, 56)	> (62, 3E, 76)	N (78, 4E, 116)	^ (94, 5E, 136)	n (110, 6E, 156)	~ (126, 7E, 176)
1 1 1 1	SI (15, F, 17)	US (31, 1F, 37)	/ (47, 2F, 57)	? (63, 3F, 77)	O (79, 4F, 117)	_ (95, 5F, 137)	o (111, 6F, 157)	DEL (127, 7F, 177)

The IACHAR intrinsic function returns for a given character the position of this character in the ASCII collating sequence; IACHAR('X') returns the integer value 88. And the ACHAR intrinsic function returns for a given nonnegative integer value the character that corresponds to this given value in the ASCII collating sequence; ACHAR(88) returns the character 'X'.

The intrinsic functions LGE, LGT, LLE, and LLT perform lexical comparisons of character strings on the basis of the ASCII collating sequence.

B MODELS FOR NUMBERS

The bit manipulation intrinsic subprograms, the numeric manipulation intrinsic functions, and the inquiry intrinsic functions are defined in terms of models for the representation of each kind of integer, real, or double precision real data implemented by the processor. These models are used to describe the characteristics and the behaviour of the corresponding number sets of a particular type. A model has parameters which are chosen by the Fortran processor such that the model best fits the hardware.

Note that these models do not dictate to the Fortran processor how it has to implement numbers and how they should behave, but the internal representation and the behaviour of the actually implemented numbers are described in terms of these (abstract) models.

B.1 Models for Integers

The models for integer data i are defined by:

$$i = s \times \sum_{k=1}^{q} w_k \times r^{k-1}$$

with

	Value	Description
s	$+1$ or -1	sign
r	integer > 1	base
q	integer > 0	max. number of digits
w_k	each an integer with $0 \leq w_k < r$	digit

The parameters r and q determine the set of values of the integer model numbers.
Example: For model $i = s \times \sum_{k=1}^{31} w_k \times 2^{k-1}$ is $q = 31$ and $r = 2$.

B.2 Models for Reals

The models for real data x are defined by:

$$x = \begin{cases} 0 & (= b^0 \times \sum_{k=1}^{p} 0 \times b^{-k}) \\ s \times b^e \times \sum_{k=1}^{p} f_k \times b^{-k} \end{cases}$$

with

	Value	Description
s	$+1$ or -1	sign
b	integer > 1	base
p	integer > 1	max. number of digits
f_k	each an integer with $0 \leq f_k < b$, but $f_1 > 0$	digit
e	integer with $e_{min} \leq e \leq e_{max}$	exponent

The parameters b, p, e_{min}, and e_{max} determine the set of values of the (real) floating point model numbers.

Example: For model $x = \begin{cases} 0 \\ s \times 2^e \times (1/2 + \sum_{k=2}^{24} f_k \times 2^{-k}), \; -126 \leq e \leq 127 \end{cases}$
is $b = 2$, $p = 24$, $e_{min} = -126$, and $e_{max} = 127$.

B.3 Models for Bit Manipulation

Bit manipulations are performed on integer data. Therefore, a **bit** is defined as a binary digit w at position k of a nonnegative integer datum. The models for these integers are defined by:

$$j = \sum_{k=0}^{s-1} w_k \times 2^k$$

with

	Value	Description
s	integer > 0	maximal number of digits
w_k	each either 0 or 1	digit, bit

The parameter s determines the set of values.

Example: For model $j = \sum_{k=0}^{31} w_k \times 2^k$ is $s = 32$. It defines a model for integer data with 32 bits.

The models for bit manipulation define that the integer datum is a sequence of s bits, which are numbered from the right to the left beginning with 0 for the rightmost bit and ending with $(s-1)$ for the leftmost bit. These models apply only to the intrinsic subprograms for bit manipulation and bit inquiry. For all other purposes, the "normal" models for integer data must be used.

C DECREMENTAL FEATURES

C.1 Deleted Language Features

The following **deleted language features** are not included in the normative part of the standard document but in its informative appendix. They are considered as redundant and largely unused. Fortran 95 is only a partly upward compatible extension to Fortran 90. Any standard-conforming Fortran 90 that does not include one of the deleted language features is also standard-conforming under Fortran 95. Since the list of deleted features is very small, migration from Fortran 90 to Fortran 95 will cause hardly problems.

Real and double precision real DO variables
Real and double precision real DO variables and loop parameters, are not supported by Fortran 95. Noninteger loop control is processor-dependent and makes programs nonportable. Thus a design fault introduced with FORTRAN 77 and for consistency ported to Fortran 90 is removed.

Branching to ENDIF
In Fortran 95, branching to an ENDIF statement is allowed only from a block inside the corresponding IF construct. Branching from the outside to an ENDIF was erroneously allowed in FORTRAN 77 and for consistency also in Fortran 90.

PAUSE statement
The anachronistic PAUSE statement, present in FORTRAN 77 and for consistency also in Fortran 90, has been deleted.

ASSIGN, assigned GOTO, and assigned format specifier
The ASSIGN statement, the assigned GOTO statement, and the ability to use an assigned integer variable as a format specifier, present in FORTRAN 77 and for consistency also in Fortran 90, have been deleted.

H edit descriptor
The H edit descriptor, present in FORTRAN 77 and for consistency also in Fortran 90, have been deleted.

C.2 Obsolescent Language Features

The following language features are described as **obsolescent language features** within the standard document. They are considered to have been redundant in Fortran 90, are still frequently used, but are candidates for removal from Fortran 2000, the next revision of the language. Therefore, programmers should use the better methods in Fortran 95 in new programs and convert existing programs to these methods.

Arithmetic IF
  Use the logical IF statement or the IF construct.

Shared DO termination for nested DO loops
  Use an ENDDO or a CONTINUE statement corresponding to each DO statement.

DO termination other than ENDDO or CONTINUE
  Use an ENDDO or a CONTINUE statement as DO termination.

Alternate return from subroutine
  Let the subroutine compute a return code that is used in a CASE construct immediately after the CALL statement on return from the subroutine.

Computed GOTO
  Use a CASE construct.

Statement function
  Use an internal function.

DATA statement amongst executable statements
  Position DATA statements within the specification part of a program unit or subprogram only.

Assumed length character functions
  Use a subroutine whose arguments correspond to the function result and the function arguments.

Fixed source form
  Use only free source form. There are tools available to convert existing programs from fixed source form to free source form.

Assumed size array
  Depending on the context, use an automatic array, an assumed-shape array, an allocatable array, or an array pointer.

CHARACTER *length
  Use the form CHARACTER ([LEN =] length).

INDEX

A edit descriptor 12-8
ABS function 14-3, 14-11, 14-13
access method 11-3
ACCESS= specifier 11-42, 11-43, 11-45
accessible
    → host association
    → USE association
ACHAR function 14-3, 14-13
ACOS function 14-3, 14-11, 14-13
ACTION= specifier 11-42, 11-44
actual argument list 13-44, 13-42
additional entry point 13-37
ADJUSTL function 14-3, 14-13
ADJUSTR function 14-3, 14-13
ADVANCE= specifier 11-10
advancing input/output 11-20
AIMAG function 14-3, 14-11, 14-14
AINT function 14-3, 14-11, 14-14
ALL function 14-3, 14-14
allocatable array 6-14
ALLOCATABLE attribute 9-3
ALLOCATABLE statement 9-19
ALLOCATE statement 5-2, 6-14
ALLOCATED function 14-3, 14-14
allocation status 6-15
ALOG function 14-6, 14-11
ALOG10 function 14-6, 14-11
alternate return 13-22, 13-40, 13-41, 13-44
alternate return target 13-22
AMAX0 function 14-2, 14-3, 14-11
AMAX1 function 14-2, 14-6, 14-11
AMIN0 function 14-2, 14-3, 14-11
AMIN1 function 14-2, 14-6, 14-11
AMOD function 14-7, 14-11

ANINT function 14-3, 14-11, 14-14
ANY function 14-3, 14-15
argument association 4-12, 13-45
argument keyword 13-44, 14-10
argument list 13-42
arithmetic IF statement 10-3
arithmetic → numeric
array 4-1, 4-5, 6-1
array assignment statement 6-21, 8-1, 8-11
array bound 6-1
array component 2-9
array constructor 6-18
array declaration 6-1
array element 4-1, 4-6, 6-5
    order, position 4-8, 4-9
array expression 6-19, 7-2
array function 6-21
array operand 6-19
array operation 6-19
array pointer 6-16
array processing 6-1
array reference 6-4
array section 4-1, 4-6, 6-6
array section of substrings 6-12
array specification 6-1
array subprogram 6-21
array variable 4-3
ASCII 7-10, A-1
ASIN function 14-3, 14-11, 14-15
assignment block 8-14
assignment interface block 8-7
assignment statement 8-1
assignment subroutine 8-7, 13-23

assignment symbol 3-3
ASSIGNMENT(=) 8-7
ASSOCIATED function 14-3, 14-15
association 4-12
association status (of a pointer) 5-3
assumed length 9-18, 13-14, 13-50
assumed-shape array 6-2, 13-55
assumed-size array 6-3, 13-55
assumed type parameter
    → assumed length
asterisk dummy argument 13-43, 13-57
ATAN function 14-3, 14-11, 14-16
ATAN2 function 14-4, 14-11, 14-16
attribute 4-1, 9-1, 9-2, 9-19
attribute specification statement 9-19
automatic array 6-13
automatic variable 4-11

B edit descriptor 12-9
BACKSPACE statement 11-52
binary constant 3-8
BIT_SIZE function 14-4, 14-16
blank
    in input 11-38, 11-43, 11-45, 12-7
    in source program 1-2, 1-4, 1-5
blank common 9-31
blank control edit descriptor 12-10
blank padding character 8-5, 11-14,
    11-23, 11-44, 11-45, 11-51, 12-8
BLANK= specifier 11-42, 11-43, 11-49
block 10 1, 10 5, 10 10
block data program unit 13-6
BLOCK DATA statement 13-6
BN edit descriptor 12-10
body of DO construct 10-15
BOZ constant 3-8
branching 10-1
branch target statement 10-1
BTEST function 14-4, 14-16
BZ edit descriptor 12-10

C comment line 1-3

CABS function 14-3, 14-11
CALL statement 13-21
carriage control (character) 11-41
CASE construct 10-10
case expression 10-10
case index 10-10
CASE statement 10-10, 10-11
case value 10-11, 10-12
case-block 10-10, 10-13
CCOS function 14-4, 14-11
CEILING function 14-4, 14-16
CEXP function 14-5, 14-11
CHAR function 14-2, 14-4, 14-11,
    14-17
character assignment statement 8-4
character constant edit descriptor
    12-11
character expression 7-12
character format specification 12-2
character length 3-7, 4-4, 9-17, 9-26,
    13-49
character operator 7-13
character position 2-4
character relational expression 7-9
character sequence type 2-7
character set
    ASCII A-1
    Fortran 1-1
    processor-dependent 7-9, A-1
CHARACTER statement 9-16
character storage unit 4-15
character string 2-4
character string edit descriptor 12-5,
    12-6
character substring 4-4
character type 2-4
CLOG function 14-6, 14-11
close (a file) 11-46
CLOSE statement 11-46
CMPLX function 14-4, 14-17
collating sequence 7-9, A-1
colon edit descriptor 12-11

Index _____ D-3

comment 1-2, 1-3, 1-5
common block 4-15, 9-31, 13-59
COMMON statement 9-31
complex type 2-3
COMPLEX statement 9-15
component
   array component 2-9
   structure component 4-9
   type component 2-5, 2-7
component attribute 2-8
component definition 2-7
computed GO TO statement 10-2
concatenation 7-13
condition
   end-of-file cond. 11-10, 11-11, 11-13
   end-of-record cond. 11-10, 11-11, 11-12
   error condition 11-11, 11-12
conformable 6-19, 7-1
CONJG function 14-4, 14-11, 14-17
connection of a file (to a unit) 11-6
constant 3-3, 4-1, 4-2
constant array constructor 6-19
constant structure constructor 2-14
constant expression 7-24
construct 1-7, 10-1
construct entity 3-3
constructed structure object 2-13
constructor 2-13, 6-18
CONTAINS statement 13-2, 13-3,
      13-11, 13-14, 13-19,13-20
continuation (line) 1-2, 1-4
CONTINUE statement 10-23
control edit descriptor 12-5, 12-6
control flow 10-1
control information list 11-8
control mask 8-12
COS function 14-4, 14-11, 14-17
COSH function 14-4, 14-11, 14-18
COUNT function 14-4, 14-18
count loop 10-14, 10-16, 10-18
CSHIFT function 14-4, 14-18
CSIN function 14-8, 14-11

CSQRT function 14-8, 14-11
current record 11-5
CYCLE statement 10-20

D edit descriptor 12-12
DABS function 14-3, 14-11
DACOS function 14-3, 14-11
DASIN function 14-3, 14-11
DATA attribute 9-4
data edit descriptor 12-5, 12-6
data entity, data object 4-1
DATA statement 9-19
data transfer statement 11-19
data type 2-1
DATAN function 14-3, 14-11
DATAN2 function 14-4, 14-11
DATE_AND_TIME subroutine
      14-9, 14-19
DBLE function 14-4, 14-19
DCOS function 14-4, 14-11
DCOSH function 14-4, 14-11
DDIM function 14-4, 14-11
DEALLOCATE statement 5-5, 6-15
deallocation 5-5, 6-15
declaration 9-1
declaration statement
      → specification statement
DEFAULT 10-11
default type 2-1
default implicit typing 4-3, 9-12, 9-40
default initialization 2-5, 2-8, 2-10
defined
   assignment 8-6
   elemental assignment 8-6
   elemental operation 7-15
   expression 7-14
   operator 7-14
defined 4-3, 4-16
definition, definition status 4-16
DELIM= specifier 11-42, 11-44
delimiter 3-1
derived type 2-5

derived type assignment stmt. 8-5
derived type definition 2-5
DEXP function 14-5, 14-11
DIGITS function 14-4, 14-19
DIM function 14-4, 14-11, 14-20
dimension (of an array) 4-5
DIMENSION attribute 9-4
DIMENSION statement 9-23
DINT function 14-3, 14-11
direct access 11-3
direct input/output 11-3
DIRECT= specifier 11-47, 11-48
disassociated 5-3
DLOG function 14-6, 14-11
DLOG10 function 14-6, 14-11
DMAX1 function 14-2, 14-6, 14-11
DMIN1 function 14-2, 14-6, 14-11
DMOD function 14-7, 14-11
DNINT function 14-3, 14-12
DO construct 10-14
DO statement 10-15
do-termination statement 10-15
DO variable 6-18, 9-22, 10-15, 11-17
DOT_PRODUCT function 14-4, 14-20
double precision complex 9-15
double precision real type 2-2
DOUBLE PRECISION statement 9-14
DPROD function 14-4, 14-12, 14-20
DSIGN function 14-8, 14-12
DSIN function 14-8, 14-12
DSINH function 14-8, 14-12
DSQRT function 14-8, 14-12
DTAN function 14-8, 14-12
DTANH function 14-8, 14-12
dummy argument list 13-42
dummy array 13-51
dummy pointer 13-52
dummy subprogram 13-28, 13-56

E edit descriptor 12-14
edit descriptor 12-1, 12-5
element sequence 13-53

elemental array assignment 8-2
elemental assignment 8-6
elemental function 13-10, 13-15, 13-39, 14-1
elemental operation 6-20, 7-1
ELEMENTAL prefix 13-11, 13-15, 13-19, 13-20, 13-21
elemental reference 13-18, 13-23, 14-10
elemental subroutine 13-10, 13-21, 13-39
else-block 10-5, 10-7
ELSE statement 10-5
elsewhere-block 8-14
ELSEWHERE statement 8-12
elseif-block 10-5, 10-7
ELSE IF statement 10-5
empty input list 11-16
empty record 11-16
empty value list 6-19
EN edit descriptor 12-14
end of the record 11-27, 11-36
END statement 1-7, 1-8
end-of-file condition 11-11, 11-13
end-of-record condition 11-10, 11-11
END= specifier 11-13
endfile record 11-2
ENDFILE statement 11-52, 11-53
ending position 4-4
END BLOCK DATA statement 13-6
END FORALL statement 8-19
END DO statement 10-15
END FUNCTION statement 13-11, 13-14
END IF statement 10-5
END INTERFACE statement 13-33
endless loop 10-14, 10-16, 10-20
END MODULE statement 13-3, 13-4
END PROGRAM statement 13-2, 13-3
END SELECT statement 10-10, 10-12

END SUBROUTINE statement 13-19, 13-20
END TYPE statement 2-6
END WHERE statement 8-12
entry point 13-37
ENTRY statement 13-37
EOR= specifier 11-10
EOSHIFT function 14-5, 14-20
EPSILON function 14-5, 14-21
equality of derived types 2-7
equals 3-3
equivalence set 9-36
EQUIVALENCE statement 9-36
equivalent expressions 7-3
ERR= specifier 11-12, 11-42, 11-43, 11-46, 11-47, 11-52
error condition 11-11, 11-12
ES edit descriptor 12-15
evaluation of expressions 7-21
executable statement 1-7
execution condition 10-15, 10-16, 10-17, 10-19
execution control statement 10-1
execution part 13-2, 13-11, 13-19
EXIST= specifier 11-47, 11-48, 11-51
EXIT statement 10-18, 10-20
EXP function 14-5, 14-12, 14-21
explicit initialization 9-13
explicit-shape array 6-2
EXPONENT function 14-5, 14-21
exponent part 12-12, 14-21
exponent range 2-2, 14-35
expression 7-1
extent of a dimension 4-7, 6-6
extended assignment 8-9
extended operator 7-18
extended Fortran 95 features 13-25
EXTERNAL attribute 9-5
external file 11-2
EXTERNAL statement 9-24
external subprogram 13-23

F edit descriptor 12-16
field width 12-6, 12-7
file 11-2
file attribute 11-2
file name 11-3
file position 11-5
file positioning statement 11-52
file status statement 11-42
FILE= specifier 11-42, 11-43, 11-45, 11-47, 11-48
FLOAT function 14-2, 14-7, 14-12
FLOOR function 14-5, 14-21
FMT= specifier 11-9
FORALL construct 8-19
FORALL construct statement 8-19
FORALL statement 8-17
form (of a file) 11-5
FORM= specifier 11-42, 11-43
format control 12-3
format item 12-1
format specification 12-1
FORMAT statement 12-1
formatted
    file 11-5
    input/output 11-21
    record 11-1
FORMATTED= spec. 11-47, 11-49
Fortran character set 1-1
Fortran processor vii
FRACTION function 14-5, 14-21
fractional part 14-21
function 13-8, 13-10
    definition 13-11
    reference, invocation 13-16
function result 13-12
FUNCTION statement 13-11

G edit descriptor 13-16
generic function 14-1
generic identifier 13-4, 13-35
generic interface block 7-15, 8-7, 13-32

generic name 13-32, 13-34, 14-1
global entity, global name 3-2
global scope 3-1
GO TO statement 10-2
group-name 9-43, 11-9, 11-33, 11-35
group object 11-35

hexadecimal constant 3-8
high precedence (defined) expression 7-3, 7-10, 7-12, 7-19
host association 4-13
HUGE function 14-5, 14-22

I edit descriptor 12-17
IABS function 14-3, 14-12
IACHAR function 14-5, 14-22
IAND function 14-5, 14-22
IBCLR function 14-5, 14-22
IBITS function 14-5, 14-22
IBSET function 14-5, 14-23
ICHAR function 14-2, 14-5, 14-12, 14-23
IDIM function 14-4, 14-12
IDINT function 14-2, 14-5, 14-12
IDNINT function 14-7, 14-12
IEOR function 14-5, 14-23
if-block 10-5, 10-7
IF construct 10-5
IF THEN statement 10-5
IF statement 10-3
IFIX function 14-2, 14-5, 14-12
implicit initialization 4-16
implicit reference 7-16, 8-8
IMPLICIT statement 9-40
implicit type declaration 9-12, 9-41
implied-DO 6-18, 9-22, 11-17
IN → INTENT attribute
INCLUDE line 1-6
INDEX function 14-5, 14-12, 14-23
index variable 8-17
indexed assignment 8-17
initial association status 9-13

initial line 1-2, 1-4
initial point (of a file) 11-5
initial value 4-16, 9-4, 9-12, 9-13, 9-19
initialization
    default 4-16
    explicit 9-4, 9-12, 9-13, 9-19
    multiple 2-11
initialization expression 7-23, 7-25, 9-12, 9-13, 9-20
initially nullified 9-19
INOUT → INTENT attribute
input arg. 9-5, 13-42, 13-58, 14-13
input field 12-12
input list 11-14
input/output 11-1
input/output arg. 9-6, 13-42, 13-59
input/output list 11-14
input/output list item 11-14
input/output specifier 11-8
input/output statement 11-8
input/output unit → unit
INQUIRE statement 11-47
inquiry (by unit, by file, by output list) 11-47
inquiry function 14-1, 14-10
instance (of a subprogram) 13-17, 13-23
INT function 14-2, 14-5, 14-12, 14-24
integer type 2-1
integer division 7-7
INTEGER statement 9-14
intended use of argument 9-5, 9-24, 13-58
INTENT attribute 9-5, 9-24, 13-58
INTENT statement 9-24
interface → subprogram interface
interface block 13-32
interface definition 7-15, 8-7, 13-32
INTERFACE statement 7-15, 8-7, 13-33
internal file 11-2, 11-30, 11-31
internal input/output 11-30

Index _____ D-7

internal subprogram 13-25
interpretation of expression 7-20
intrinsic
    assignment statement 8-1
    data type 2-1
    expression 7-2
    function 14-1
    operator 6-20, 7-1,7-3,7-8
    subprogram 14-1
    subroutine 14-9
INTRINSIC attribute 9-6
INTRINSIC statement 9-24
invalid operation 7-7
invocation 13-16, 13-21
IOLENGTH= specifier 11-47
IOR function 14-5, 14-24
IOSTAT= specifier 11-11, 11-42,
    11-46, 11-47, 11-52
ISHFT function 14-5, 14-24
ISHFTC function 14-5, 14-24
ISIGN function 14-8, 14-12
iteration count 9-23, 10-17, 10-18

keyword 3-1, 13-44, 14-10
keyword argument 13-44, 14-10
KIND function 14-5, 14-25
kind type parameter 2-1, 9-11,
    14-25, 14-37
kind type param. conversion 7-5, 7-12
KIND= 9-11

L edit descriptor 12-18
label → statement label
LBOUND function 14-5, 14-25
leading array specification 9-11
leading length specification 9-16
left tab limit 12-23
LEN function 14-6, 14-12, 14-25
LEN= → length specification
length → character length
length specification 9-12, 9-16, 9-17,
    9-18

LEN_TRIM function 14-6, 14-25
lexical comparison 7-10, 14-2, 14-25,
    14-26
lexical token 1-4, 3-1
LGE function 14-2, 14-6, 14-12, 14-25
LGT function 14-2, 14-6, 14-12, 14-26
line
    INCLUDE line 1-6
    printed line 11-41
    program line 1-1
list-directed formatting 11-9, 11-25
list-directed input/output 11-25
literal constant 3-3
LLE function 14-2, 14-6, 14-12, 14-26
LLT function 14-2, 14-6, 14-12, 14-26
local entity, local name 3-2
local scope 3-1
LOG function 14-6, 14-27
LOG10 function 14-6, 14-27
logical type 2-4
logical assignment statement 8-3
logical expression 7-10
LOGICAL function 14-6, 14-27
logical IF statement 10-4
logical operator 7-10
LOGICAL statement 9-16
loop
    DO construct 10-14
    implied-DO 6-18, 9-22, 11-17
    nested loops 10-21
loop control 10-17
loop parameter 9-22, 10-15, 11-17
loop termination 10-18
low precedence (defined) expression
    7-19
lower array bound 6-1

main entry point 13-37
main program 13-1
mask expression 8-11, 8-12, 8-17,
    8-18
MASK= argument 14-2

masked array assignment 8-11
MATMUL function 14-6, 14-27
MAX function 14-6, 14-28
MAXEXPONENT function 14-6, 14-28
maximum record length 11-43,
    11-45, 11-49
MAXLOC function 14-6, 14-28
MAXVAL function 14-6, 14-29
MAX0 function 14-2, 14-6, 14-12
MAX1 function 14-2, 14-3, 14-12, 14-29
memory management 6-13
MERGE function 14-6, 14-29
MIN function 14-6, 14-30
MINEXPONENT function 14-7, 14-30
MINLOC function 14-7, 14-30
MINVAL function 14-7, 14-31
MIN0 function 14-2, 14-6, 14-12
MIN1 function 14-2, 14-3, 14-12, 14-31
MOD function 14-7, 14-12, 14-31
model number B-1
module 13-3
module reference 13-3, 13-4
MODULE statement 13-3
module subprogram 13-4, 13-27
MODULE PROCEDURE statement
    7-16, 8-7, 13-33
MODULO function 14-7, 14-31
multiple initializations 2-11
multiple statements 1-2
MVBITS subroutine 14-9, 14-32

name 3-2
name association 4-12
NAME= specifier 11-47, 11-52
named common block 4-15, 4-18,
    9-29, 9-31
named constant 4-2, 9-7, 9-25
NAMED= specifier 11-47
namelist comment 11-37
namelist group 9-43, 9-44
namelist group object 11-35
namelist input/output 11-33

NAMELIST statement 9-43
NEAREST function 14-7, 14-32
NEQV → .NEQV.
nested constructs 8-15, 8-20, 8-21,
    10-9, 10-21, 10-22
next record 11-5
NEXTREC= specifier 11-47
NINT function 14-7, 14-12, 14-32
NML= specifier 11-9
nonadvancing input/output 11-38
NONE → IMPLICIT statement
noneffective list items 11-14
nonexecutable statement 1-7
nonrepeatable edit descriptor 12-6
NOT function 14-7, 14-32
NULL function 2-8, 2-10, 9-12, 9-13,
    14-7, 14-32
null value 11-28, 11-37
nullification 5-6
NULLIFY statement 5-6
NUMBER= specifier 11-47
numeric assignment statement 8-3
numeric editing 12-7
numeric expression 7-3
numeric operator 7-3
numeric relational expression 7-8
numeric sequence type 2-7
numeric storage unit 4-14

O edit descriptor 12-19
object → data object
octal constant 3-8
ONLY → USE statement
only-list 13-4, 13-6
open (a file) 11-6, 11-42
OPEN statement 11-42
OPENED= specifier 11-47
operand 7-1, 7-3, 7-10, 7-12, 7-14,
    7-21
operation 2-2, 2-3, 2-4, 2-5, 7-7
OPERATOR
    → operator interface block

operator 3-3, 7-1, 7-3, 7-8, 7-10, 7-13, 7-14, 7-19
operator function 7-15, 13-18
operator interface block 7-15
operator precedence 7-19
OPTIONAL attribute 9-7, 13-57
optional dummy argument 13-57
OPTIONAL statement 9-25
ordering of statements 1-8
OUT → INTENT attribute
output → input/output
output argument 9-6, 13-42, 13-58
output list 11-16
output statements 11-8
overloaded subprogram name 13-36
own array specification 9-12
own length specification 9-12, 9-16, 9-17

P edit descriptor 12-20
PACK function 14-7, 14-33
PAD= specifier 11-42, 11-45, 11-47
padding character 8-5, 11-14, 11-23, 11-44, 11-45, 11-51, 12-8
PARAMETER attribute 9-7
PARAMETER statement 9-25
parent object vii
parenthesized expression 7-20
partially associated 4-16
pending control mask 8-14
pointer 4-1, 5-1, 9-8, 13-52, 9-26
pointer assignment statement 8-10
pointer association 4-14
POINTER attribute 9-8
pointer component 2-9, 2-10
pointer function 13-13
pointer nullification 5-6
POINTER statement 9-26
pointer target 5-2
position
    array element 4-8, 4-9
    character substring 4-4
    file 11-5

POSITION= specifier 11-42, 11-45, 11-47, 11-51
positional argument 13-45
positioning 11-5, 11-39, 11-52
precedence 7-4, 7-11, 7-19
preceding record 11-5
precision 2-2, 14-33
PRECISION function 14-7, 14-33
preconnected file or unit 11-6, 11-7
prefix 13-11, 13-19
present (argument) 13-57, 14-33
PRESENT function 14-7, 14-33
PRINT statement 11-19
printing 11-41
private 2-12, 9-8, 9-27
PRIVATE attribute 9-8
PRIVATE statement 9-27
procedure → subprogram
processor → Fortran processor
processor time 14-18
processor-dependent character set A-1
PRODUCT function 14-7, 14-34
program 1-1, 13-1
program line 1-1
PROGRAM statement 13-2
program unit 1-1, 13-1
pseudorandom number 14-34
public 2-12, 9-9, 9-27
PUBLIC attribute 9-9
PUBLIC statement 9-27
pure function 13-10, 13-15, 13-39
PURE prefix 13-11, 13-19
pure subroutine 13-10, 13-20, 13-39

RADIX function 14-7, 14-34
RANDOM_NUMBER subroutine 14-9, 14-34
RANDOM_SEED subroutine 14-9, 14-34
range 9-22, 10-14, 10-15, 11-17, 14-35
RANGE function 14-7, 14-35

rank 4-1, 4-7, 6-7
READ statement 11-19
READ= specifier 11-47
reading 11-1
READWRITE= specifier 11-47
real type
REAL function 14-2, 14-7, 14-12, 14-35
REAL statement 9-14
real-time clock 14-40
REC= specifier 11-10
RECL= specifier 11-42, 11-45, 11-47
record 11-1
record length 11-4, 11-42, 11-45, 11-47
record number 11-4, 11-10
RECURSIVE prefix 13-11, 13-14, 13-19, 13-20
recursive function 13-14
recursive subroutine 13-20
reference → vii
    assignment subroutine 8-8
    function 13-16
    module 13-4
    operator function 7-16
    statement function 13-31
    subroutine 13-21
relational expression 7-8
relational operator 7-8
repeat factor 11-29, 11-37
REPEAT function 14-8, 14-36
repeat specification 12-4
repeatable edit descriptor 12-6
representable character 3-7
restricted expression 7-26
RESHAPE function 14-8, 14-36
RESULT 13-11, 13-12, 13-14, 13-37
result variable 13-11, 13-38
return 13-39
RETURN statement 13-39
reversion (of format control) 12-4
REWIND statement 11-52
RRSPACING function 14-8, 14-36

S edit descriptor 12-22
SAVE attribute 9-9
SAVE statement 9-28
saved variable 4-18, 9-9
scalar 4-1, 4-4
scale factor 12-20
SCALE function 14-8, 14-36
SCAN function 14-8, 14-37
scope 3-1
scoping unit 3-1
section subscript 6-6
SELECTED_INT_KIND function 14-8, 14-37
SELECTED_REAL_KIND function 14-8, 14-37
selector 10-10
SELECT CASE statement 10-10
separator
    format specification 12-1
    internal input 11-32
    list-directed input 11-26
    namelist input 11-35
    statement 1-2
sequence association 13-52
SEQUENCE statement 2-6
SEQUENCE attribute 2-6
sequence type 2-7
sequential access 11-3
SEQUENTIAL= specifier 11-47
set of values 2-1, 2-2, 2-3, 2-4, 2-5
SET_EXPONENT function 14-8, 14-37
shape 4-1, 4-7, 6-7
SHAPE function 14-8, 14-38
side effect 6-5, 6-7, 7-21, 8-16, 8-22
sign control edit descriptor 12-22
SIGN function 14-8, 14-12, 14-38
significand 12-13
simple executable statement 1-7
SIN function 14-8, 14-12, 14-38
SINH function 14-8, 14-12, 14-38

Index _____ D-11

SIZE function 14-8, 14-38
size of an array 4-7, 6-6, 14-38
SIZE= specifier 11-14
slash edit descriptor 12-22
SNGL function 14-2, 14-7, 14-12
source form 1-1
SP edit descriptor 12-22
SPACING function 14-8, 14-39
special character 1-1
special expression 7-23
special name 3-3
specific function 14-1, 14-10
specific interface block 13-32, 13-34
specific name 13-11, 13-19, 14-1, 14-10
specification expression 7-26
specification part 13-2, 13-3, 13-6, 13-11, 13-19
specification statement 9-1
SPREAD function 14-8, 14-39
SQRT function 14-8, 14-12, 14-39
SS edit descriptor 12-22
standard unit 11-7
starting position 4-4
STAT= 5-2, 5-5, 6-14
statement function 13-29
statement label 1-1, 1-2, 1-4, 3-3
statement ordering 1-8
statement separator 1-2
status variable 5-2, 5-5, 6-14
status
    allocation status 6-15
    association status (of a pointer) 5-3
    definition status 4-16
    input/output status information 11-11, 11-42, 11-44, 11-46, 11-47, 11-52
STATUS= specifier 11-42, 11-45, 11-46
STOP statement 10-23
storage association 4-14, 9-33
storage sequence 2-6, 4-14, 4-15
storage unit 4-14
stride 6-9, 6-18, 9-22, 10-15, 11-17

structure component 4-9
structure constant 2-14
structure constructor 2-13
structure object 2-13, 4-1
subobject 4-2, 4-3, 4-4
subprogram 13-1, 13-8
subprogram definition 13-11, 13-19, 14-10
subprogram interface 13-9
subroutine 13-18
    reference, invocation 13-21
    definition 13-19
SUBROUTINE statement 13-19
subscript expression 6-5, 6-7
subscript list 6-5
subscript value sequence 6-6, 6-9
subscript-triplet 6-9
substring 4-4
SUM function 14-8, 14-39
SYSTEM_CLOCK subroutine 14-9, 14-40

T edit descriptor 12-23
tabulator edit descriptor 12-23
TAN function 14-8, 14-12, 14-40
TANH function 14-8, 14-12, 14-40
target 5-1
TARGET attribute 9-10
TARGET statement 9-30
temporary file 11-45
terminal point (of a file) 11-5
THEN → IF THEN statement
TINY function 14-9, 14-40
TL edit descriptor 12-23
token → lexical token
totally associated 4-16
TR edit descriptor 12-23
TRANSFER function 14-9, 14-41
transfer of control 10-1
transformational function 14-1
TRANSPOSE function 14-9, 14-41
TRIM function 14-9, 14-41

type 2-1
type component definition 2-7
type concept 2-1
type conversion 7-5, 14-2
type declaration statement 9-11
type definition 2-5, 2-12
type specification 1-4, 9-11, 13-11
TYPE statement
    type declaration 9-18
    type definition 2-6

UBOUND function 14-9, 14-41
ultimate component 2-7
unconditional GO TO 10-2
undefined 4-3, 4-16, 4-18, 5-3
unformatted
    file 11-5
    input/output 11-24
    record 11-1
UNFORMATTED= specifier 11-47
unit 11-6
unit number 11-9
UNIT= specifier 11-9, 11-42,
    11-44, 11-46, 11-47
UNPACK function 14-9, 14-42
unspecified storage unit 4-15
upper array bound 6-1
USE association 4-12, 13-5
USE statement 13-4

value list 6-18
value separator 11-26
variable vii, 4-1, 4-3
variable (dummy) array 4-7, 6-1, 13-51
variable format specification 12-2
vector subscript 6-11
VERIFY function 14-9, 14-42
visible 2-12, 9-9, 9-28

where-block 8-13
WHERE construct 8-12
WHERE construct statement 8-12

WHERE statement 8-11
WHILE loop 10-14, 10-16, 10-19
whole array 6-4
WRITE statement 11-19
WRITE= specifier 11-47
writing 11-1

X edit descriptor 12-24

Z edit descriptor 12-25
zero-length character string 2-4,
    7-22
zero-size array 6-2, 7-22

.AND.	7-10
.EQ.	7-8
.EQV.	7-10
.FALSE.	3-6
.GE.	7-8
.GT.	7-8
.LE.	7-8
.LT.	7-8
.NE.	7-8
.NEQV.	7-10
.NOT.	7-10
.OR.	7-10
.TRUE.	3-6
==	7-8
>=	7-8
>	7-8
<=	7-8
<	7-8
/=	7-8
=>	8-10